Organometals and Organometalloids

Organometals and Organometalloids

Occurrence and Fate in the Environment

F. E. Brinckman, EDITOR
National Bureau of Standards

J. M. Bellama, EDITOR
University of Maryland

Based on a symposium sponsored
by the Division of Inorganic Chemistry
at the 175th Meeting of the
American Chemical Society, Anaheim,
California, March 13–17, 1978.

ACS SYMPOSIUM SERIES 82

AMERICAN CHEMICAL SOCIETY
WASHINGTON, D. C. 1978

Library of Congress CIP Data

Organometals and organometalloids.
(ACS symposium series; 82 ISSN 0097-6156)

Includes bibliographies and index.

1. Organometallic compounds—Environmental aspects—Congresses.
I. Brinckman, F. E. II. Bellama, Jon M., 1938- III. American Chemical Society. Division of Inorganic Chemistry. IV. Title: Organometalloids: occurrence and fate in the environment. V. Series: American Chemical Society. ACS symposium series; 82.

QH545.074073 574.5′2 79-24316
ISBN 0-8412-0461-6 ACSMC8-82 1-447 1979

PRINTED IN THE UNITED STATES OF AMERICA

ACS Symposium Series

Robert F. Gould, *Editor*

FOREWORD

The ACS SYMPOSIUM SERIES was founded in 1974 to provide a medium for publishing symposia quickly in book form. The format of the Series parallels that of the continuing ADVANCES IN CHEMISTRY SERIES except that in order to save time the papers are not typeset but are reproduced as they are submitted by the authors in camera-ready form. Papers are reviewed under the supervision of the Editors with the assistance of the Series Advisory Board and are selected to maintain the integrity of the symposia; however, verbatim reproductions of previously published papers are not accepted. Both reviews and reports of research are acceptable since symposia may embrace both types of presentation.

CONTENTS

To Frederick Challenger

EMERITUS PROFESSOR OF ORGANIC CHEMISTRY
UNIVERSITY OF LEEDS

DEDICATION

The significance of the metal–carbon or metalloid–carbon bond to that rather vague area of descriptive science variously termed "environmental chemistry," "bioinorganic chemistry," or "biogeochemistry," is now perceived as basic to the assessment of man's activities and future on our planet, and there is one individual who stands out as the pioneer in this field that encompasses so many disciplines. More than 40 years ago, long before chemists recognized the implications of sterically constrained donor sites or enjoyed spectrometric means for virtually nondestructive characterization of trace materials, this researcher singlehandedly applied all available chemical forces to a unified study of the biogenesis of organometalloids. His work still stands as a beacon in the field, and even today a report rarely appears without citing one or more of the many papers of Professor Frederick Challenger—papers that span 60 years of personal research.

Today technological, social, and political pressures give urgency to expanded studies on the occurrence and fate of organometals and their implications for mankind: this volume is an attempt to highlight such work. It is obvious that no attempt to give a topical perspective on our present state of knowledge concerning problems in the field could succeed without the most direct recognition of Professor Challenger's contributions and views.

The editors deem themselves and all readers most fortunate that Professor Challenger, now enjoying more than 90 summers of health, could and would give this symposium volume a keynote paper. Most happily, all contributors to this volume take this opportunity as fitting and proper to recognize his past and present primal role in the field of environmental organometallic chemistry. We dedicate this book to him and to his work.

F. E. Brinckman
National Bureau of Standards
Washington, DC

J. M. Bellama
University of Maryland
College Park, MD

PREFACE

The inorganic chemist today is confronted with an intriguing and challenging prospect which will significantly advance our understanding of chemical reactions and transport of toxic elements in the environment. The opportunity lies partly in extending fundamental studies of bioactive sigma carbon–metal bonds or other similar toxic heteroatomic combinations that are formed in polar transport media such as water. The challenge also exists for inorganic chemists to inject their ideas and findings into that diverse interdisciplinary field loosely described as "environmental chemistry" and thereby to profit from such an interchange by finding exciting new problems to solve.

Our current perception of mobilization and transport of certain combinations of organometals and organometalloids is that certain toxic moieties can be and are relocated on a global scale. Recent unambiguous findings show that even simple, labile methylmetal species occur in the environment as a result of biogenic processes. Moreover, anthropogenic inputs of organometals parallel growing technological and agricultural demands. Matching these increased demands are amplified requirements to understand the potential, or lack of it, for wastewater treatment of refractory organometals and organometalloids by chlorination or ozonation, a neglected study area. Similarly, study and discussion are necessary for us to understand the means by which man introduces organometals and organometalloids into the environment, and the pathways by which the environment can pass on (return) such bioactive metal species to the human organism. Consequently, not only from pragmatic environmental quality or public health perspectives, but also from basic needs to understand biogeochemical cycles, participation of inorganic chemists in the dialogue is timely and is needed urgently.

The symposium on inorganic chemical problems in the environment could not deal with the entire range of such identifiable problems amenable to inorganic research in a brief two-day meeting. Therefore, the symposium focused on two main objectives: a) definition, by example, of the latest research concerned with organometal and organometalloid chemistry relevant to environmental concerns, with particular emphasis on aqueous reactions and transport mechanisms; and b) dialogue, with interested and competent colleagues outside the inorganic community, who can best transmit the current needs and consequences of alternative courses for future research relevant to biological implications of organometallic chemistry.

Therefore, this symposium comprises a series of invited research and review papers by researchers active internationally in both biological and organometallic chemistry. At the welcome suggestion of reviewers, additional papers were solicited from highly competent colleagues to achieve better balance and topical presentation for this volume. The editors assume full responsibility for the fact that additional areas in the title field are not discussed. Such omissions were dictated by our personal biases, constraints of publication costs, and the obvious need to produce similar symposia in the near future.

We have not attempted to arrange papers in chronological order, e.g. in the symposium format. Rather, we have exercised editorial prerogative and (with apologies to authors) have assembled the volume into three broad categories of needed and ongoing research: biogenesis of organometals and organometalloids; underlying abiotic aquatic chemistry and mechanisms; and the nature of entry, transport, or uptake of organometals and organometalloids into environmental compartments. Clearly, many papers overlap these classifications, just as expected for such an interdisciplinary dialogue. Just as clearly, some readers will find useful a structuring which aids them in placing their own interests into perspective with others.

The editors also chose to adopt a symposium format which devoted approximately 30% of available meeting time to discussion. (Authors were asked to complete their manuscripts after the symposium to exploit this feedback feature.) We have recorded and edited these additional verbal interactions between symposium participants, and have attached these discussions to the end of each paper. Again, rigorous, and hopefully equitable, editing was imposed to provide the best presentation of each commentator's remarks while adhering to minimum length. The editors hope that readers will find the discussions to be useful supplements to the formal papers, particularly from the standpoint of highlighting untouched research topics in this fascinating field.

In addition to the authors who have steadfastly supported the attempts of the editors to produce a volume of excellent papers, and the referees (those from the American Chemical Society (ACS) as well as those who critically read the individual papers), we owe debts of gratitude to many others. We are deeply grateful to S. Sisk (University of Maryland) and to C. Lamb (National Bureau of Standards (NBS)) who unstintingly provided the secretarial and professional typing support critical to a book in this format. We especially thank our colleagues, particularly Professors J. M. Wood, M. L. Good, J. S. Thayer, J. K. Kochi, and W. R. Cullen, who sparked and sustained our effort for so many months. In the final analysis, these colleagues gave their talents to conducting and pacing symposium sessions, without which realization of this volume would have been impossible.

Finally, we are grateful to our sponsors who both financially and spiritually supported our effort. We thank the ACS for many valuable suggestions and aids in producing the symposium. We are indebted to the National Measurements Laboratory, NBS, and the Division of Inorganic Chemistry, ACS, for assistance in organizing and defraying costs for invited authors.

F. E. Brinckman
Center for Materials Science
National Bureau of Standards
Washington, DC 20234
August, 1978

J. M. Bellama
Department of Chemistry
University of Maryland
College Park, MD 20742

1

Biosynthesis of Organometallic and Organometalloidal Compounds

FREDERICK CHALLENGER

19 Elm Avenue, Beeston, Nottingham, NG9 1BU, U. K.

I feel greatly honoured to have been asked to send to the Symposium an introductory communication which shall form a keynote to the proceedings. The programme covers almost every aspect of the biological methylation of a large number of elements and the properties of several other organo-metallic compounds e.g. those of tin. Clearly, only a brief survey of certain areas of this wide field is possible, and this paper will be confined to those aspects of the Symposium which approach most closely to the work of my students and myself, especially on the biological methylation of compounds of arsenic, selenium, and tellurium (1931-1953) at the University of Leeds.

A strictly historical treatment takes us back to the old laboratories at University College, Nottingham (now merged in the Trent Polytechnic) where between 1900 and 1944 Kipping laid the foundations of the organic chemistry of silicon. It was my privilege to work with him for 2½ years and to succeed in the optical resolution of D,L-dibenzylethylpropylsilane monosulphonic acid by means of brucine in December 1909. The study of the Si-C bond (though strictly speaking silicon is neither metal nor a metalloid) gave me a budding interest in the metal-carbon link in general, and particularly in organo-derivatives of bismuth which later was extended to arsenic.

It is impossible here for me to emphasize sufficiently my deep indebtedness to Professor Frederick Stanley Kipping, which I have endeavoured partially to discharge in an appreciation of the man and his work, published in 1950 and 1951 (1-2). His exacting laboratory methods were based on those of von Baeyer (in whose laboratory he had worked) and required little but beakers, flasks, test tubes and glass rods. They were at once the despair and the inspiration of his research students.

It was not solely an interest in organo-metallic compounds that led to our work on the arsenical Gosio-gas but an equally keen attraction, microbiological chemistry, of which I gained some rudimentary knowledge in the laboratory of Professor Alfred Koch in Göttingen (1910-1912) and maintained in Manchester. In 1931 it

0-8412-0461-6/78/47-082-001$05.00/0

became possible, with the assistance of Constance Higginbottom and Louis Ellis (3), to combine these two interests. I was attracted by the early work on the unknown composition of Gosio-gas which was evolved from arsenical wallpaper containing the mineral pigments Scheele's green and Paris green.

Later on, this vapour was found by Gosio (4) to be evolved from pure cultures of the mould *Penicillium brevicaulis* (now designated *Scopulariopsis brevicaulis*) containing arsenious oxide. At the time it appeared that the possible identification of Gosio-gas would, while interesting, relate to a small area of mycological chemistry with, possibly, a quite restricted significance. When we found that Gosio-gas was pure trimethylarsine, we were suddenly ejected from our small corner and pitchforked directly into the growing field of Transmethylation, then in the early stages of its development in animals by the fundamental work of Vincent du Vigneaud (5). There we have remained ever since. I realize that my metaphors are very mixed here, but the personal excitement of the investigations which followed may perhaps be accepted as an excuse!

Biological Methylation of Arsenite and Arsenate

The earliest attempts to identify Gosio-gas were made by Biginelli who passed the gases from aerated cultures of *S. brevicaulis* containing As_2O_3 through $HgCl_2$ in dilute hydrochloric acid. He regarded the resulting precipitate as $(CH_3CH_2)_2AsH{\cdot}2HgCl_2$, but a study in my laboratory at the University of Leeds showed it to be a mixture of the mono- and dimercurichlorides of trimethylarsine. The first clue to the identity of Gosio-gas was obtained, however, by passage through alcoholic benzylchloride, formation of trimethylbenzylarsonium picrate, and comparison with an authentic specimen prepared some years previously by Ingold, Shaw, and Wilson at Leeds in research (6) on the orienting influence of positive poles in aromatic substitution. The identity was confirmed by the formation of several other derivatives, see below.

Sodium methylarsenate $CH_3AsO(ONa)_2$ and sodium cacodylate $(CH_3)_2AsO(ONa)$ when added to the mould cultures on bread crumbs also gave trimethylarsine. This reaction was, however, ambiguous owing to the possibility of the fission of the As-C link in these acids and formation of As_2O_3.

However, when several mono- or dialkylarsonic acids $RAsO(OH)_2$ and $RR'AsO{\cdot}OH$ or their sodium salts were added to bread cultures of *S. brevicaulis*, methylation occurred giving ethyldimethylarsine, *n*-propyldimethylarsine, allyldimethylarsine and methylethyl-*n*-propylarsine. These were characterized by formation of various derivatives such as the mercurichloride, the benzyltrialkyl arsonium picrate and the hydroxytrialkylarsonium picrate, and comparisons with authentic specimens. It was therefore clear that the methyl group was supplied by the mould (7) as summarized in the scheme below.

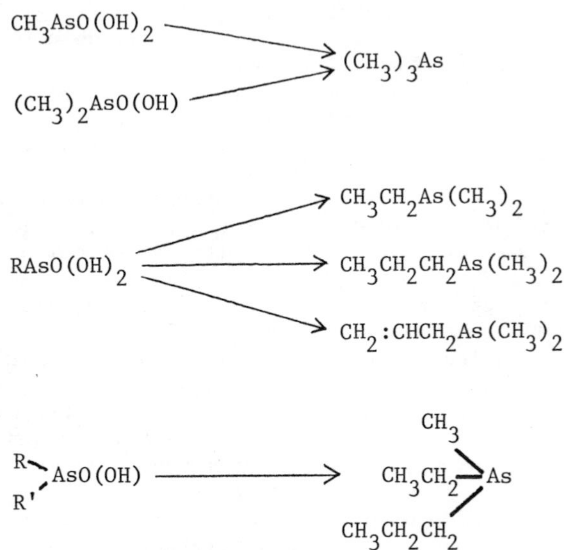

Methylation of Selenate, Selenite, and Tellurite

Several workers had already drawn attention to the odourous products evolved from cultures of *S. brevicaulis* containing various oxy-acids of selenium and tellurium, but without identifying them. Using similar methods to those employed for arsenic compounds, the odours were identified by Harry North (8) and Marjorie Bird (9) as due to dimethylselenide and dimethyltelluride. These were characterised by various derivatives such as the mercurichloride, mercuribromide, platinochloride ($(CH_3)_2Se \cdot PtCl_2$), hydroxyselenonium nitrate $(CH_3)_2Se(OH) \cdot NO_3$, and the picrates prepared using benzylchloride as before. Dimethylselenide and -telluride were also evolved from various cultures of *Penicillium chrysogenum*, *P. notatum*, and a mould closely allied to *P. notatum*. *Aspergillus niger* gave $(CH_3)_2Se$ with selenate.

Dimethyltelluride could only be detected (except by its powerful odour) when the cultures were grown in test tubes in series to minimise atmospheric oxidation and the minimum of absorbent was used.

The Metabolism of Some Sulphur Compounds in Mould Cultures

Strictly speaking, this subject is not within the purview of this Symposium, but as the results now to be summarised are of rather general application, some account of them may be justified.

S. brevicaulis was also shown by Alan Rawlings (10), Philip Charlton (11), and Stanley Blackburn (12) to cause fission of dimethyl, diethyl, di-*n*-propyl and di-*n*-butyl disulphides, RS-SR, giving RSH and $RSCH_3$. No fission was observed with dibenzyl- and diphenyldisulphides, in accord with the results obtained with phenyl-and benzylarsonic acids which give no methylated arsines in the mould cultures. Another interesting fission reaction was observed by Charlton (11). *S. brevicaulis*, in bread cultures, produces alkylthiol and alkylmethylsulphide from S-methyl-, S-ethyl- and S-*n*-propylcysteines, $RSCH_2CHNH_2COOH$. The sulphur compounds were characterised by formation of well-known authentic compounds. These fission reactions have well-established analogies in animal biochemistry.

Addition of sulphur, Na_2SO_3, $Na_2S_2O_3$, thiourea, sodium ethanesulphinate and sulphonate, and sodium thiodiglycollate, $S(CH_2COONa)_2$, to cultures of two strains of *S. brevicaulis* gave no dimethylsulphide. These negative results are interesting, and at present the author knows of only one mould which will methylate simple inorganic compounds of sulphur, namely *Schizophyllum commune*, a higher fungus which destroys wood and which was shown by Birkinshaw, Findlay, and Webb (13) to give methanethiol and traces of hydrogen sulphide when grown on a glucose medium containing sulphate. We found at Leeds that dimethylsulphide and -disulphide are also evolved, and that cultures on wort or bread without addition of sulphate evolve methanethiol due to sulphur compounds in the medium. When the cultures are almost odourless, addition of sodium selenate gave dimethylselenide in small amount. Only faint odours were observed on adding arsenite, tellurite, or methyl- or *n*-propylarsonic acids to similar cultures.

Fission to the corresponding thiol, RSH, was observed with dimethyl, diethyl, and di-*n*-butyl disulphides. With dimethyldisulphide some dimethylsulphide was detected. A careful study of the metabolism of this fungus might yield interesting results.

Selective Methylation

Bird *et al*. (14) studied the capacity of a number of other moulds to methylate inorganic and organic compounds of arsenic, selenium, and tellurium. *A. niger*, *P. notatum* and *P. chrysogenum* in bread cultures gave no trimethylarsine with arsenite, but with methyl- and dimethylarsonic acids as sodium salts, trimethylarsine was evolved. *A. niger* gave ethyldimethylarsine with sodium ethylarsonate. The same mould with selenate and the two *Penicillia* with selenite gave dimethylselenide. *A. versicolor* and *A. glaucus* gave trimethylarsine with arsenite and sodium methylarsonate, but no dimethylselenide or -telluride with selenite or tellurite.

Attempts at an explanation of selective methylation were made by Bird *et al*. (14), but no definite conclusions were reached.

The Development of Ideas on The Mechanism of Biological Methylation

In 1887 it was shown by His (15) that pyridine acetate administered to dogs and turtles is excreted as methylpyridinium acetate. An analogous reaction occurs with quinoline in dogs. A strong garlic odour was observed by Gmelin (15) and by Hansen (16) in animals and man, respectively, after doses of potassium tellurite. This odour was regarded by Hofmeister (15) as being due to dimethyltelluride, though without proof. In his article we can recognize the first rather vague conception of a possible transfer of a methyl group. He suggested that "the methyl group is already present in the tissues which possess the capacity for methylation. In presence of pyridine and tellurium these are methylated, whereas under normal conditions methyl derivatives such as choline and creatine are produced." No particular compound was suggested as the source of the methyl group. This conception was carried much further by Riesser in 1913 (17) who considered that methyl groups of the (assumed) dimethyltelluride, formed in the animal body after tellurite administration, probably arose from choline or betaine. He based this suggestion partly on his observation that, when tellurite was heated with sodium formate and either choline chloride or betaine hydrochloride, an odour resembling dimethyltelluride was evolved. Challenger *et al*. (18) extended this reaction to include sodium selenite and sulphite, which on heating with betaine free from hydrochloride and without formate, gave dimethylselenide and -sulphide which were identified by formation of derivatives. Under similar conditions pure betaine, when heated with primary aromatic amines, gave $RNHCH_3$. Phenol and β-naphthol gave the corresponding methyl ethers, an imitation at high temperatures of many biological methylations.

In 1935 Challenger and Higginbottom (19) stated "It is not impossible that some ingredient of the cell substance containing a methylated nitrogen atom may, under the special conditions obtaining in the cell, lose a methyl group which if it be eliminated with a positive charge could be easily co-ordinated by the unshared electrons of tervalent arsenic or quadrivalent selenium and tellurium." The underlining indicates the degree to which this suggestion extends those of Hofmeister and Riesser. Challenger, in several publications, discussed this conception in detail and proposed a series of reactions for the methylation of arsenate or arsenite, selenate or selenite which for brevity may be summarised on the next page.

The scheme involves ionisation, reduction, and the coordination of a positive methyl group. The suggested intermediate arsenic compounds were not found in the medium but were all converted to trimethylarsine in bread cultures of *S. brevicaulis*. Methylarsonate, however, was found by McBride and Wolfe (20) as an intermediate compound in the biological formation of dimethylarsine by a methanobacterium from canal mud.

The potassium salts of methane-, ethane-, and *n*-propaneseleninic acids, $RSeO_2H$, readily gave the corresponding methylalkylselenides in bread cultures of S. brevicaulis. The potassium salts of methane-, ethane- and n-propaneselenonic acids, $RSeO_2OH$, unexpectedly gave only dimethylselenide due to hydrolysis to selenite. Dimethylselenone, $(CH_3)_2SeO_2$, has not been prepared, but methylselenoxide nitrate, $(CH_3)_2Se(OH)(ONO_2)$, also gave dimethylselenide.

Biological Methylation of Arsenite and Selenate

1. $$As(OH)_3 \longrightarrow CH_3AsO(OH)_2 \longrightarrow (CH_3)_2AsO(OH) \longrightarrow (CH_3)_3AsO \longrightarrow (CH_3)_3As$$

2. $$SeO_2(OH)_2 \longrightarrow CH_3SeO(OH) \longrightarrow (CH_3)_2SeO_2 \longrightarrow (CH_3)_2SeO \longrightarrow (CH_3)_2Se$$

The reactions set out above may be represented as the additions of a methylcarbonium ion, CH_3^+, to a neutral molecule, followed by expulsion of a proton:

$$CH_3^+ + :As(OH)_3 \longrightarrow \overset{+}{C}H_3As(OH)_3 \longrightarrow CH_3AsO(OH)_2 + H^+$$

In recent years much attention has been directed towards methionine. In liver or kidney enzyme systems which can effect methylation, Cantoni (21) found in 1952 that added methionine forms a sulphonium compound, S-adenosylmethionine, or "active methionine", as it was first called. Its formula and involvement in biological methylation are so well-known as to need no comment. If it is represented as $RR'\overset{+}{S}CH_3$, and if we assume that methionine is similarly "activated" in moulds the biomethylation of arsenite could be represented:

$$RR'\overset{+}{S}CH_3 + :As(OH)_3 \longrightarrow [RR'S \longleftarrow \overset{+}{C}H_3:As(OH)_3] \longrightarrow$$

$$RSR' + [CH_3\overset{+}{As}(OH)_3] \longrightarrow CH_3AsO(OH)_2 + H^+$$

The attraction of the positive sulphur center for the electrons of the $-S-CH_3$ link might allow nucleophilic attack on the methyl group by the arsenic atom with its unshared electrons. The resulting transition state would lead to a neutral sulphide (R·S·R'), a proton, and methylarsonic acid without formation of a free positive methyl ion at any state. This explanation is preferred by some of my colleagues. It was put forward by Challenger in 1955 and 1959 in two reviews (22, 23) and is still regarded as a satisfactory explanation.

The failure of the author and co-workers (24) in their admittedly preliminary experiments of 1935 to observe any methylation of mercuric oxide by S. brevicaulis is explicable because the Hg^{2+} ion would not behave as a nucleophile. For the later results of other workers see p. 16 of this review.

It may, perhaps, be emphasised here that the fundamental demonstration of the activation of methionine prior to the transfer of its methyl group (transmethylation) was first made in 1952 by Cantoni (21). My suggestion (22, 23) that methionine might be similarly activated in moulds and that this might precede the methylation of arsenic was an adaption of Cantoni's work at a much later date. This historical point is not, perhaps, made quite clear by Ridley, Dizikes, and Wood (25) and by Ridley, Dizikes, Cheh, and Wood (26).

Methylation of Selenium using Methionine Labelled with ^{14}C- in Liquid Cultures of *Aspergillus niger*

In an extended investigation with Philip Dransfield and Denis Lisle (27, 28), the author showed that in liquid cultures of *A. niger* containing sucrose, glycine, inorganic salts, selenate, and either DL-, L- or D-[Me^{14}C]methionine, over 90% of the methyl groups of the evolved dimethylselenide were derived from the labelled methionine. The dimethylselenide was collected in aqueous mercuric chloride and counted as the mercurichloride adduct, $(CH_3)_2Se{\cdot}HgCl_2$, m.p. 153-154°. *A. niger* does not methylate As_2O_3, but in one experiment with *S. brevicaulis* in bread cultures containing arsenious oxide and DL-[Me^{14}C]methionine, the methylation percentage was 28.3%, a very much lower figure than was obtained with *A. niger* and selenate.

Search for Methanol in Arsenical Liquid Cultures of *S. brevicaulis*

If a free positive CH_3 group is the methylating agent in *S. brevicaulis* cultures, it might be expected to react with the water of the medium to give methanol. Attempts were made to detect the methanol in 2-3 litres of arsenical medium upon which the mould had grown for 46 days. After careful fractional distillation, the first runnings were tested for methanol by Wright's method (29). This depends on oxidation to formaldehyde with potassium permanganate, and its detection by Schiff's reagent. The results obtained by Douglas Barnard (30) suggested the presence of not more than 0.001 mL per litre of medium. This figure was quite insufficient to allow any conclusion to be drawn as to the mechanism of formation or even of the actual presence of methanol.

Later experiments by Fernand Kieffer (31) were equally inconclusive. More delicate methods employed by Axelrod and Daly in 1965 (32, 33) have shown that an enzyme occurring in the pituitary gland of several mammals can form methanol from [Me^{14}C]-S-adenosylmethionine. In parallel experiments using the same enzyme preparation and [2-^{14}C]S-adenosylmethionine, the demethylated product [2-^{14}C]S-adenosylhomocysteine was formed. The methanol and the homocysteine derivative were formed enzymically and in stoichiometrical proportions.

The methanol was detected by conversion to its 3,5-dinitrobenzoate which was radioactive and was identified by thin layer chromatography after dilution with an inactive specimen and crystallisation to constant specific activity.

The authors state "this reaction proceeds by methylation of water or by hydrolysis of S—adenosylmethionine".

McBride and Wolfe (20) showed that arsenate is reduced and methylated under anaerobic conditions (in a hydrogen atmosphere) by cell extracts and by whole cells of a *Methanobacterium* (strain MoH) in present of methylcobalamin (methyl-B_{12}) as a methyldonor. The organism was isolated from the mud of a canal near Delft. The gaseous product was regarded as dimethylarsine, and methylarsonic acid $CH_3AsO(OH)_2$ was identified by electrophoresis. Dimethylarsonic (cacodylic) acid was not found in the medium, but both it and methylarsonic acid gave dimethylarsine in cell extracts of the bacterium. The reaction was shown to be enzymic. The authors pointed out that a mixture of CH_3AsH_2 and $(CH_3)_3As$ would give the same ratio of CH_3 to As as was found by analysis of the volatile product. However, variation in the concentration of the methyl donor did not affect this ratio.

In another paper of this symposium Dr. McBride and his colleagues discuss their considerable later work in which dimethylarsine is produced by anaerobic organisms, and trimethylarsine is produced by aerobic organisms. All the biologically methylated arsines described by Challenger were produced in well-aerated mould cultures which nevertheless exhibited a strong reducing action. The authors state that they are not aware of any other references to the bacterial synthesis of arsine or its alkylated derivatives. This recalls the negative results of Challenger and Higginbottom (19) with As_2O_3 and several bacteria, a point emphasised by Challenger (34) at the Brussels Biochemical Congress in 1955.

Attempts to Methylate Derivatives of Elements other than Arsenic

Antimony In Vienna a case of chronic antimony poisoning occurred in a house containing silk curtains mordanted with a compound of antimony. A series of experiments were performed in 1913 by Knaffl-Lenz (35) to detect the possible formation of volatile compounds of antimony. Thus, *S. brevicaulis* was grown on media containing one percent of tartar emetic (potassium antimonyl tartrate, O=Sb-O-CO·CHOH·CHOH·CO·OK), with the volatile products being aspirated through concentrated nitric acid which was later tested for antimony with negative results. Later, similar work by

Tiegs (36) and also by Smith and Cameron (36) was also unsuccessful. The last named authors state that antimony compounds do not interfere with the biological test for arsenic using S. brevicaulis. Challenger and Ellis (36) found that bread or liquid cultures of S. brevicaulis containing tartar emetic gave no odour or any precipitate in acidified mercuric chloride solution. When the cultures were left for 9 months, a large amount of antimony trioxide Sb_2O_3 was deposited in the mycelium. An attempt to isolate trimethylstibine oxide as the picrate from these mould cultures failed. Tartar emetic and Sb_2O_3 contain antimony as a cation. Methylation by nucleophilic mechanism analogous to that exerted by S. brevicaulis in arsenical media would therefore not be expected. In experiments by Barnard (37) with S. brevicaulis and P. notatum cultures containing phenylstibonic acid $C_6H_5SbO(OH)_2$ or potassium meta-antimoniate $KSbO_3$, in which Sb is an anion, methylations might be expected but none was observed. Aspiration of volatile products from the cultures into concentrated nitric acid, evaporation of the acid and application of the Gutzeit or Marsh test to the residue gave varying results. The amount of volatile antimony compound was, however, far too small to allow any conclusions as to its nature or origin.

The position does not seem to have altered since these experiments were performed in 1933 and 1947. Parris and Brinckman (38) state "At this time it has not been demonstrated that methylstibines are metabolites of microorganisms acting on inorganic antimony compounds, but the extensive similarity of the chemistry of arsenic and antimony gives reasons to believe that antimony can be biologically methylated."

In a later paper (39) these authors say "There is no obvious thermodynamic or kinetic barrier to biomethylation and the chemical similarities between Sb and Sn, Pb, As, Se, and Te, which literally surround Sb in the Periodic Table, and all of which have been shown to be subject to biomethylation, would suggest biomethylation pathways for antimony." The authors then refer to the use of inorganic and organic compounds of antimony along with halogenated hydrocarbons in fire retardant systems. Should biomethylation occur, the antimony in various commercial products would become more soluble in water and become a potential hazard. The occurrence of arsenious oxide in antimony preparations should also be borne in mind.

An unusual outbread of antimony poisoning (not due, however, to biomethylation) occurred several years ago in a large shop in Newcastle-upon-Tyne. On a very hot day the employees were given lemonade, presumably "synthetic" lemonade, in large enamelled jugs. The enamel, as in many other products, contained Sb_2O_3 which dissolved in the tartaric (or citric) acid of the lemonade to give tartar emetic or an analogous compound. The result can be imagined.

Parris and Brinckman (38) also refer to the atmospheric oxidation of a solution of trimethylarsine in methanol and the formation of cacodylic acid and trimethylarsine oxide, $(CH_3)_3AsO$, detected by NMR. This was also observed by Ellis (36) in ether. The cacodylic acid was separated and the oxide isolated as the picrate. Dimethylallylarsine behaved in a similar manner.

Metabolism of Selenium Compounds

Any study of the biochemistry of selenium compounds will involve not only their methylation but also their uptake by plants and transfer to animals. This has been discussed in very many publications and need not be dealt with in detail here. Many selenium analogues of biologically important amino acids containing sulphur have been found in plants growing on seleniferous soils.

The literature on almost every aspect of selenium metabolism has been admirably discussed by Diplock (40) in a review of 57 closely printed pages with 430 references.

Mention may here be made of a statement by Cerwenka and Cooper (41) that the odours produced in animals by administration of selenite and of tellurite are similar. The odour of dimethyltelluride resembles that of trimethylarsine which is, however, quite different from that of dimethylselenide. Here it may be said that the well-known methylation of inorganic tellurium compounds in man and animals, so much more pronounced at similar concentrations than that of selenium, will not be dealt with further in this review. Its history and chemistry have been discussed previously by the author (22, 23). The exhalation of dimethylselenide after injection of radioactive (^{75}Se) sodium selenate into rats has been reported by McConnell and Portman (42). Much attention is now being paid to the effect of compounds of a second element on the biogenesis of organometallic or -metalloidal derivatives. Diplock (40) discusses in detail the effect of arsenic, cadmium, mercury, silver, and thallium on the toxicity of selenium compounds in animals. Results up to the present are not always easy to interpret and further advances will be awaited with interest.

An important step forward was made by Byard (43) in the course of a study of the metabolism of sodium selenite containing traces of $H_2{}^{75}SeO_3$ after oral administration to rats. He found that the urine contained the trimethylselenonium ion Me_3Se^+. This ion was isolated after ion exchange chromatography as the reineckate and converted to the chloride $(CH_3)_3SeC\ell$. The identity of this salt was established by paper chromatography, NMR, and mass spectrometry. It was also detected in the bladder 30 minutes after injection of selenite and was therefore not a product of bacterial action. A second compound of selenium was isolated but not identified.

Almost simultaneously, Palmer *et al.* (44) isolated trimethylselenonium chloride from the urine of rats after injection with [^{75}Se] selenite by methods very similar to those employed by

Byard. They used the reineckate, and by co-crystallisation with the authentic trimethylselenonium salt, identified their specimen as the corresponding chloride, a conclusion confirmed by the NMR spectrum.

In later studies, Palmer et al. (45) injected rats intraperitoneally with sodium selenate containing $H_2{}^{75}SeO_4$. Several seleno-amino acids were also used.

Amino-Acids Containing Selenium

$HOOCCH(NH_2)CH_2Se\text{-}SeCH_2CH(NH_2)COOH$

Selenocystine

$HOOCCH(NH_2)CH_2SeCH_3$

Selenomethylselenocysteine

$HOOCCH(NH_2)CH_2CH_2SeCH_3$

Selenomethionine

Some of the selenomethionine was labelled with ^{75}Se. Co-crystallisations of the reineckates were again used and all the selenoamino acids gave the same trimethylselenium chloride, $(CH_3)_3SeCl$, which was also formed on feeding the rats with seleniferous wheat grown on soil containing selenite or selenate.

Co-Enzyme M (2-Mercaptoethanesulphonic Acid)

Recent work by McBride, Wolfe, and their colleagues (46, 47, 48, 49, 50) has provided much information about a heat-stable co-factor for methyl transfer prior to methane formation in cell extracts of *Methanobacterium* strain M.O.H. grown in hydrogen. The co-factor is acidic and dialysable. It also occurs in rumen fluid and has been designated Co-enzyme M. It contains phosphate when first isolated, and this is removable by prolonged acid hydrolysis. Its relation to the rest of the molecule has not been established. After prolonged purification of cell extracts by ion exchange, or on a smaller scale by dialysis, chemical analysis and numerous spectroscopic studies showed that the product was 2:2'-dithiodiethanesulphonic acid (I), a structure verified by comparison with a synthetic specimen prepared from sodium 2-bromoethanesulphonate and H_2S in ammoniacal solution. This procedure gave the corresponding -SH compound (II), which was converted to the disulphide by gaseous oxygen. By using methanethiol instead of H_2S, the S-methylated product (III) was obtained which was identical with the product formed biologically from Co-enzyme M

Co-Enzyme M and Derivatives

$HOSO_2CH_2CH_2SSCH_2CH_2SO_2OH$ (I) $HSCH_2CH_2SO_2OH$ (II)

$CH_3SCH_2CH_2SO_2OH$ (III) $(CH_3)_2\overset{+}{S}\cdot CH_2CH_2SO_2\bar{O}$ (IV)

(thiol form) and methyl-B_{12} in cell extracts of the Methanobacterium. The enzyme responsible was named methylcobalamin Co-enzyme M methyl transferase, and was purified 100 times.

By reaction with methyl iodide in methanol, the $-SCH_3$ derivative gave the corresponding dimethylsulphonium ethanesulphonate (IV). This was found to be inert as a methyl donor in the methyl reductase-catalyzed reaction in incubation periods up to 70 minutes. No methane production was observed when the onium compound (IV) and the thiol forms of Co-enzyme M (II) were present together in the reaction medium. The onium compound (IV) would not methylate the thiol to yield biologically active methylCo-enzyme M (III). This species could not be further methylated by the enzyme methyltransferase.

It is probable that the disulphide form of Co-enzyme M (I) is an artifact arising during the purification process, and that the effective Co-enzyme M is the thiol form (II). In biological systems which obtain in the mud of swamps, rivers, and lakes, this is methylated to the $-SCH_3$ compound (III) which by enzymatic reduction gives methane.

Speculations on the Biosynthesis of Co-enzyme M

A compound of closely related structure is isethionic acid $HOCH_2CH_2\cdot SO_2OH$. Some speculations on the possible biochemical significance of this sulphonic acid were made by the author (51, 52) in 1970 in another connection.

By reaction with H_2S, isethionic acid might give the thiol form of Co-enzyme M, a reaction analogous to the enzymatic conversion of serine to cysteine in yeast by H_2S, and the enzyme formerly known as serine sulfhydrase. Dagley and Nicholson (53) name the enzyme "L-serine hydrolyase, adding H_2S". The modern name for this enzyme (because of one of its chief reactions) is cystathionine β-synthase[L-serine-hydro-lyase (adding homocysteine) E.C.4.2.1.21.] (54).

Similarly, O-phosphohomoserine with H_2S and a sulfhydrase gives homocysteine (55). Isethionic acid is connected with cysteine by a series of reactions set out below. Cysteine is, no doubt, present in the cell protein of the methanobacterium. It can be biologically converted to taurine, $H_2N\cdot CH_2\cdot CH_2\cdot SO_2OH$, which yields isethionic acid by enzymatic oxidative deamination followed by reduction. Some of these reactions were cited by Challenger in 1970 (51, 52). I have not access to all my files at present, but I think the conversion of cysteine to taurine is well established; I forget the circumstances. I think it is described in Meister's "Amino-acids"(2nd edition).

$$HSCH_2CH(NH_2)COOH \longrightarrow HO_2SCH_2CH(NH_2)COOH \xrightarrow{-CO_2} HO_2S{\cdot}CH_2CH_2NH_2 \longrightarrow$$

Cysteine — Cysteinesulphinic Acid — Hypotaurine

$$HOSO_2CH_2CH_2NH_2 \longrightarrow [HOSO_2CH_2CHO] \longrightarrow HOSO_2CH_2CH_2OH \xrightarrow{H_2S}$$

Taurine — Isethionic acid

$$\longrightarrow HOSO_2{\cdot}CH_2CH_2SH$$

Co-enzyme M

Taurine is a product of the hydrolysis of the cell walls of a B. subtilis mutant (56). This is interesting in view of the formation of Co-enzyme M in extracts of sonicated cells of Methanobacterium. When taurine is present as sole source of sulphur in A. niger cultures, isethionic acid is formed (57).

Another possible (though not established) source of isethionic acid and so of Co-enzyme M is 3-L-sulfolactic acid, $HO_3SCH_2{\cdot}CHOH{\cdot}COOH$, which has been provisionally identified in 1969 as a major sulphur compound in B. subtilis spores by Bonsen et al. (58). By a well recognized biochemical reaction this might give formic acid and $HO_3S{\cdot}CH_2{\cdot}CHO$ which by reduction would yield isethionic acid. I have not followed the further work (if any) on sulfolactic acid, but the clear association of sulphonic acids with cell walls, disrupted cells, and spores suggests further study of these relationships.

I have always thought that isethionic acid is worthy of more attention by biochemists, and the facts set out above may be said to strengthen this view. This is, I think, true for sulfonic acids in general.

Ability of Metals for Incorporation into Living Systems

In a most useful publication (59) summarising the results of six or seven years' work, Wood discusses the availability of numerous metals for incorporation into living systems. He points out that titanium and aluminum are not so available because of the insolubility of their hydroxides; and that nickel and chromium are almost absent from biological systems, owing to the stability of these cations in octahedral sites in silicates. These last two elements do not form complexes with proteins for geometrical reasons. The limits imposed by space and the subject of his review, "Biological Cycles for Elements in the Environment", unfortunately do not permit him to discuss these interesting observations in detail. He refers, however, to ten metals which, after transport into the cell by suitable chelating agents, bind to various ligands, and mentions the molybdenum-phosphorus link. This recalls the ready formation of the yellow phosphomolybdate precipitate in qualitative analysis. The requirements of many enzyme systems for magnesium, zinc, and manganese are well-recognized.

Having digressed slightly into the realm of inorganic biochemistry, detailed reference may now be made to a somewhat similar investigation into some aspects of the behaviour of toxic metals in soils. The results were published in 1975 and only very recently came to the author's notice.

Much valuable work was published in 1975 (60, 61) as the result of a survey of the metal content of the soil of the site for a satellite town at Beaumont-Leys, two miles from Leicester. Prior to 1964 the site was used as a sewage farm, after which it was let to farmers. In 1970, in view of its impending use, an investigation by the National Agricultural Advisory Service revealed a high concentration of heavy metals, chiefly zinc, copper, and nickel, on the site. A value, based on the content of these three metals, known as the *zinc equivalent* was used for assessing their content in the soil, and 250 parts per million (ppm) was regarded as permissible. Maximum values of 1000-6000 ppm were obtained for soils from Beaumont-Leys. Moreover, the zinc content of grain grown on the estate was found to be 115 ppm, the maximum permissible value being 50 ppm.

The study was then extended to include other more poisonous metals such as cadmium, arsenic, and lead. The results were of particular interest because of the imminent development of the site which was known to be contaminated (a) with sewage sludge and (b) by sewage effluent which flowed off after deposition of the sludge. The results of this investigation, representing much extensive and detailed work by E. R. Pike, the Leicestershire County Analyst, Miss L. C. Graham (his deputy), and M. W. Fogden, are set out (60, 61) in two publications. Part I deals with zinc, copper, and nickel, Part II with lead, cadmium, arsenic, and chromium. Some of the conclusions will be discussed later.

In consequence, a re-distribution of the proposed sites for houses and gardens, relative to other non-domestic building, was found necessary. The question of the uptake of toxic elements by vegetables also had to be considered. I am greatly indebted to Mr. Pike and his colleagues for sending me reprints of their two papers and a list of references. The microbiological and biomethylation aspects of the subject were not investigated.

The main lines of study pursued by Pike *et al*. included (1) determination of total metal pollution in the soil by (a) sewage sludge deposits and (b) sewage effluent; (2) determination of "available" metal in the soil by extraction with 0.5 M acetic acid. The "non-available" metal is probably held by reaction with the organic matter of the soil. Most of the analyses were carried out by atomic absorption spectrophotometry. A third line of investigation involved the determination of toxic metal content in vegetables grown on both (a) and (b) types of soil, both at Beaumont-Leys and on sites remote therefrom. For details regarding most metals, reference must be made to the original papers, but because of work on the uptake of cadmium by an American oyster (62) and by certain bacteria (63), results obtained for this metal may be given in some detail. At Beaumont-Leys, however, zinc appears to be present in the greatest concentration.

The symptoms of the Itai-Itai disease (Japan, 1939-1945) are described and attributed to the cadmium content of river water that received waste from a zinc, lead, and cadmium mine and which was used for drinking purposes and for irrigation of rice fields. The symptoms are probably due to an inhibition of enzymes requiring zinc. Cadmium hazards in Great Britain have arisen from fumes produced during the welding of cadmium in badly ventilated places. It has been found in foodstuffs and in human tissues, where it may increase with age, causing high blood pressure and cardiac disease. The cadmium content of normal soils has been variously given as 0.01-5.00 ppm. In spite of the high zinc content of soil at Beaumont-Leys, the total cadmium ranged from less than 5 ppm (effluent area) to 50 ppm in the sludge areas. The corresponding figures for available cadmium were less than 5 ppm, rising to 25 ppm, on sludge ground. Those figures were based on a very large number of samples (sludge areas over 1200 samples, effluent areas over 1700 samples).

The ratio of available to total cadmium (30%), and the total cadmium do not alter greatly with the depth of the soil. Lead presents a different picture with a possible available percentage of only 1.0.

The mode of fixation of compounds of metals to protein and humus of the soil and to various acids, e.g. phosphoric, or to silicates would present an interesting subject for a leisurely study beginning with the simplest analogues.

The cadmium content of many garden and allotment soils remote from Beaumont-Leys is little different from that of the soils which have received effluent, but much lower than that in those receiving sludge. As regards the uptake of cadmium by vegetables, a careful study of seven garden varieties grown in soils containing cadmium revealed a content of 0.05 to 0.5 ppm, but usually nearer the lower limit. The figures obtained at Beaumont-Leys and elsewhere may represent a normal cadmium content. Arsenic does not appear to be a problem at Beaumont-Leys. The highest observed figure is 59 ppm and then only in the areas heavily polluted by sludge. Its concentration in the soil of most of the area is similar to that in the gardens and allotments already mentioned as remote from the site. On careful analysis of 10 types of vegetables grown on plots at Beaumont-Leys only traces of arsenic were found in lettuces and radishes and no detectable amounts in the others.

The results at Beaumont-Leys confirm those cited by Monier-Williams (64) that only negligible amounts of arsenic are taken up by plants even from highly arsenical soils.

Lead

The analytical results of a study of the distribution of lead on the Beaumont-Leys site are illustrated by two maps for (1) total and (2) available lead. Land receiving only effluent treatment had a lead content not greatly different from the 200 ppm given by Swaine (65) for natural soil. His studies on the metal content of soils are frequently quoted by Pike *et al*. It seems that it is the content of "available" lead which is significant in the uptake of this metal by plants, much lead becoming insoluble due to reaction with protein and other compounds present in the soil after sludge deposition. The presence of lead appears to be confined to the first two feet of soil, below which the figure approaches that of the normal lead content of the area. The "available" lead can, however, increase with depth, probably owing to a decrease in the organic matter of the soil.

The lead content of the soil has little effect on its uptake by the usual vegetables employed in this study, and in all cases was much less than the 2 ppm allowed for lead in Food Regulations (48, 53, 66). The paper presented to the Symposium by Dr. G. K. Pagenkopf on the transport of ions of transition and heavy metals, mediated by fulvic and humic acids, is very relevant to the question of "available" and "non-available" metals ions to which Pike *et al*. make such frequent reference in their discussion of the Beaumont-Leys soil. Although Pagenkopf appears to be concerned mainly with reactions in aqueous media, his full paper will be read with much interest by all agriculturists.

The Biological Degradation of Organo-derivatives of Mercury

Wood (59), when referring to his well-known work on the methylation of compounds of mercury by methyl-B_{12}, points out that methylation proceeds more rapidly than degradation by other organisms. Nelson, Brinckman *et al*. (67), working in a closed system, find that benzene and mercury vapours are produced from phenylmercury ($C_6H_5HgOCOCH_3$) by cultures of mercury-resistant bacteria. Under natural conditions the mercury so produced may possibly, owing to its volatility, escape into the atmosphere. Details are given of the cultivation of mercury-resistant bacteria on media containing mercuric chloride and phenylmercury acetate, and also references to their identification as to genus. The phenylmercury acetate was labelled with ^{203}Hg and full details of the analyses for Hg and benzene are provided and the apparatus described. References are also given to the conversion of $C_6H_5HgOCOCH_3$ to mercury and diphenylmercury by aerobic bacteria, which also convert CH_3Hg^+ and $C_2H_5Hg^+$ to methane, ethane, and mercury.

The work of Spangler *et al*. (68) has shown that, in lake sediments containing Hg^{++}, formation of CH_3Hg^+ occurs followed by a fall in concentration and formation of metallic mercury. Four purified cultures of what appeared to be a *Pseudomonas* were isolated from the sediment. These also converted CH_3Hg^+ to methane and mercury. These products were identified in the head-space gases by flame ionisation gas chromatography and mass spectrometry, respectively, thus explaining why CH_3Hg^+ has been so difficult to identify in natural sediments or rivers.

Methane formation was only observed with cultures that degraded CH_3Hg^+ under identical conditions.

Somewhat similar results were obtained by Edwards and McBride (69) who studied the biosynthesis and degradation of CH_3Hg^+ in human feces. Formation of methane is mentioned several times but an extended discussion of its origin would have been interesting.

The Natural Occurrence of Methylated Compounds of Arsenic:

Braman and Foreback (70) have reported the occurrence of arsenate, arsenite, methylarsonate, and dimethylarsonate (cacodylate) in many environmental samples including sea-shells, egg shells, natural waters and human urine. The methods of analysis devised for the purpose are discussed and also earlier methods suitable only for total arsenic. Arsenite is reduced to AsH_3 by sodium borohydride at pH 4-9. Arsenate is stable to this reagent and must first be reduced to arsenite by sodium cyanoborohydride at pH 1-2, followed by reduction with $NaBH_4$ at pH 1-2. Methylarsonate and cacodylate gave the corresponding arsines with $NaBH_4$ at pH 1-2.

The AsH_3, CH_3AsH_2, and $(CH_3)_2AsH$ were collected in glass beads cooled in liquid nitrogen and then passed after fractional volatilisation through an electrical discharge giving arsenic emission lines which were examined photometrically.

Methylarsonic acid and/or cacodylic acid were similiarly found in human urine, and the authors suggest that inorganic arsenic is methylated by methylcobalamin or methionine in the body. All these arsenic compounds were found in nanogram quantities.

The methylation of inorganic arsenic to volatile di- or trimethylarsine in the human body has, to the author's knowledge, never been rigidly established. Methylation may, of course, cease at the cacodylic acid state, or the methylarsines may be oxidised in the body. The evidence was discussed in 1945 by Challenger (71), and later work may have escaped his notice. The author would appreciate any references which his colleagues might send him. At the most the amount volatilised must be minute as medicinal doses of arsenite produce hardly any odour, whereas that arising from absorption of traces of tellurite is intense.

Acknowledgements

I need hardly say how much I am indebted to my research collaborators, the results of whose labours for over 20 years, often with unpleasant compounds, have added so greatly to my enjoyment of University life and work.

Dr. D. Barnard has kindly allowed me to mention some of his unpublished work on antimony compounds, and his research for methanol in mould cultures [Ph.D. Thesis, University of Leeds]. My thanks are also due to Dr. P. A. Briscoe and the late Dr. J. W. Baker for valuable discussions on theoretical points. Discussion in this review on the development of ideas on biological methylation is based on parts of pages 164 and 170-173 of the author's book "Aspects of the Organic Chemistry of Sulfur". I am indebted to my publishers, Messrs. Butterworths, London, for permission to use this material.

In conclusion, may I once more thank the organisers of this Symposium for asking me to write this review, and particularly Dr. Brinckman, Professor Wood and Professor Thayer for the encouragement and help they have accorded me during its preparation.

References

1. Challenger, F., Obituary Notices of Fellows of the Royal Society, (1950) 7, 183.
2. Challenger, F., J. Chem. Soc. (1951) 849.
3. Challenger, F., Higginbottom, Miss C., and Ellis, L., J. Chem. Soc. (1933) 95.
4. Gosio, B., Arch. Ital. Biol., (1901) 35, 201.
5. duVigneaud, V., "A Trail of Research". Cornell Univ. Press, Ithaca, New York.
6. Ingold, C.K., Shaw, F.R., and Wilson, C., J. Chem. Soc. (1928), 1280.
7. Challenger, F., "Aspects of the Organic Chemistry of Sulphur". 165. Butterworths, London, 1959.
8. Challenger, F., and North, H.E., J. Chem. Soc. (1934) 68.
9. Bird, M.L., and Challenger, F., J. Chem. Soc. (1939) 163.
10. Challenger, F., and Rawlings, A.A., J. Chem. Soc. (1937) 868.
11. Challenger, F., and Charlton, P.T., J. Chem. Soc. (1947) 424.
12. Blackburn, S., and Challenger, F., J. Chem. Soc. (1938) 1872.
13. Birkinshaw, J.H., Findlay, W.P.K., and Webb, R.A., Biochem. J. (1942) 36, 526.
14. Bird, M.L., Challenger, F., Charlton, P.T., and Smith, J.O., Biochem. J. (1948) 43, 78.
15. Challenger, F., Quart. Rev. Chem. Soc. (1955) 9, 255.
16. Hansen, A., Ann. der Chemie (1853) 86, 213.
17. Riesser, O., Z. physiol. Chem. (1913) 86, 440.
18. Challenger, F., Taylor, P., and Taylor, B., J. Chem. Soc. (1942) 48.
19. Challenger, F., and Higginbottom, C., Biochem. J. (1935) 29, 1757.
20. McBride, B.C., and Wolfe, R.S., Biochemistry (1971) 10, 4312.
21. Cantoni, G.L., J. Amer. Chem. Soc. (1952) 74, 2942.

22. See reference 7.
23. See reference 15.
24. Challenger, F., J. Soc. Chem. Ind., London (1935) 54, 657.
25. Ridley, W.P., Dizikes, L.J., and Wood, J.M., Science (1977) 197, 329.
26. Ridley, W.P., Dizikes, L.J., Cheh, A., and Wood, J.M., Environ. Health Perspectives (1977) 19, 43.
27. Dransfield, P.B., and Challenger, F., J. Chem. Soc. (1955) 1153.
28. Challenger, F., Lisle, D.B., and Dransfield, P.B., J. Chem. Soc. (1954) 1760.
29. Wright, J., Ind. & Eng. Chem., (1927) 19, 750.
30. Barnard, D., Thesis, Univ. of Leeds, (1947) 92-93.
31. Kieffer, F., Unpublished observations.
32. Axelrod, J., and Daly, J., Science (1965) 150, 892-3.
33. Axelrod, J., in "Transmethylation and Methionine Biosynthesis", (Shapiro, S.R. and Schlenk F., eds.) Univ. of Chicago Press, Chicago, 1965, pp. 71-83.
34. Challenger, F., Conférences et Rapports, 3ème Congrès Intern. de Biochimie, Bruxelles (1955) 238.
35. Knaffl-Lenz, Arch. exp. Path. Pharm., (1913) 72, 224.
36. Challenger, F., and Ellis, L., J. Chem. Soc. (1935) 396.
37. Barnard, D., Thesis, Univ. of Leeds, (1947) 16-29.
38. Parris, G.E. and Brinckman, F.E., J. Org. Chem. (1975) 40, 3801.
39. Parris, G.E., and Brinckman, F.E., Environ. Sci Technol. (1976) 10, 1128.
40. Diplock, A., Critical Review in Toxicology, Chemical Rubber Co., Cleveland, Ohio (1976) 271.
41. Cerwenka, E.A., and Cooper, W.C., Archiv. Environ. Health (1961) 3, 189.
42. McConnell, K.P., and Portman, O.W., J. Biol. Chem. (1952) 195, 277.
43. Byard, J.L., Arch. Biochem. Biophys. (1969) 130, 556.
44. Palmer, I.S., Fischer, D.D., Halverson, A.W., and Olson, O.E. Biochem. Biophys. Acta (1969) 177, 336.
45. Palmer, I.S., Gunsalus, R.P., Halverson, A.W., and Olson, O.E., Biochem. Biophys. Acta (1970) 208, 260.
46. McBride, B.C., and Wolfe, R.S., Biochemistry (1971) 10, 2317.
47. Taylor, C.D., and Wolfe, R.S., J. Biol. Chem. (1974) 249, 4879.
48. Taylor, C.D., and Wolfe, R.S., J. Biol. Chem (1974) 249, 4886.
49. Taylor, C.D., McBride, B.C., Wolfe, R.S., and Bryant, M.P., J. Bacteriol. (1974) 120, 974.
50. Frick, T., Francia, M.D., and Wood, J.M., Biochem. Biophys. Acta (1976) 428, 808.

51. Challenger, F., Biochem. J. (1970) 117, 65P.
52. Challenger, F., Komrower, G.M., and Robins, A.J., Quart. Rep. Sulfur Chem. (1970) 5, 91.
53. Dagley, S.J., and Nicholson, D.E., "An Introduction to Metabolic Pathways", Blackwell, Oxford (1970), p.212.
54. Tallan, H.H., Sturman, J.A., Pascal, T.A., and Gaull, G.E., Biochem. Med. (1974) 9, 90.
55. Datko, A.H., Mudd, S.H., Giovanelli, J., J. Biol. Chem. (1977) 252, 3436.
56. Kelly, A.P., and Weed, L.L., J. Biol. Chem. (1965) 240, 2519.
57. Braun, R., and Fromageot, P., Biochem. Biophys. Acta (1962) 62, 548.
58. Bonsen, P.P.M., Spudich, J.A., Nelson, D.L., and Kornberg,A., J. Bacteriol. (1969) 98, 62.
59. Wood, J.M., Naturwissensch. (1975) 62, 357 (Note especially page 358.)
60. Pike, E.R., Graham, L.C., and Fogden, N.W., J. Assoc. Publ. Analysts (1975) 13, 19.
61. Pike, E.R., Graham, L.C., and Fogden, N.W., J. Assoc. Publ Analysts (1975) 13, 48.
62. Zaroogian, G.E., and Cheer, S., Nature (1976) 261, 408.
63. Doyle, J.J., Marshall, R.T., and Pfander, W.H., Appl. Microbiol. (1975) 29, 562.
64. Monier-Williams, G.W., "Trace Elements in Foods", Chapman and Hall, London (1949), p. 168.
65. Swaine, D.N., Comm. Bur. Soil Sci., Tech. Bull. No. 48, (1955).
66. The Lead in Food Regulations, S.I. 1961, No. 1931, H.M.S.O. (London) 1961.
67. Nelson, J.D., Blair, W., Brinckman, F.E., Colwell, R.R., and Iverson, W.P., Appl. Microbiol. (1973) 26, 321.
68. Spangler, W.J., Spigarelli, J.L., Rose, J.M., and Miller, H.M., Science (1973) 180, 192.
69. Edwards, T., and McBride, B.C., Nature (1975) 253, 462.
70. Braman, R.S., and Foreback, C.C., Science (1973) 182, 1247.
71. Challenger, F., Chem. Rev. (1945) 36, 326.
72. Thayer, J.S., "Organometallic Chemistry: A Historical Perspective." Advances in Organo-Metallic Chemistry, (Stone, F.G.A., and West, R., eds.), vol. 13, Academic Press, New York (1975), p. 1.
73. Thayer, J.S., "Organometallic Compounds and Living Organisms" J. Organometal. Chem. (1974) 76, 265.
74. Thayer, J.S., "Teaching Bio-Organometal Chemistry. I. The Metalloids." J. Chem. Ed. (1977) 54, 604.
75. Thayer, J.S., "Teaching Bio-Organometal Chemistry. II. The Metals. J. Chem. Ed. (1977) 54, 662.

Discussion

J. S. THAYER (University of Cincinnati): I would like to add just a bit of historical perspective in pointing out that Challenger's work had origin in the real problem that "Gosio-gas" was emitted in poorly ventilated housing, and quite a number of people died or suffered from arsenic poisoning. In England, in 1930, there was a rather notorious case that led directly to Challenger's interest in this area. He tried to methylate mercury using this same route and unfortunately failed by a very narrow margin; otherwise, the situation that developed in the 50's and later might have at least been anticipated.

J. M. WOOD (University of Minnesota): In 1945, Challenger wrote a review [Chem. Rev. 36, 315 (1945)] which is a masterpiece, not only in terms of the chemistry, but also in terms of the history of the Gosio-gas poisoning cases. [It described the] history of children in the Forest of Dean who were poisoned by trimethylarsine when a farmer put some arsenic compounds in a particular area in the forest. He gives details of the clinical history as well as the chemistry. That review is almost as long as his present paper, but it's worth the effort to sit down and read it.

W. R. CULLEN (University of British Columbia): Yes, I can add that the reference to arsenic in concrete was that the house the children were living in had arsenic in the fly ash used for the concrete. This is what the molds worked on to produce trimethylarsine.

F. E. BRINCKMAN (National Bureau of Standards): From Challenger's history and commentary on his nucleophilic reaction shown in Figure 2, I'd like to address a simple question that's of great interest to us. In the final step he points out formation of trimethylarsine, which is the observed volatile product. There is a trimethylarsine oxide intermediate step and this of course involves the two electron reduction to form ultimately the observed volatile species. Dr. Parris and I, several years ago, looked at the rates of oxidation of trimethylarsine in aqueous media and in air [Environ. Sci. Technol. (1976), 10, 1128]. It was fairly clear that while comparatively slow as a chemical reaction, it proceeded rapidly compared to biological rates of formation, at least our estimates of biological rates of formation. I'm rather curious if anybody has a comment on that final reduction step, whether that is an abiotic or whether it is a biogenic step.

CULLEN: I think I can answer that question. We have actually used trimethylarsine oxide as a substrate for *Candida humicola* and have obtained trimethylarsine from it. So there is some fungal reduction anyway. I think we can comment further on that when Professor McBride speaks later today.

F. CHALLENGER (*in absentia*): Regarding the matter of possible biomethylation of antimony, the question of its charge in tartar emetic bears further note. Reihlen and Hezel [Ann. (1931) 487, 213], after much experimental work, have proposed a formula for tartar emetic

$$\left[H_2O \ldots Sb \begin{matrix} \diagup O \text{———} C = O \\ \text{———} O \text{———} \overset{|}{C}H \\ \diagdown O \text{———} \overset{|}{C}HCOO^- \end{matrix} \right] \qquad K^+$$

in which antimony appears in the anion. This work escaped my notice until recently. The few experiments of Barnard were insufficient to allow for a discussion of the implications of this formula for the possible methylation of antimony. [cf. paper by G.E. Parris in this volume for additional discussion of charge on Sb and potential for methylation - Eds.]

F. CHALLENGER (*in absentia*): Regarding methylation of inorganic arsenic in the body, a further comment is appropriate. In a series of important studies, Crecelius [Environmental Health Perspectives (1977) 19, 147] demonstrated, for example, that human ingestion of As^{3+}-rich wine results within 5-10 hours in about a 5-fold increase in urinary levels of As^{3+} and As^{5+}, methylarsonic acid (MAA), and dimethylarsonic acid (DMAA). Most (~ 80%) of the ingested arsenic is excreted in urine over several days. Other experiments reveal that both As^{3+} and As^{5+} are excreted from the body, chiefly in methylated forms. Ingestion of "marine arsenic" in crab meat results in urinary elimination, but the organically-bound As entering the body apparently undergoes no methylation. In all of these studies, the reductive volatilization method of Braman and Forelock [cf. reference 70] was used. Consequently, my earlier statement that methylation of inorganic arsenic to volatile di- or trimethylarsine in the human body is not established cannot yet be revised.

RECEIVED September 15, 1978.

2

Biotransformations of Sulfur as Evolutionary Prototypes for Metabolism of Metals and Metalloids

GEORGE E. PARRIS

Division of Chemical Technology HFF-424, Food and Drug Administration, Washington, DC 20204

Sulfur, like phosphorus, is an abundant element in the earth's crust and sulfur compounds became necessary nutrients early in the process of evolution. Some non-metals and metalloids (e.g., Se, Te, As) have chemical properties similar to metabolicly active forms of sulfur and/or phosphorus. There is always a possibility that an organism will encounter an element, compound or ion in its environment which will act as a pseudo-sulfur or pseudo-phosphorus nutrient. Pseudo-nutrients may or may not be actively toxic, but they can inhibit cell functions as they compete with true nutrients in respiration and assimilation.

It is observed that pseudo-nutrients are often transformed *in vivo* into products which no longer encumber the cell's function. Ideally, such a transformation will produce a material which has few labile valences capable of combining with enzymes, nucleic acids or constituents of the cytoplasm, is not itself a mimic of a normal metabolite, and has low polarity so that it can passively diffuse directly out of the cytoplasm through the cell membrane. Whether these transformations are merely serendipitous or are evolved specifically for the purpose of dealing with pseudo-nutrients may be debatable.

Because the analogy between the chemistry of S, Se and Te is rather obvious and has been reviewed at length (1), attention in this manuscript is focused on arsenic and other elements where current understanding is relatively incomplete.

Evolution of Respiration Chains

It is convenient to first consider biotransformation of sulfur in the respiration of cells which obtain energy from the oxidation of sulfur compounds. In 1945, Horowitz (2) stated a hypothesis which is summarized as follows: The primordial organism is assumed to be capable of a simple form of respiration

0-8412-0461-6/78/47-082-023$05.00/0

involving one enzyme. Within the environment of the organism

$$\text{substrate a} \xrightarrow{\text{Enzyme A}} \text{products + energy}$$

there are other substances such as substrate "b" which could be converted to substrate "a", but the primordial organism cannot utilize substrate "b". There may be, however, spontaneous mutants of the primordial strain which possess the genetic blueprint for "Enzyme B" which could convert "b" to "a" intracellularly, perhaps even with a small release of useful energy. Under the stress of Malthusian growth, the primordial organisms become so numerous that substrate "a" is consumed as fast as it enters the organisms' environment. When the availability of substrate "a" becomes critical, the basic strain of the primordial organism becomes extinct and is replaced by a new organism capable of the biochemical transformations. Thus, a chain of respiratory enzymes is

$$\text{b} \xrightarrow{\text{B}} \text{a} \xrightarrow{\text{A}} \text{products + energy}$$

gradually built up. Generally, the primeval enzyme systems will be retained in more highly evolved and specialized organisms. However, a new group of enzymes, F, G, H, etc., may evolve that is capable of carrying out these or other energy-releasing reactions without resorting to use of the primeval enzyme system A, B, C, etc. In such a case, an entirely new respiratory system may evolve as the now-superficial primeval system is lost piecemeal in subsequent mutations (3,4,5). Unlike the development of a

$$\text{h} \xrightarrow{\text{H}} \text{g} \xrightarrow{\text{G}} \text{f} \xrightarrow{\text{F}} \text{products + energy}$$

respiration sequence in an evolving family of organisms, the introduction of a defensive detoxification system can logically spring from any of the existing enzymes. Also, the organisms will seldom be so lucky as to be able to derive any net useful energy from detoxification and it is likely that detoxification will actually amount to a considerable tax on the normal energy pools of the organism.

There are two types of biochemical oxidation reactions: dehydrogenations and oxygen insertions. Dehydrogenations are believed to have evolved in the time before molecular oxygen was available in quantity in the atmosphere. Oxygen insertion reactions could not have been of much importance until after the evolution of photosynthetic blue-green algae. The dehydrogenation process of most organisms are similar. Typically, an organism has a respiratory chain which facilitates dehydrogenation of some substrate (DH_2) (6). Each time the hydride ($H:^-$ or $H\cdot + e^-$) is

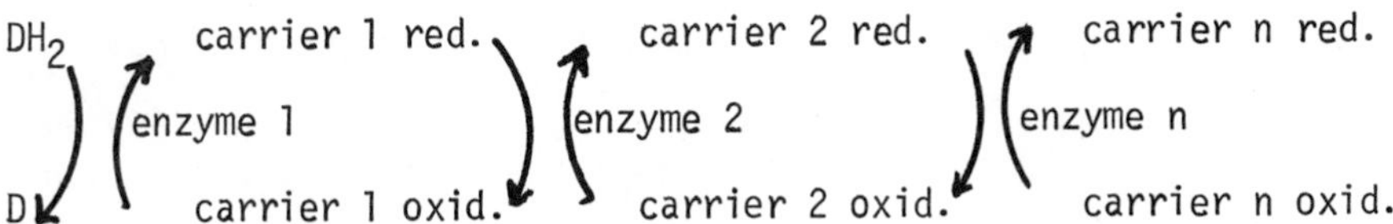

transferred, chemical potential is made available to the cell. This chemical potential is utilized by coupling the oxidation steps to phosphorylation steps which ultimately form ATP from ADP + P_i (inorganic phosphate).

The first electron carriers in the respiration chain are adapted to the type of substrate being oxidized (e.g., dehydrogenated). Nicotinamide adenine dinucleotide (NAD^+, Figure 1) is a common receptor which functions as a coenzyme in the oxidation of many organic compounds. NAD^+ can be reduced to NADH by transfer of a hydride ($H:^-$) to the 4-position of the nicotinamide. A very similar coenzyme $NADP^+$ has a third phosphate group at the C-2 hydroxyl group of the ribose moiety of the adenine portion of the ion. There are different enzymes and coenzyme receptors for oxidation of different sulfur species (e.g., sulfur, sulfide, thiosulfite, sulfite). Once the electrons have been removed from the substrate, they can be passed into a carrier chain which is made up of a series of cytochromes. Cytochromes are complexes of metals (such as iron or copper) which have convenient oxidation potentials and favorable electron transfer kinetics for enzymatic coupling to phosphorylation reactions producing ATP. As discussed above, the major elements of this chain may be common to many species. Complexes of

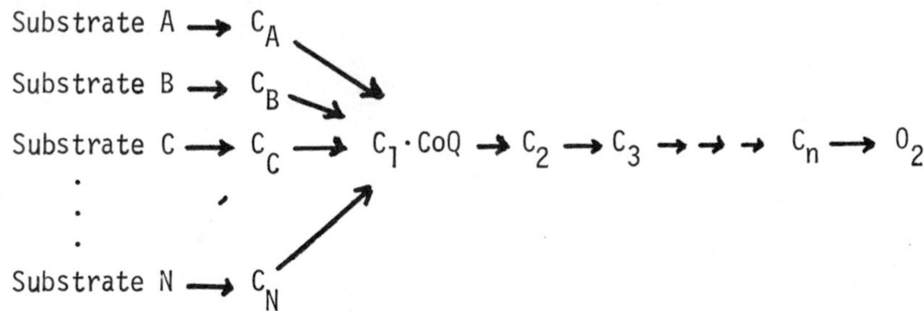

ubiquinones (coenzyme Q) with cytochromes are known to be the focal point for electrons arriving from different substrate branches of respiratory chains (7).

Respiration in Thiobacilli

There are several genera of bacteria which oxidize inorganic sulfur compounds. The genus *Thiobacillus* has been the most thoroughly studied. *Thiobacilli* obtain all their carbon from CO_2 and all their energy from reaction of various reduced forms of sulfur (S_8, S^{2-}, SO_3^{2-}, $S_2O_3^{2-}$) with oxygen or an oxygen donor (NO_3^-) to form oxidized forms of sulfur (SO_4^{2-})(8).

Thiobacillus thiooxidans derives its energy autotrophicly by oxidizing elemental sulfur to sulfate. The efficiency with which

$$1/8\ S_8 + 3/2\ O_2 + H_2O \rightleftharpoons H_2SO_4$$

$$\Delta G = -596\ kJ$$

this energy fixes carbon to form cellular material has been calculated to be about 6%. Nearly 32 gm of sulfur are oxidized for every gram of carbon assimilated into cell components. Cytochromes and coenzyme Q_8 have been identified in T. thiooxidans and it has been observed that low concentrations of cyanide, azide or carbon monoxide (all potent cytochrome inhibitors) block sulfur oxidation in this organism. The amount of CoQ present was shown to be comparable to that in bacteria which are not specialized sulfur oxidizers. Thus, the cytochrome-linked phosphorylation was concluded to be an integral part of the respiratory machinery of the organism (9,10).

Thiobacillus thioparus, unlike T. thiooxidans, does not appear capable of using native sulfur as a substrate since when it is grown on thiosulfate, it accumulates elemental sulfur and produces sulfate. A cytochrome (cyt s, tentatively classified as a c-type cytochrome) has been isolated from T. thioparus. Nonetheless, evidence seems to indicate that the classical cytochrome-linked oxidative phosphorylation chain may have been supplemented, in the course of evolutionary specialization, by a direct oxidation sequence which involves formation of high energy AP∿S (APS, adenosine 5'-phosphosulfate). Notice that in

$$(1)\quad S_2O_3^{=} + 2\ H^+ + 2\ e^- \xrightarrow[\text{reductase}]{\text{thiosulfate}} SO_3^{=} + H_2S$$

$$(2)\quad 2\ H_2S + O_2 \xrightarrow{\text{sulfide oxidase}} 2\ S + 2\ H_2O$$

(3) $SO_3^{=}$ + AMP —2 e⁻→ APS-reductase → $APS^{=}$; cyt (ox) —2 e⁻→ cyt (red)

reaction 3 a high energy sulfate bond has been formed in $APS^{=}$ analogous to ADP . The hydrolysis of APS to AMP + S_i is said to be about 42 kJ more exothermic than the hydrolysis of ADP to AMP + P_i. While for some energy requiring functions of the cell, $APS^{=}$ may be used directly, a more conventional view indicates that $APS^{=}$ is primarily used as a substrate to form ADP . Ultimately,

$$(4)\quad APS^{=} + P_i \xrightarrow{\text{ADP-sulfurylase}} ADP + S_i^{=}$$

the conventional cellular energy source ATP is formed. Peck and

$$(5)\quad 2\ ADP \xrightarrow[\text{kinase}]{\text{adenylic}} AMP + ATP$$

Fisher and Stulberg noted that reactions 3 and 4 may be capable of supplying the necessary ATP for all cellular energy requirements (11,12). However, it is unlikely that cytochrome-linked phosphorylation (initiated in reaction 3) is completely supplanted (13,14).

Involvement of Arsenic in Respiratory Processes

Arsenite is not only a pseudo-sulfite nutrient, it is very toxic to cells because it combines indiscriminately with thiol groups causing mass disruption in metabolism. Arsenate is a pseudo-phosphate nutrient, but since it does not readily react with thiols, it only "clogs-up" certain phosphate-processing enzymes rather than rendering enzymes nonfunctional. Thus, the oxidation of arsenite to arsenate can be viewed as a defensive mechanism.

The first report of bacterial oxidation of arsenite to arsenate dates to 1918 when an organism known as Bacillus arsenoxydans was observed in cattle dipping preparations based on arsenite. However, the organism was never studied systematicly and has since been lost. In 1949 Turner (15,16) reinvestigated organisms responsible for the oxidation of arsenite to arsenate in sheep dip and isolated fifteen strains of heterotrophic bacteria which both tolerated up to 0.1 M arsenite and accomplished the oxidation to arsenate. The bacteria were believed to be strains of Pseudomonas. The biochemical functions of one strain were studied. An optimal pH for cell growth was found to be between 6.0 and 6.5 where arsenite is protonated (pK_a 9.12). The optimum oxidation rate was 90 μM arsenite per hour per mg dry weight of cells. Oxidation was completely inhibited outside pH 3-11. The arsenite oxidation was inhibited by cyanide, azide and carbon monoxide. The inhibition by CO was only in the dark. These observations are typical of cytochrome electron transfer systems. In particular, they are suggestive of the so-called P-450 cytochrome system. The reversible inhibition by CO is due to the equilibrium below. The active site of P-450 is believed to

$$\text{P-450}(Fe^{2+}) \underset{h\nu}{\overset{CO}{\rightleftharpoons}} \text{P-450}(Fe^{2+})\cdot CO$$

involve a pentacoordinate iron moiety which has been recently discussed (17,18,19). While electrons given up by As(III) in this process are apparently funneled into the normal cytochrome respiration chain, there was no evidence that any useful energy is derived from arsenite oxidation as measured by growth rate of cultures. Turner regarded the transformation as purely defensive.

In a study related to biological oxidation of arsenite to arsenate, Johnson and Pilson (20) have examined oxidation of arsenite in sterilized sea water. The sterilization was accomplished by filtration (0.1 micron) to remove live cells. Thus, the sea water can be regarded as a dilute cell-free extract since rupture of nonviable cells in a population under natural conditions undoubtedly frees a whole range of enzymes. The observed abiotic (i.e., extra-cellular) oxidation rate of As(III) was sufficient to account for most of the arsenic in sea water each year (0.023 micromoles As(III)/l·yr).

At this time no organism has been found which oxidizes arsenite for release of useful energy. Arsenate is an uncoupler of oxidative phosphorylations. It is believed that in (unadapted) cells arsenate is mistaken for phosphate and unstable ADP∿As products are formed which are hydrolyzed before they can be used as ATP (ADP∿P) analogues. In the thiosulfate oxidizing bacteria *T. thioparus*, arsenate was believed to function in place of phosphate in the liberation of $SO_4^=$ from APS (see Reaction 4).

$$[Ado\text{-}OPO_2OSO_3]^= + AsO_4 \xrightarrow{\text{ADP-sulfurylase}} AMP{\sim}As + SO_4^=$$

There has apparently been no investigation of the possible oxidation of arsenite by *T. thioparus* in analogy to sulfite oxidation. Thermodynamicly the oxidation of sulfite is slightly more favorable than the oxidation of arsenite.

$$H_2SO_3 + 1/2\ O_2 \longrightarrow H_2SO_4 \qquad \Delta G = -205\ kJ$$

$$H_3AsO_3 + 1/2\ O_2 \longrightarrow H_3AsO_4 \qquad \Delta G = -130\ kJ$$

Reduction of Sulfate

Adenosine-5'-phosphosulfate (APS) also plays a central role in the assimilation of sulfate (S_i) (14). Reaction 6 lies very

$$(6)\quad ATP + S_i \underset{\text{ATP-sulfurylase}}{\rightleftharpoons} APS + PP_i$$

much in favor of ATP + S_i ($\Delta G \simeq +46$ kJ). Some of this energy can be recouped *in vivo* by hydrolysis of PP_i. Regardless, APS is

$$(7)\quad PP_i^= + H_2O \xrightarrow[\text{pyrophosphatase}]{} 2\ P_i^= + 2\ H^+$$

trapped in a usable form for assimilatory reactions by phosphorylation of the 3' hydroxyl. Reaction 8 is favored by

$$(8)\quad APS + ATP \underset{\text{APS-kinase}}{\rightleftharpoons} PAPS + ADP$$

about 21 kJ. Reactions 6 and 8 are found in all cells which utilize S_i. 3'-Phosphoadenosine-5'-phosphosulfate (PAPS, Figure 2) can be reduced by NADPH to sulfite.

$$(9)\quad PAPS + NADPH \xrightarrow[\text{PAPS-reductase}]{} PAP + SO_3^{=} + NADP^+ + H^+$$

PAPS is also the sulfur source in formation of sulfate esters.

$$R\text{-}OH + PAPS \underset{\text{sulfokinase}}{\rightleftharpoons} PAP + ROSO_3^- + H^+$$

An inorganic reduction of sulfite to sulfide has been postulated by Woolfork (21):

$$(10)\quad 2\,HSO_3^- \rightleftharpoons {}^-O{-}S(=O)_2{-}S(=O){-}O^- + H_2O$$

(metabisulfite)

$$(11)\quad S_2O_5^{2-} + H_2 \longrightarrow {}^-O{-}S(=O){-}S(=O){-}O^- + H_2O$$

(dithionite)

$$(12)\quad S_2O_4^{2-} + H_2 \longrightarrow S_2O_3^{2-} + H_2O$$

(thiosulfate)

$$(13)\quad S_2O_3^{2-} + H_2 \longrightarrow HSO_3^- + HS^-$$

Similar pathways involving trithionate $S_3O_6^{2-}$ have also been proposed (22). An enzymatic scheme for sulfate reduction to enzyme bound -SH via PAPS has been outlined (23).

Reduction of Arsenate

Little if any biochemical experimentation has been directed at biological reduction of arsenate, or formation of arsenic hydrogen bonds. Many sulfate reducing bacteria (Desulfovibrio) are known. The production of hydrogen sulfide is usually observed under anaerobic conditions. Reduction of arsenate to arsenite commonly occurs in aerobic media, though formation of $(CH_3)_3AsH$ has only been reported under anaerobic conditions (24,25,26). Johnson and Pilson (20,27) have pointed out that the presence of arsenite in sea water is undoubtedly due to activity of arsenate reducers. The absence of any organism which produces AsH_3 is probably due to the fact that this compound is one of the most potent biochemical toxins known.

Alkylation of Metals and Metalloids

With regard to the criteria listed in the introduction for an ideal detoxification process, biological methylation is highly desirable for bacteria and fungi. Not only are carbon-metal (metal = Sn,Pb,Sb,Hg) and carbon-metalloid (metalloid = Se,Te,As, Si) bonds usually unreactive towards -OH, -SH and $-NH_2$ groups in the cytoplasm of cells, organometal and organometalloids are frequently lipophilic enough to diffuse through the cell's membrane and escape more or less permanently from the cell's environment. In complex organisms, lipophilicity leads to undesirable toxic effects since fat-soluble compounds breach the body's compartmentalization of sensitive systems.

The non-methyl carbon-sulfur bonds in proteins are formed by reactions in which carbon-oxygen bonds are displaced. The sulfur nucleophile is generally sulfide ($S^=$) or hydrosulfide (HS^-)(14). These powerful nucleophiles are produced intracellularly by enzymatic reduction of sulfite. NADPH is the most common reducing coenzyme. Little is known about the details of the reaction because the sulfur intermediates are usually protein bound. Thiosulfate may be involved. Reactions 11 and 12 account for the introduction of sulfide into proteins.

(14) HO-protein + acetylCoA $\xrightarrow[\text{enzyme}]{}$ AcO-protein

(15) AcO-protein + H_2S $\xrightarrow[\text{enzyme}]{}$ HS-protein + AcOH

The formation of methyl-sulfur bonds is a common biochemical transformation and it appears that biomethylation of metals can be traced to the methylation of sulfur. The biosynthesis of methionine from aspartate by way of cysteine has been outlined by Mahler and Cordes (28). For the current discussion, only the last major transformation in which homocysteine is methylated to form methionine is important.

$HSCH_2CH_2CHNH_3^+CO_2^-$ — homocysteine

$CH_3SCH_2CH_2CHNH_3^+CO_2^-$ — methionine

There are at least two enzymatic pathways by which this methylation occurs. In species of organisms which lack vitamin B_{12}, there are relatively inefficient transmethylases which bind methyl donor poly-L-glutanate-N^5-methyltetrahydrofolate (N^5-methyl-H_4-folate (Glu_1,Glu_2, etc.) Figure 3) and homocysteine and promote methyl transfer to form methionine and tetrahydrofolate (H_4-folate). Most enzyme systems of this type do not function with the mono-L-glutamate-N^5-methyl-H_4-folate. The main function of the glutanate moieties appears to be to provide a suitable binding site for the methyl donor to the enzyme. There is little if any information available concerning the labilization of the methyl group in this transfer. The turnover rate for one of these direct transmethylase reactions is only 14 moles of N^5-methyl groups per minute per mole of enzyme (0.25 mole/sec·mole enzyme)(29).

The second and more efficient transmethylase enzyme system depends upon vitamin B_{12}. Pratt has provided an extensive review of vitamin B_{12} chemistry (30). In vitamin B_{12}, cobalt is complexed by a porphyrin-like corrin ring. The fifth coordination site is occupied by a 5,6-benzimidazole (Bz) ring and the sixth site can be occupied by a methyl group or a variety of other ligands. Unlike the direct methyltransferase described above, the CH_3-B_{12} mediated methyl transfer process requires a reducing (anaerobic) medium. The CH_3-B_{12}-enzyme system transfers methyl groups from N^5-methyl-H_4-folate (Glu_1, etc.) to homocysteine at a steady rate of about 800 moles per min per mole of enzyme (13 mole CH_3/sec·mole enzyme)(31).

Weissbach and Taylor (32) summarized the work of a number of others to describe the role of vitamin B_{12} in the synthesis of methionine. The scheme which evolved from a variety of experimental observations is outlined in Figure 4.

Here "ENZ" stands for the "N^5-methyltetrahydrofolate-homocysteine cobalamin methyltransferase" which has a molecular weight of about 140,000 and has usually been isolated from strains of E. coli. The enzyme itself may or may not be involved in any other biomethylation reactions; but the overall scheme, where "ENZ" stands for some other enzyme and different methyl donors and receivers are involved, appears to be adaptable to many situations where the methyl group is donated as an incipient carbon cation producing a reduced form of the B_{12}-enzyme complex (B_{12s}). The mode of B_{12}-ENZ binding in the methionine synthetase system has been discussed by Law and Wood (33) who point out that the photostability of the CH_3-Co bond in this CH_3-B_{12}-ENZ complex weighs against a sulfur ligand replacing the benzimidazole base in spite of the fact that CH_3-B_{12} will not bind to the enzyme if all the enzyme's sulfhydryl groups have been previously blocked. The role of S-adenosylmethionine (SAM), in this scheme, is to provide a relatively high-energy methyl donor which, in the presence of a suitable reducing agent, will methylate the oxidized, inactive, enzyme-bound B_{12}. SAM will also methylate the reduced-reactive, enzyme-bound B_{12} moiety but it is more than 10^2 times slower than

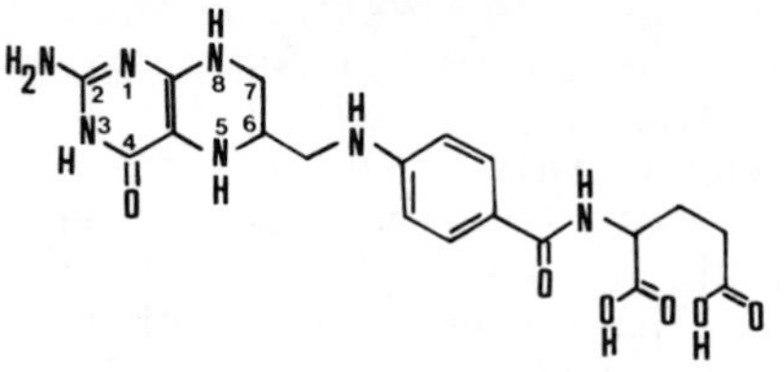

Figure 1. Nicotinamide adenine dinucleotide (NAD^+)

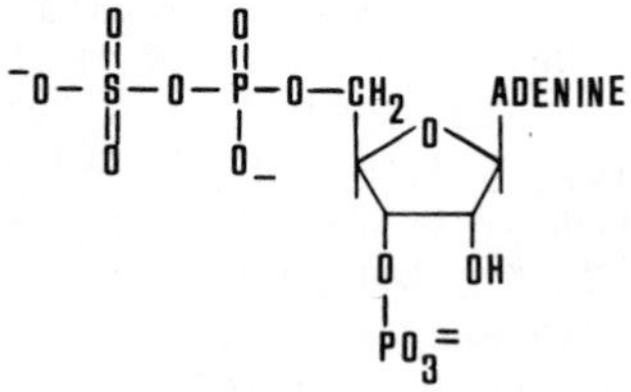

Figure 2. 3′-Phosphoadenosine-5′phosphosulfate (PAPS)

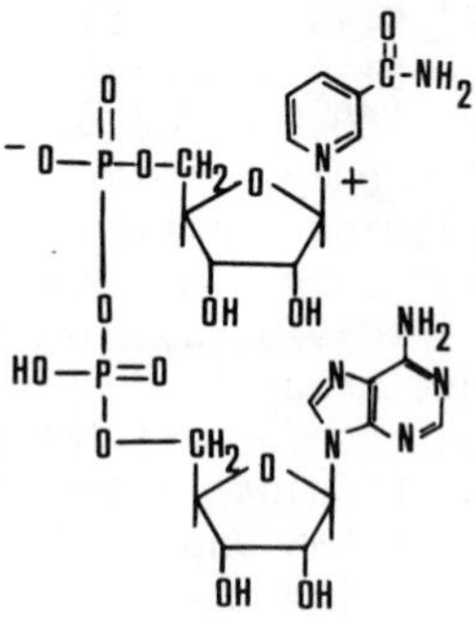

Figure 3. Tetrahydrofolic acid

N^5-methyl-H_4-folate in this reaction. Thus, in the presence of N^5-methyl-H_4-folate, little SAM is consumed except to regenerate the active methyl-B_{12}-ENZ from the oxidized form. It is noteworthy that SAM can be replaced by other alkylating agents such as CH_3I in the presence of a suitable reducing agent (34).

Formation of Carbon-Metal and Carbon-Metalloid Bonds

Alkylcobalamins are attractive models for biological alkyl donors where the alkyl receiver is a metal or metalloid. Depending upon the ability of the ligands (e.g., enzyme) bound to cobalt to stabilize the product oxidation state (Co^I, Co^{II} or Co^{III}), the alkyl-cobalt bond can be broken in three ways (35,36, 37):

$$R\text{-}B_{12} \longrightarrow \begin{cases} R^+ + B_{12s}(Co^I) \\ R\cdot + B_{12r}(Co^{II}) \\ R^- + B_{12}(Co^{III}) \end{cases}$$

The ability of CH_3-B_{12} to yield an incipient methyl anion may mean that cobalamin plays a unique role as a coenzyme in the biomethylation of metals in high oxidation states (e.g., Hg^{2+}, Sn^{4+}, Sb^{5+}, Pt^{4+}, Pb^{4+}).

In vitro transmethylations involving CH_3-B_{12} have been studied extensively. In these cases the methyl is transferred as a carbon anion in an S_E2 type mechanism. The overall reaction

$$M'\text{-}CH_3 + M^+ \rightleftharpoons \left[M \cdots CH_3 \cdots M\right]^+ \rightleftharpoons M'^+ + CH_3\text{-}M$$

has an interesting complication in that the metals which have been studied as methyl receptors compete with the cobalt for the benzimidazole ligand. This reaction has an equilibrium constant

$$\underset{\text{base-on}}{CH_3\text{-}B_{12}\text{-}Bz} + M \rightleftharpoons \underset{\text{base-off}}{CH_3\text{-}B_{12}(\text{-}Bz\text{-}M)}$$

on the order of 10^2 and the "base-off" complex is formed more rapidly than methyl transfer from the "base-on" complex to the metal M. The "base-off" complex is probably a cul-de-sac since, with the electron donating benzimidazole group gone, the cobalt would develop a very high effective positive charge density if the methyl group left as an anion. Thus, the ultimate productive reaction is hampered by the diversion of the reactive substrate (CH_3B_{12} base-on)(38). Hughes has discussed the observation that the alkylcobalamins behave as though they are Co(II) rather than

Co(III) since ligand exchange at the fifth coordination site is rapid and there appears to be an equilibrium between 5 and 6 coordinate species in the alkylcobalamins. Normally, Co(III) has very slow ligand exchange and it is strictly 6 coordinate (39).

There is a host of potential coenzymes for transfer of incipient alkyl cations to metals and metalloids in low oxidation states (e.g., As^{3+}, Sb^{3+}, Sn^{2+}, Pt^{2+}, Hg^0, Pb^{2+}, divalent S, Se and Te, and I^-). The mechanism most likely to account for alkyl transfer in these cases can be called "oxidative addition" and presumably is related to the well-known non-enzymatic S_N2 mechanisms (40). In this reaction a leaving group (X) on carbon is displaced with its electrons as a nucleophile (N) forms a new

$$N: \quad R_3C\text{—}X \rightleftharpoons \left[N \cdots CR_3 \cdots X \right]^{\neq} \rightleftharpoons \overset{+}{N}\text{—}CR_3 + X^-$$

bond to carbon. The transition state is stabilized by polarizable N and X groups which can form relatively strong bonds even when the N-C and X-C bond distances are rather long. Bulky substituents (R) on carbon inhibit the reaction by steric crowding.

S-Adenosylmethionine (Figure 5) comes to mind in this class of CH_3^+ donating coenzymes. There has been some work which suggests that the enzymatic transfer of methyl from S-adenosylmethionine to 3,4-dihydroxyacetophenone by rat liver extract involves a rate limiting S_N2 step. It is not known whether or not the methyl is transferred to the enzyme at any point during the reaction (41). ATP alkylates (adenosylates) methionine with PPP_i acting as a leaving group (42). Other phosphate esters may have similar alkylating potential. Acetate has been mentioned as a potential leaving group (reaction 15 above). Sulfate esters, $ROSO_3^-$, would appear to be potentially very good R^+ donor coenzymes. N^5-methyl-H_4-folate (Figure 3) has been mentioned as a methyl donor in non-B_{12} formation of methionine (29). Ethylation has not been seriously considered since Challenger refuted Gosio's claim for diethylarsine formation by fungi (43,44). Gosio's proposal might not have been so readily overturned if the production of ethionine by *E. coli* (45) and the formation of S-adenosylethionine (46) had been recognized in 1930.

The nature of the metal or metalloid nucleophile is as important as the leaving group in alkyl transfers. March (40) and Pocker and Parker (47) should be consulted for general discussions about nucleophilicity. Sulfur seems to be bioalkylated only after reduction to a powerful RS^- or HS^- nucleophile (Equation 12 and Figure 5). Selenium and tellurium may also be reduced to the divalent state prior to displacing a leaving group from carbon. The reduction of arsenite to arsenide does not seem to be likely. It is noteworthy that sulfite is a nucleophile which is capable

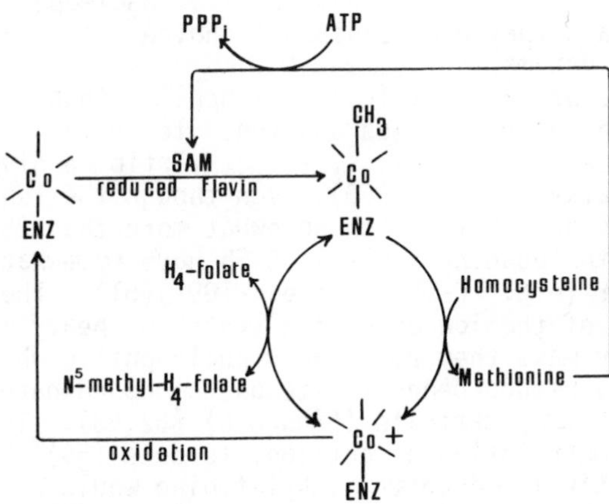

Figure 4. B_{12} System for methylation of homocysteine

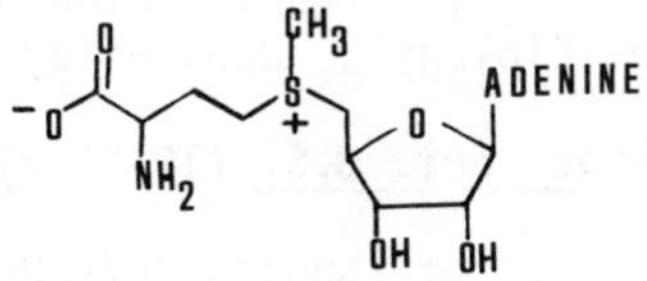

Figure 5. S-Adenosylmethionine

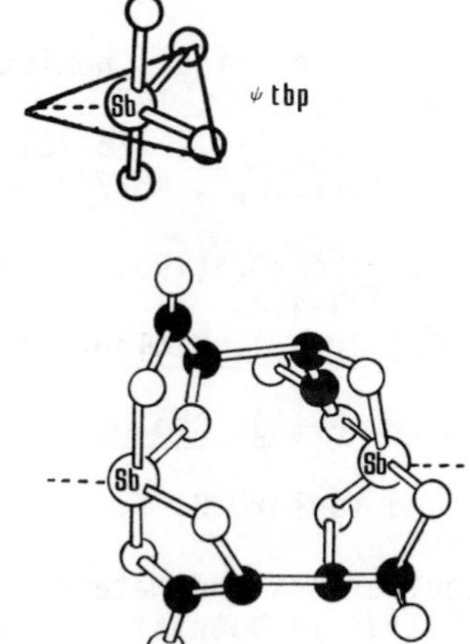

Figure 6. Structure of antimony(III) tartrate showing pseudotrigonal bipyrimid (ψ-tbp) coordination

of displacing halides from alkyl halides (48,49). Selenite, tellurite and arsenite should have similar nucleophilic capabilities, and they are reasonable choices as alkyl acceptors in biological systems.

Antimonite probably is less nucleophilic than arsenite because its lone pair of electrons tends to be in a more s-type molecular orbital. Nonetheless, trimethylstibine displaces halogens from alkyl halides (50), even though the lone pair of electrons is in an orbital with somewhat more than 25% s character and the bonding orbitals C-Sb have somewhat more than 75% p character (e.g., C-Sb-C angles<<109°)(51). The polarizability of the low oxidation states of heavy metals (e.g., Pb^{2+}, Sb^{3+}) may make them reasonably nucleophilic when the lone pair occupies a pseudo-pendent site on the coordinated metal ion, e.g., as in antimony tartrate (Figure 6) (52,53). The inability to observe biomethylation of antimony to date (54) should not be taken too negatively since trimethylstibine would not survive long enough in the aerobic systems investigated to allow detection by the usual technique of headspace gases analysis (55).

Literature Cited

1. Klayman, D. L. and Gunther, W. H. H., "Organic Selenium Compounds: Their Chemistry and Biology", John Wiley and Sons, New York, N.Y., 1973.
2. Horowitz, N. H., Proc. Natl. Acad. Sci. U.S. (1945) 31, 153-157.
3. Foster, J. W., "Chemical Activities of Fungi", 207-208, Academic Press, New York, N.Y., 1949.
4. Calvin, M., Amer. Scientist (1975) 63, 169-177.
5. Lock, D. M., "Enzymes-The Agents of Life", 177-207, Crown Publishers, New York, N.Y., 1969.
6. Mahler, H. R. and Cordes, E. H., "Biological Chemistry, 2nd Ed.", 632, Harper and Row, New York, N.Y., 1971.
7. Mahler, H. R. and Cordes, E. H., loc. cit., 778-781.
8. Zajic, J. E., "Microbial Biogeochemistry", 46-78, Academic Press, New York, N.Y., 1969.
9. Cook; T. M. and Umbreit, W. W., Biochem. (1963) 2, 194-196.
10. London, J., Science (1963) 140, 409-410.
11. Peck, H. D., Jr. and Fisher, E., Jr., J. Biological Chem. (1962) 237, 190-197.
12. Peck, H. D., Jr. and Stulberg, M. P., J. Biological Chem. (1962) 237, 1648-1652.
13. Bowen, T. J., Happold, F. C. and Taylor, B. F., Biochim. Biophys. Acta (1966) 118, 566-576.
14. Peck, H. D., Jr., Sulfur Requirements and Metabolism of Microorganisms,"Symposium: Sulfur in Nutrition", Muth, O. H. and Oldfield, J. E., editors, 61-79, Avi Publishing Co., Westport, Conn., 1970.
15. Turner, A. W., Nature (1949) 164, 76-77.

16. Turner, A. W., Austral. J. Biological Sci. (1954) 7, 452.
17. Collman, J. P., Sorrell, T. N. and Hoffman, B. M., J. Amer. Chem. Soc. (1975) 97, 913-914.
18. Koch, S., Tang, S. C., Holm, R. H. and Frankel, R. B., J. Amer. Chem. Soc. (1975) 97, 914-916.
19. Koch, S., Tang, S. C., Holm, R. H., Frankel, R. B. and Ibers, J. A., J. Amer. Chem. Soc. (1975) 97, 916-918.
20. Johnson, D. L. and Pilson, M. E. Q., Environ. Lets. (1975) 8, 157-171.
21. Woolfork, C. A., J. Bacteriol. (1962) 84, 659.
22. Findley, J. E. and Akagi, J. M., Biochem. Biophys. Res. Commun. (1969) 36, 266.
23. Roy, A. B. and Trudinger, P. A., "The Biochemistry of Inorganic Compounds of Sulfur", 256, Cambridge at the University Press, London, 1970.
24. Zajic, J. E., loc. cit., 79-95.
25. Johnson, D. L., Nature (1972) 240, 44-45.
26. McBride, B. C. and Wolf, R. S., Biochemistry (1971) 10, 4312-4317.
27. Pilson, M. E. Q., Limnol. Oceanog. (1974) 19, 339-341.
28. Mahler, H. R. and Cordes, E. H., loc. cit., 778-781.
29. Whitfield, C. D. and Weissbach, H., J. Biol. Chem. (1970) 245, 402.
30. Pratt, J. M., "Inorganic Chemistry of Vitamin B_{12}", Academic Press, New York, N.Y., 1972.
31. Taylor, R. T. and Hanna, M. L., Arch. Biochem. Biophys. (1970) 137, 453.
32. Weisbach, H. and Taylor, R. T., Roles of Vitamin B_{12} and Folic Acid in Methionine Synthesis, "Vitamins and Hormones", Harris, R. S., Munson, P. L. and Piczfalusy, E., editors, 415-440, Academic Press, New York, N.Y., 1970.
33. Law, P. Y. and Wood, J. M., J. Amer. Chem. Soc. (1973) 95, 914-919.
34. Penley, M. W., Brown, D. G. and Wood, J. M., Biochem. (1970) 9, 4302-4310.
35. Brodie, J. D., Proc. Natl. Acad. Sci. U.S. (1969) 62, 461.
36. Schrauzer, G. N., Accounts Chem. Res. (1968) 1, 97-103.
37. Silverman, R. B. and Dolphin, D., J. Amer. Chem. Soc. (1974) 96, 7094-7096.
38. Scovell, W. M., J. Amer. Chem. Soc. (1974) 96, 3451-3456.
39. Hughes, M. N., "The Inorganic Chemistry of Biological Processes", 186-187, John Wiley and Sons, New York, N.Y., 1972.
40. March, J., "Advanced Organic Chemistry", 251-375, McGraw-Hill Book Co., New York, N.Y., 1968.
41. Hegazi, M. F., Borchardt, R. T. and Schowen, R. L., J. Amer. Chem. Soc. (1976) 98, 3048-3049.
42. Mahler, H. R. and Cordes, E. H., loc. cit., 382.
43. Gosio, B., Arch. Ital. Biol. (1893) 18, 253.

44. Challenger, F., Higginbottom, C. and Ellis, L., J. Chem. Soc. (1933), 95-101.
45. Fisher, J. F. and Mallette, M. F., J. Gen. Physiol. (1961) 45,1.
46. Farber, E., Adv. Cancer Res. (1963) 7, 383
47. Pocker, Y. and Parker, A. J., J. Org. Chem. (1966) 31, 1526.
48. March, J., loc. cit., 330.
49. Gilbert, E. E., "Sulfonation and Related Reactions", 136-163, Interscience Publishers, Inc., New York, N.Y., 1965.
50. Parris, G. E. and Brinckman, F. E., J. Org. Chem. (1975) 40, 3801-3803.
51. Brill, T. B., Parris, G. E., Long, G. G. and Bowen, L. H., Inorg. Chem. (1973) 12, 1888-1891.
52. Poore, M. C. and Russell, D. R., J. Chem. Soc. A (1971), 18.
53. Cotton, F. A. and Wilkinson, G., "Advanced Inorganic Chemistry; 3rd Ed.", 392-393, John Wiley and Sons, New York, N.Y., 1972.
54. Challenger, F. and Ellis, L., J. Chem. Soc. (1935), 396.
55. Parris, G. E. and Brinckman, F. E., Environ. Sci. Tech. (1976) 10, 1128-1134.

RECEIVED August 22, 1978.

3

Occurrence of Biological Methylation of Elements in the Environment

Y. K. CHAU and P. T. S. WONG

Canada Centre for Inland Waters, Burlington, Ontario, Canada

Biological methylation of elements is one of the most intriguing biochemical processes involved with living systems. This phenomenon was observed in the early nineteenth century (1815) when several cases of arsenical poisoning occurred in Germany due to the use of domestic wallpapers containing arsenic pigments. In 1839 Gmelin (1) noted that a garlic odor was present in rooms associated with the incident. The mystery was not thoroughly unveiled until 1893 when Gosio (2) observed the evolution of a garlic odor gas from a mould-infected sample of mashed potato on exposure to air. The gas was called Gosio gas which was later identified by Challenger as trimethylarsine.

The term "biological methylation" was first used by Challenger (3) to describe the replacement of the oxy-groups of arsenic, selenium and tellurium compounds by methyl groups through the action of moulds, resulting in the formation of organometalloids or organometallic compounds. It has since been shown that biological methylation is a general process for living organisms (4). Available information indicates that microorganisms, especially bacteria and fungi play an important role in the transformation. While the biological function of methylation is not clearly known, it has been proposed to be a detoxification process. Alternately, it may be energetically preferable for some organisms to transmethylate metal rather than to synthesize methane (5).

Studies on the environmental impacts of biological methylation have gained much momentum since the discovery that microorganisms in a natural lake sediment were able to methylate mercury to a highly-neurotoxic methylmercury species (6). Because methylmercury is produced at a rate faster than organisms can accomplish its degradation, it may accumulate in fish and so poses a threat to public health (7). Indeed, several incidents of environmental catastrophies caused by mercury have been documented (8,9).

Methylation in the environment results in the formation of organometalloids or organometals which are generally more toxic and easily bioaccumulated; it plays an important role in the mobilization of elements from sediment to water; it may cause

0-8412-0461-6/78/47-082-039$05.00/0

transmethylation of other elements. Attempts have been made to predict the possibility of methylation of other elements by the relative ease of formation of metal-carbon bonds (7) and by the reduction potential of the elements (10). The metals Hg, Sn, Pd, Pt, Au and Tl and metalloids As, Se, Te and S have been postulated to accept methyl groups from methyl-cobalamin in biological systems. However, Pb, Cd and Zn have been predicted not to be methylated because of the extreme instability of their monoalkyl derivatives in aqueous systems.

Methylation studies pertaining to environmental impact began when Jensen and Jernelov (6) investigated the transformation of $HgCl_2$ in bottom sediments from freshwater aquaria. Then Cox and Alexander (11) studied the transformation of methylated arsenic and selenium compounds from their inorganic salts by fungi isolated from raw sewage and grown on agar plates. Similarly, Huey *et al.* (12,13) investigated methylation of Sn, Cd, by a *Pseudomonas* species isolated from Chesapeake Bay.

All these investigations were carried out with environmentally originated microorganisms grown in laboratory media. More realistic approaches were adopted by Bramen (14), Langley (15), and Wong *et al.* (16) who used systems containing natural waters and sediments to investigate respectively the methylation of As, Hg, and Pb. All these investigations, however, bear certain relevance and significance to the environment and can be extrapolated to living ecosystems. Methylation of mercury in the environment has been well established and documented by other workers.

Experimental

A sediment-lake water system is used for the investigation of methylation of metals and metalloids in the aquatic environment. In each study, 50 g of sediment and 150 ml of lake water were placed in a 250 ml filter flask. Nutrient broth (0.5%), glucose (0.1%), and yeast extract (0.1%) were added to stimulate microbial growth. The compounds to be tested for methylation were added to the sediment (5 mg/ℓ) and the flasks were capped and incubated at 20°C for 7-10 days. The headspace gas was analyzed for volatile methylated compounds and the lake water was analyzed for the presence of methylated species of the element.

A specially-developed Gas Chromatograph-Atomic Absorption technique was used for the analyses of volatile alkyllead compounds (17), methyl selenides (18), methylarsines and methylarsenic acids (19).

Sediments from several lakes in Ontario were used for the studies. In all these experiments, appropriate controls were prepared either by omitting the test compounds or by sterilizing the medium by autoclaving.

The toxicity of the volatile alkyl metals (Pb) and metalloids (As, Se) on algal growth was investigated by using the biologically generated alkyls since most of the methylated products are highly volatile and insoluble in water, making the dosing of the compounds to biota difficult. The biological generator (Figure 1) consists of a 4 ℓ culture flask containing a bacterial inoculum (Aeromonas sp. 150 ml) and 1350 ml of the fresh nutrient medium with and without the desired compound (at 5 ppm level as the element) for generation of the volatile methylated products. When sediment was used, 500 g sediment and one litre of lake water with and without the compound were incubated with glucose (0.1%). After about 10 days incubation, the headspace gases were analyzed for the presence of the methyl derivatives before the culture flask was connected to a test flask (4 litre) containing 1.4 ℓ of fresh CHU-10 medium and 100 ml of algal inoculum. The biologically-generated methylated product was sucked through the algal culture flask by a peristaltic pump. The setup without the addition of the compound was used as a control. The effects of a particular methylated product on algal growth and primary productivity were determined.

Results and Discussion

Methylation of lead. Wong et al. (16) presented the first evidence that under laboratory conditions microorganisms in sediments from several Canadian lakes would transform certain inorganic and organic lead compounds into a volatile and highly toxic tetramethyllead. It was also observed that incubation of certain lake sediments produced Me_4Pb even without the addition of extraneous lead compound. There was no direct relationship between lead concentrations in the sediment and the amount of Me_4Pb produced (Table I).

The conversion of Me_3PbOAc to Me_4Pb was observed in all experiments but that of lead nitrate was only sporadic. Subsequently, Jarvie et al. (20) confirmed the methylation of Me_3PbOAc to Me_4Pb and proposed a mechanism involving hydrogen sulfide complexed with Me_3PbOAc followed by decomposition to Me_4Pb. These workers suggested that lead methylation was a chemical process, without considering the microbial production of sulfide in sediment as being a biological process. Later Schmidt and Huber (21) demonstrated that not only trimethyllead, but also lead acetate, could be methylated to tetramethyllead in aquarium water. The mechanism of methylation is yet to be established.

Chemical disproportionation reactions of Me_3Pb^+ salts are known to produce Me_4Pb. The experiments set up to determine the portion of Me_4Pb due to chemical disproportionation reactions consisted of a series of culture of Aeromonas species in nutrient broth with addition of 0-100 mg Pb/ℓ of Me_3PbOAc. An identical set of samples was sterilized for determining the Me_4Pb produced by chemical disproportionation reactions. After three weeks incubation, the bacteria growth was measured and the amounts of Me_4Pb produced in the headspace of the chemical and biological systems were quantified. The amounts of Me_4Pb generated

Table I. Methylation of lead in lake sediments.

	Lake	Total Pb conc. in sediment (mg/kg dry wt.)	μg Me_4Pb generated from sediment supplemented with: No addition	$Pb(NO_3)_2$	Me_3PbOAc
1.	Mitchel Bay	110	1.20	2.20	4.7
2.	Erieau Harbor	60	0	0	2.9
3.	Port Stanley	69	0	0	5.2
4.	Long Lake	116	0	0.09	8.6
5.	Kelly Lake	285	1.80	1.10	4.7
6.	Lunch Lake	47	0.16	0.14	4.0
7.	Robinson Lake	48	0	2.10	21.0
8.	Dill Lake	47	0.71	0.55	7.6
9.	Norway Lake	48	0	0	2.4
10.	Babine Lake	43	0	0	1.4
11.	Hamilton Harbor	273	0.01	0.13	6.4

50 gm sediment (wet wt.), 150 ml lake water, 0.5% nutrient broth and 0.1% glucose with and without the addition of 1 mg Pb as Pb $(NO_3)_2$ or Me_3PbOAc. Final Pb concentration 5 ppm.

biologically and chemically at different concentrations of Me_3PbOAc and their relationships to bacterial growth are illustrated in Figure 2. At any concentration of Me_3PbOAc where the microorganisms were actively growing, the Me_4Pb generated chemically only represented about 15-20% of the total Me_4Pb produced in the biological system. When growth was inhibited at 100 ppm of Me_3PbOAc, the Me_4Pb generated in the system was solely due to chemical disproportionation. Ultraviolet irradiation did not cause further chemical conversion of Me_3PbOAc to Me_4Pb in the absence of microorganisms.

Direct chemical synthesis of methyllead compounds through alkylation of inorganic lead is very difficult because of the extreme instability of the postulated first intermediate monoalkyllead salt ($MePb^{+3}$). The difficulties have also been explained in terms of oxidation-reduction reactions involved in biomethylation (10). However, from a biochemical point of view, it is not entirely unreasonable to envisage ligand systems which could form stable monomethyllead complexes before the successive methylation steps occur. Such may have been the case in the observations of methylation of lead nitrate and lead chloride to Me_4Pb in some sediments.

The toxicity of Me_4Pb on an alga (*Scenedesmus quadricauda*) was studied by bubbling the biologically-generated Me_4Pb into the culture medium (22). It was estimated that less than 0.5 mg (as Me_4Pb) had passed through the culture medium. The primary productivity (^{14}C technique) and cell growth (dry weight) decreased by 85% and 32%, respectively, as compared with the controls without exposure to Me_4Pb. Furthermore, cells exposed to Me_4Pb tended to clump together. Similar results were obtained with *Ankistrodesmus falcatus*. To obtain similar effects, twice as much lead in the form of Me_3PbOAc, and twenty times as much lead nitrate would be required.

Lead methylation is analogous to mercury in several aspects. It is dependent on temperature, pH and microbial activities of the medium but independent of the concentration of lead in the sediment. So far there are only three reports of laboratory experiments on lead methylation. Evidence of its occurrence in the environment is still lacking. However, existence of significantly high ratios of tetraalkyllead to total lead in certain fishery products indicates the possibility of methylation in sediment or in fish tissues (23). The factors and the occurrence of *in situ* lead methylation in the environment and the mechanisms of methylation are being further investigated.

Methylation of Selenium. The production of volatile selenium compounds by microorganisms has been acknowledged for decades (24). It is known that volatile selenium compounds (Me_2Se and Me_2Se_2) are produced through methylation by fungi, bacteria, rats, and higher plants (25). Not much is known about the methylation of selenium in the aquatic environment. Under laboratory conditions, Chau *et*

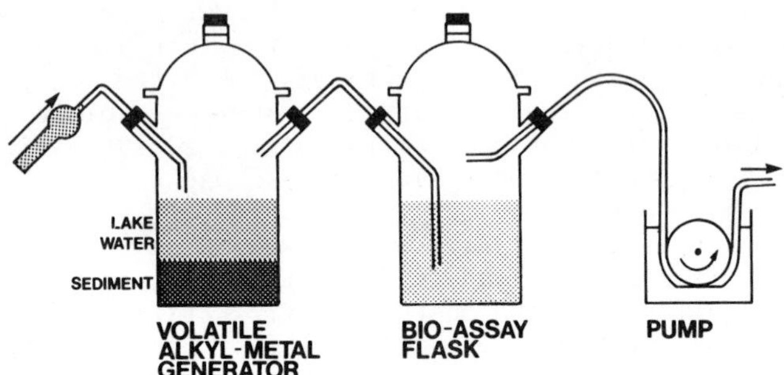

Figure 1. Biological generation of volatile methyl alkyls for algal toxicity testing

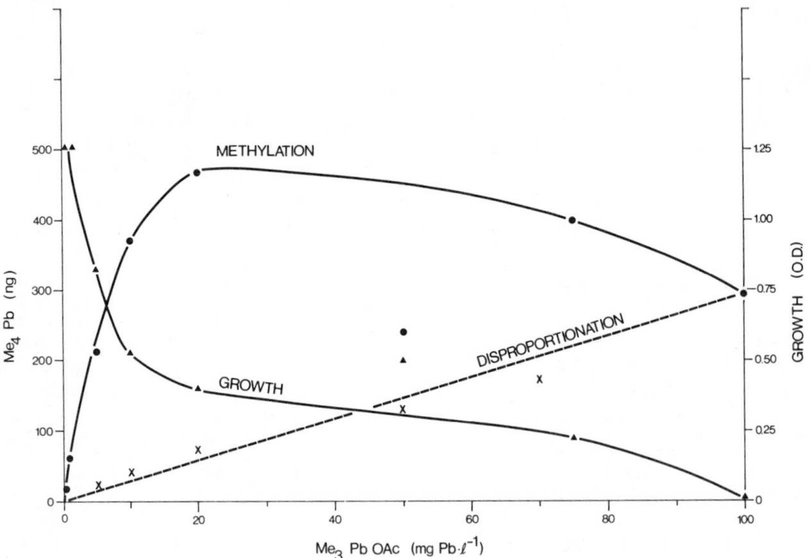

Figure 2. Production of Me_4Pb by chemical disproportionation and by biological methylation and their relationships to growth of Aeromonas *species. Growth was measured by optical density.*

al. (26) observed the production of Me_2Se and Me_2Se_2 from several sediment and soil samples with and without enrichment with the following selenium compounds: sodium selenite, sodium selenate, selenocystine, selenourea, and selenomethionine. In many cases, an unidentified volatile selenium compound was produced with a retention time between that of Me_2Se and Me_2Se_2 in the gas chromatogram. The production of volatile selenium compounds was observed to be associated with microbiological growth and was temperature dependent. A summary of the investigations with lake sediments with and without addition of sodium selenite and selenate is given in Table II.

It has been shown that significant concentrations of selenium exist in both fresh and salt water fish (27). In studying the relative toxicity of organic and inorganic selenium compounds to fish, Niimi and LaHam (28) observed a noticeable daily decrease of selenium levels in the test solution. The loss was probably due to the microbial methylation of the inorganic selenium to a volatile organoselenium compound which was more toxic to fish.

Selenium is of particular interest as a potential environmental toxicant because of the small safety margin between the levels necessary in the diet and the concentrations hazardous to man (29). Organic selenium compounds are more toxic and have longer retention times than the inorganic selenium salts (30). Thus it is of considerable interest to study the methylation of selenium in the aquatic environment.

Methylation of arsenic. Methylation of arsenic by fungi and bacteria has been known for several decades. Challenger and co-workers (3) extensively investigated the ability of *Scopulariopsis brevicaulis* to methylate organic and inorganic arsenic compounds. Cox and Alexander (11) found 3 sewage fungi that would methylate various aresenic compounds used as pesticides to form trimethylarsine. McBride *et al.* (31) also showed that aerobic microorganisms produced trimethylarsine whereas the anaerobic methanogenic bacteria produced dimethylarsine when incubated in the presence of pentavalent and trivalent arsenic derivatives.

In our experiments with lake water and sediment, we have demonstrated that in certain sediments containing high arsenic levels, such as those from the Moira River area in Ontario, non-volatile methylated arsenic compounds are detected in the overlaying lake water without the addition of extraneous arsenic compounds (Table III). In other sediments with low arsenic levels, addition of arsenic compounds was required for methylation. However in most of these experiments, no volatile methylated arsines were detected in the headspace of the incubation flasks. In several experiments with the sediments from the Moira River, volatile methylated arsines (Me_2AsH, Me_3As) were detected. Factors which control arsenic methylation are not completely understood. Adenosine triphosphate and hydrogen were found to be essential for the formation of dimethylarsine by cell extracts of

Table II. Methylation of selenium compounds in 12 Sudbury area lake sediment samples.

Lake	Se in Sediment µg/g dry wt.	No Addition			Na selenite			Na selenate		
		$(CH_3)_2Se$	$(CH_3)_2Se_2$	Unknown	$(CH_3)_2Se$	$(CH_3)_2Se_2$	Unknown	$(CH_3)_2Se$	$(CH_3)_2Se_2$	Unknown
1. Elbow	0.48	34	0	3.3	33.3	0	0	34	0	3.3
2. Ramsey	1.64	0	0	0	0	0	0	33.6	0	0
3. Kelley	20.48	2.7	3.3	2.3	14	0	0	20	3.3	5.3
4. Long	0.90	1.7	0	0	27.3	0	0	24.7	0	0
5. Simons	16.28	0	0	0	18.7	0	18	20	4.7	19.7
6. Vermillion	0.52	0	0	0	20.3	0	0	9.7	8	5.3
7. Windy	0.65	0	0	0	12.3	0	0	0	0	0
8. Moose	0.67	0	0	0	5	0	0	10.5	0	2
9. Kukagami	0.44	0	0	0	33.3	0	0	23.4	0	0
10. Nepewassi	0.53	0	0	0	25.3	0	0	28.7	0	0
11. Johnnie	0.55	6.3	0	3.7	43.3	16.5	0	94.6	19.4	54.8
12. George	0.67	0	0	0	56.8	0	8.7	94.7	11	7.3

Table III. Methylations of arsenic compounds in sediment samples from Moira River area, Ontario.

Sample	Arsenic addition (5 mg/1)	Arsenic in medium (μg/1)			
		As(III) As(V)	methylarsonic acid	dimethylarsinic acid	trimethyl arsine oxide
Lake sediment	none	1.92	2.14	3.25	-
	arsenate	- *	1.03	9.57	-
	arsenite	0.80	0.80	7.84	-
River sediment	none	1.16	3.75	2.29	-
	arsenate	0.45	0.50	3.27	-
	arsenite	-	-	0.52	-
Pond sediment	none	6.03	0.63	2.93	7.59
	arsenate	8.89	-	7.19	17.92
	arsenite	3.96	-	3.73	20.40

* - below 0.1 ng detection limit

Table IV. Methylation of arsenic compounds by pure bacteria.

Bacterial species	Arsenic addition (10 mg/ℓ)	As in medium (mg/ℓ)				Volatile As in headspace (ng)	
		As(III) As (V)	methyl-arsonic acid	dimethyl-arsinic acid	trimethyl-arsine oxide	Arsine	Trimethyl-arsine
<u>Aeromonas</u> sp	arsenate	8.59	-	-	-	0.8	-
	arsenite	8.92	-	0.78	-	0.5	-
	methylarsonic acid	- *	9.63	2.88	-	0.2	-
	dimethylarsinic acid	-	-	2.72	-	0.6	-
<u>E. Coli</u>	arsenate	7.61	0.08	0.51	-	-	-
	arsenite	8.09	0.09	1.67	-	0.4	-
	methylarsonic acid	-	11.73	1.79	-	-	113.0
	dimethylarsinic acid	-	-	6.15	-	1.1	-
<u>Flavobacterium</u> sp	arsenate	4.89	-	0.40	-	1.0	-
	arsenite	9.10	-	0.47	-	0.5	-
	methylarsonic acid	-	12.60	0.14	-	-	-
	dimethylarsinic acid	-	-	3.76	1.18	-	10.9

* - below 0.1 ng detection limit

Methanobacterium (32). Phosphate, selenate and tellurate, however, inhibited the conversion of arsenate to trimethylarsine by a fungus (11).

Since sediment is a complex biological and chemical mosaic, it is difficult to understand what organisms(s) and reactions are responsible for the arsenic methylation. In an attempt to simplify the situation, we have found two pure bacterial cultures Aeromonas and Flavobacterium sp., isolated from lake water and another bacterium Escherichia coli, commonly found in the intestine of an organism and in polluted water, had the ability to methylate arsenic compounds when grown in a medium of 0.5% nutrient broth, 0.1% glucose and 10 mg/ℓ arsenic compound (as As) at 20°C under aerobic conditions (Table IV). Results show that the additions of arsenic compounds to the medium would generally result in the formation of dimethylarsenic acid in the medium. Trimethylarsine oxide was seldom detected. In these experiments, volatile arsine and trimethylarsine were also found in the headspace of the culture flasks.

The toxicity of a mixture of biologically-generated volatile arsines on an alga (Chlorella pyrenoidosa) was investigated by the previously mentioned technique. The primary productivity (^{14}C technique) and cell growth (cell count) decreased by 45% and 44%, respectively, as compared with the control algae without exposure to arsines. Chemical analyses of the algal cells revealed that the total As levels in the exposed cells were 10 times more than that in the unexposed cells.

McBride et al. (31), using cell-free extracts of the aerobic soil organism, Candida humicola, demonstrated that in the presence of S-adenosylmethionine and NADPH, the extracts could synthesize arsenite, methylarsonate and dimethylarsinate from arsenate. Whether such reactions occur in sediments requires further investigation.

Literature Cited

1. Gmelin. Karlsruhen Zeit. (1839) November.

2. Gosio, B. Arch. ital. biol. (1893) 18, 253.

3. Challenger, F. Chem. Rev. (1945) 36, 315.

4. Thayer, J.S. J. Chem. Educ. (1973) 50, 390.

5. Ferguson, J.F. and Gavis, J. Water Res. (1972) 6, 1259.

6. Jensen S. and Jernelöv, A. Nature (1969) 220, 173.

7. Wood, J.M. Science (1974) 183, 1049.

8. Irukayama, K. "The pollution of Minamata Bay and Minamata Bay disease". p. 153 in Advances in Water Pollutions Research. Proc. 3rd Int. Conf. Water Pollution Control Fed. Washington, D.C. vol. 3. (1967).

9. Niigata Report. Report on the cases of mercury poisoning in Niigata. Ministry of Health and Welfare, Tokyo, Japan (1967).

10. Ridley, W.P., Dizikes, L.J. and Wood, J.M. Science (1977) 197, 329.

11. Cox, D.P. and Alexander, M. Bull. Environ. Contam. Toxicol. (1973) 9, 84.

12. Huey, C., Brinckman, F.E., Grim, S. and Iverson, W.P. Proc. Int. Conf. "Transport of Persistent Chemicals in Aquatic Ecosystems". O.N. LeHam, Ed. (National Research Council of Canada, Ottawa 1974) pp. 73-78.

13. Huey, C., Brinckman, F.E., Iverson, W.P. and Grim, S.O. Abst. Int. Conf. Heavy Metals Environ. Toronto, Ont. (1975) Paper C214.

14. Braman, R.S. "Arsenical Pesticides", E.A. Woolson, Ed. pp. 108-123. Amer. Chem. Soc. Washington, D.C. (1975).

15. Langley, D.G. J. Water Pollut. Control. Fed. (1973) 45, 44.

16. Wong, P.T.S., Chau, Y.K. and Luxon, P.L. Nature (1975) 253, 263.

17. Chau, Y.K., Wong, P.T.S. and Goulden, P.D. Anal. Chim. Acta (1976) 85, 421.

18. Chau, Y.K., Wong, P.T.S. and Goulden, P.D. Anal. Chem. (1975) 47, 2279.

19. Wong, P.T.S., Chau, Y.K., Luxon, P.L. and Bengert, G.A. Proc. 11th Ann. Conf. "Trace Substances in Environ. Health XI" Missouri. Ed. D.D. Hemphill, pp. 100-106 (1977).

20. Jarvie, A.W.P., Markall, R.N. and Potter, H.R. Nature (1975) 225, 217.

21. Schmidt, U. and Huber, F. Nature (1976) 259, 159.

22. Silverberg, B.A., Wong, P.T.S. and Chau, Y.K. Arch. Environ. Contam. Toxicol. (1977) 5, 305.

23. Sirota, G.R. and Uthe, J.F. Anal. Chem. (1977) 49, 823.

24. Challenger, F. Chem. and Ind. (1935) 54, 657.

25. Challenger, F. Adv. Enzymology and related areas of molecular biol. (1951) 12, 429.

26. Chau, Y.K., Wong, P.T.S., Silverberg, B.A., Luxon, P.L. and Bengert, G.A. Science (1976) 192, 1130.

27. Oelschlager, V.W. and Menke, K.H. Ernahrungswiss (1969) 9, 216.

28. Niimi, A.J. and LaHam, Q.N. Can. J. Zool. (1976) 54, 501.

29. Copeland, R. Limnos (1970) 3, 7.

30. Schroeder, H.A., Frost, D.A. and Balassa, J.J. J. Chronic Disease (1970) 23, 227.

31. McBride, B.C., Reimer, M.M. and Cullen, W.R. 175th ACS National Meeting, Anaheim, Calif. (1978) Abstract of Papers, Inorganic, no. 116.

32. McBride, B.C. and Wolfe, R.S. Biochem. (1971) 10, 4312.

Discussion

J. M. WOOD (University of Minnesota): One of the things I find puzzling about the methylation of lead is what sort of complex has to be formed in biological systems to stablize a single lead-carbon bond in water? This is a tremendous problem from a thermodynamic point of view. Yet, it appears that somehow this happens because tetramethyllead is produced. I wonder whether anybody here has any idea on what sort of ligand is likey to coordinate lead to stablize a single lead-carbon bond in water.

CHAU: We are working on a number of biological ligands, namely, glutathione polycarboxylic acids, and sulfur ligands, trying to stablize the monomethyllead complex.

F. HUBER (University of Dortmund): Did you add vitamin B_{12} to your solutions when studying the production of tetramethyllead from lead(II) compounds?

CHAU: Yes, we did add B_{12}, but in our general experiments we use no B_{12}.

HUBER: So there is no influence on the tetramethyllead production when you add vitamin B_{12}?

CHAU: No, there was no enhancement.

HUBER: This is about the same result we got, especially when we tried it with thallium. You talked about the problem of the proportion where you found about 15%-20%; we find about 80% chemically produced tetramethyllead. It might be possible that the compositions of the solutions are different. We show that, depending on the composition of the solution, there is a tremendous influence on the rate of redistribution reactions of organolead compounds.

J. J. ZUCKERMAN (University of Oklahoma): In some cases you specified the acetate; I take it that these were soluble lead compounds, and that these engaged in the redistribution reaction. Did you try colloidal or insoluble lead compounds to see if they could be mobilized by the biological species in these reactions?

CHAU: We tried several insoluble compounds, such as lead hydroxide, lead chloride, and sparingly soluble lead carbonate. There was no methylation. But even with lead nitrate we had difficulty getting consistent results, so I'm not surprised that those insoluble compounds didn't react.

W. P. RIDLEY (University of Minnesota): I'm interested in your experiments on the methylation of lead in fish intestines. Have you examined the transport of the methylated arsenic species across the intestinal wall of the fish?

CHAU: No, we haven't. We used fish intestine cut into pieces and mixed it with an inorganic arsenic solution. The bacteria from the fish intestine could transform the arsenate and arsenite into methyl and dimethyl arsenic compounds.

K. IRGOLIC (Texas A & M University): You showed data on arsenic compounds which you identified as dimethylarsinic and methylarsonic acids. As you pointed out, it is not necessarily true that the dimethylarsine or monomethylarsine you detected need come from these compounds. They could come from an arsenobetaine. I would like to warn that use of dimethylarsinic acid or methylarsonic acid as a standard in analytical procedures does not mean that you have these compounds in solution. I would like to ask what kind of structure was suggested for the volatile unknown selenium intermediate in the methylation of the selenium.

WOOD: In a model system, we were looking at the methylation of selenium by dimethylmercury. Under the right conditions, you can get methyl transfer from dimethylmercury. We isolated and characterized a selenium ylid as an intermediate. Which is, in fact, the predominant product of the reaction and which is very rapidly converted to dimethylselenide by beta-elimination. Dr. Chau had isolated something similar and I sent him the mass spectral data and the nmr data. I just learned today when you talked that these things are identical.

CHAU: Yes, they are identical.

E. BEVAGE (Universal Oil Products): Is it your opinion that the methylation of these elements is given by a wide number of species of bacteria or do you feel it is limited to fairly few? I noticed you had data on *E. coli*.

CHAU: Dr. Wong has identified several bacterial species, namely *Aeromonas* and *Pseudomonas*, very common in lake sediments. They do carry out methylation. I think we have identified four species of bacteria which could carry out this methylation. It is more favorable in anaerobic situations. Sometimes methylation occurs with either aerobic or anaerobic conditions as with arsenic.

RECEIVED August 22, 1978.

4

Kinetic and Mechanistic Studies on B_{12}-Dependent Methyl Transfer to Certain Toxic Metal Ions

Y.-T. FANCHIANG, W. P. RIDLEY, and J. M. WOOD

Freshwater Biological Institute/Department of Biochemistry, College of Biological Sciences, University of Minnesota, P.O. Box 100, Navarre, MN 55392

Introduction

In the 1930's Challenger discovered the biomethylation of arsenic, and provided us with the first example of how biological systems possess the capability for synthesizing very toxic organo-arsenic compounds from less toxic inorganic substrates [1,2]. Even in those early days Challenger recognized that biomethylation could only present a local environmental hazard, if the methylated product is produced in significant concentration so as to exert its toxic effect and if the methylated product is stable to hydrolysis.

Oxidation-reduction reactions were not well understood in the 1930's; in fact, they are not too well understood to this day, especially in biological systems. For example, biochemists and chemists do not understand the fundamental processes of nitrogen fixation, hydrogen evolution and fixation, photosynthesis, sulfite and nitrite reduction, etc., etc. These processes all use electron transfer systems which contain transition metal ions such as molybdenum, iron, manganese, etc., for catalysis, and most of these systems are sensitive to molecular oxygen. A major, as yet unanswered question is: "How did such a great variety of single electron transfer reactions evolve and remain active in a global system which is bathed in molecular oxygen?" Clearly, the "redox" conditions in a specific environment must select for the preferred reaction mechanism, but also control the kinetics for environmentally significant reactions such as biomethylation.

Dynamic aspects of these reactions are of critical importance, because even though most methylated metals are thermodynamically unstable in water, many of them are kinetically stable. In fact, it can be shown that metals which are lower in their periodic groups form metal-alkyls which are kinetically more stable. For example, mercury, platinum and possibly lead offer potentially stable systems whereas palladium, chromium and cadmium do not.

In this brief report we examine the different mechanisms for

0-8412-0461-6/78/47-082-054$05.00/0

B_{12}-dependent methyl-transfer to a selected group of toxic elements placing special emphasis on "redox" conditions. Four in vitro mechanisms have been elucidated for B_{12}-dependent methyl-transfer to date: (1) heterolytic cleavage of the Co-C bond with the transfer of a carbanion to the attacking metal ion; (2) heterolytic cleavage of the Co-C bond with the transfer of a carbonium ion to an attacking nucleophile; (3) homolytic cleavage of the Co-C bond with transfer of a methyl-radical to an attacking free radical; and (4) "Redox-Switch", a mechanism where metal ions complex with the corrin macrocycle to labilize the Co-C bond to attack by weak electrophiles [3,4,5,6]. The transfer of CH_3^- or $CH_3^\cdot$ have been found to be the most predominant reaction mechanisms for a number of metals and metalloids.

(1) Electrophilic Attack on the Co-C Bond of Methyl-B_{12}. Methyl-transfer from methyl-B_{12} requires cleavage of the Co-C bond. The methyl to cobalt bond can break under different conditions to give a carbanion, a radical or a carbonium ion (Figure 1). We have shown that the carbanion and radical forms are the principal species involved in methyl-transfer to metals and metalloids [7].

The reaction between mercuric ion and methyl corrinoids is an example of carbanion methyl-transfer (Figure 2). Because mercuric ion is a good electrophile, it also coordinates to the nitrogen of the 5,6-dimethylbenzimidazole base to give a mixture of "base off" and "base on" methyl-B_{12}. The "base on" species reacts 1000 times faster than the "base off" species to give methylmercury as the product [3] (Figure 2). Other metals which are known to react with methyl-B_{12} by a similar mechanism to mercuric ion are lead (Pb^{IV}), thallium (Tl^{III}) and palladium (Pd^{II}) [8,9,10].

The reactions described above all involve the displacement of a carbanion from the cobalt atom of methyl-B_{12}. These reactions occur under aerobic conditions with rate constants in the order of milliseconds. It is apparent that metals which react by electrophilic attack on the Co-C bond (SE_2 mechanism) occur with the more oxidized state of the metal (i.e. Pb^{IV}, Tl^{III}, Hg^{II}, Pd^{II}, etc.).

(2) Free Radical Attack on the Co-C bond of Methyl-B_{12}. A second general mechanism for biomethylation involves the displacement of a methyl radical from methyl-B_{12} by the attacking metallic species. Homolytic cleavage of the Co-C bond occurs with the transfer of the methyl radical. This reaction can be viewed as one electron oxidative addition [5,7]. Reactions of this kind require the generation of a radical intermediate. This radical intermediate can be produced under anaerobic conditions in the laboratory either by adding a single electron oxidant to a metal ion in the reduced state of a redox couple (e.g. $Sn^{II} \longrightarrow Sn^{III} + e$) or by adding a single electron reductant to a metal

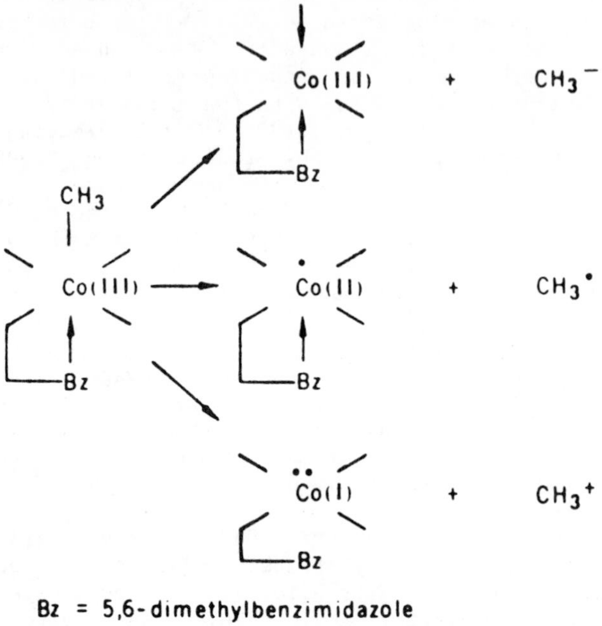

Bz = 5,6-dimethylbenzimidazole

Figure 1. Methyl transfer from methyl-B_{12} requires cleavage of the Co–C bond. The methyl-to-cobalt bold can break under different conditions giving a carbanion, a radical, or a carbonium ion.

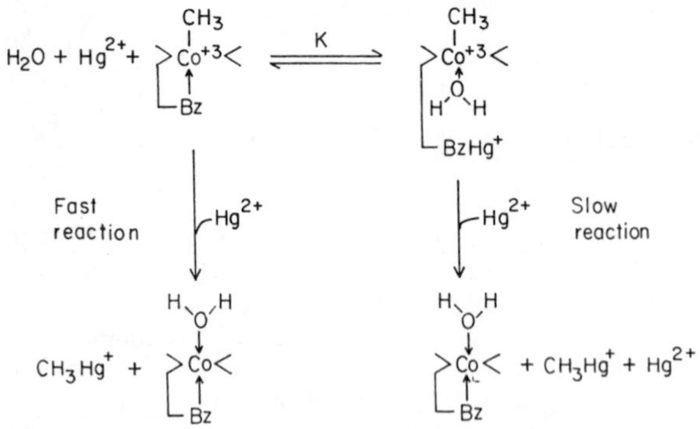

Figure 2. The reaction between mercuric ion and methyl corrinoids is an example of carbanion methyl transfer.

ion in the oxidized state of a redox couple (e.g. $Au^{III} + e \longrightarrow Au^{II}$) [11]. In the case of tin we have shown that reductive Co-C bond cleavage of methyl-B_{12} occurs by a transient Sn^{III} radical which is generated by one equivalent oxidation of Sn^{II} (Figure 3) [5,7]. In the case of gold we have shown that cleavage of the Co-C bond requires the generation of Au^{II} by single electron reduction of Au^{III} (Figure 4) [11]. For this reaction preincubation of the Au^{III} salt with the Fe^{II} salt is required before a reaction with methyl-B_{12} proceeds.

Similar reaction mechanisms have been demonstrated for methyl-transfer to sulfhydryl groups [12] and to chromium (Cr^{II}) [13]. The standard reduction potentials (E° volts) for elements known to be methylated by methyl-B_{12} are shown in Table I. It is clear that those metals with a high reduction potential (oxidizing agents) react by an electrophilic mechanism (Type I) and those with a low reduction potential (reducing agents) react by a reductive mechanism (Type II) [5,7]. This connection between standard reduction potential and mechanism for biomethylation seems highly rational, because E° describes the relative thermodynamic tendency for the metals involved to accept or donate electrons.

Table I. Relationship Between Standard Reduction Potential (E°) and the Mechanism of Methylation for Selected Elements.

Redox Couple	E° (volts)	Mechanism of Methylation
Pb(IV)/Pb(II)	+1.46	Type I [7,8]
Tl(III)/Tl(I)	+1.26	Type I [9]
Se(VI)/Se(IV) acid	+1.15	
Pd(II)/Pd(0)	+0.987	Type I [10]
Hg(II)/Hg(0)	+0.854	Type I [3,7]
Au(III)/Au(I)	+0.805	Redox Switch [14]
Pt(IV)/Pt(II)	+0.760	Redox Switch [14]
As(V)/As(III) acid	+0.559	
Au(III)/Au(II)	+0.50*	Type II [6]
Sn(IV)/Sn(II)	+0.154	Type II [5,7]
Se(VI)/Se(IV) base	+0.05	
Cys-S-S Cys/2Cys-SH	-0.22	Type II [12]
Cr(III)/Cr(II)	-0.41	Type II [13]
As(V)/As(III) base	-0.67	

* The Au(III)/Au(II) couple is estimated [19].

Those metals which react by electrophilic attack (Type I mechanism) have standard reduction potentials greater than +0.80 volts, whereas those metals which react by reductive homolytic cleavage (Type II mechanism) tend to have standard reduction potentials less than +0.56. Reactions which are not clearly defined occur close to the standard reduction potential for molecular oxygen (+0.68 volts), and this group of metals have been called "Redox

Switch" metals [14] because both oxidation states of the metal are required for biomethylation to occur. In the following section we will present recent data on this "Redox Switch" mechanism.

(3) The "Redox Switch" Mechanism for Attack on the Co-C Bond of Methyl-B_{12}. In 1972, Brown and Wood [15] showed that 2'3'-isopropylidene-5'-deoxy-β-(D)riboxyl cobinamide could exist as two stable corrin-ring isomers. This isomerization was shown to be pH dependent and solvent dependent. The yellow colored isomer (Y) had a λ_{max} at 420 nm and the Co-C bond was shown to be stable to visible light, but the red isomer (R) had a λ_{max} at 475 mμ and was shown to be photolabile. A 220 MHz NMR study of these two isomers gave markedly different spectra. For (R) the vinyl proton at C_{10} was assigned at $\sigma = 6.7$ shifting to $\sigma = 7.48$ in the (Y) form. The (Y) to (R) isomerization was shown to be reversible. This research represented the first example of the influence of corrin-ring conformation on the stability of the Co-C bond. At that time we suggested that the interaction of B_{12} with proteins could lead to corrin-ring isomerizations which may represent an important feature in understanding substrate directed labilization of the Co-C bond in the B_{12}-enzymes [15]. Certainly, it seems reasonable that the propionamide side chains on B_{12}-coenzymes could hydrogen bond to proteins to cause some distortion of the corrin macrocycle. In 1975, first Hogenkamp et al., [16] and then Cockle et al., [17] used 270 MHz NMR to demonstrate that a similar red-yellow shift to that observed by Brown and Wood [15] could be explained by corrin-ring isomerization of cobalamins. Recently, Abeles et al., [18] showed that a red to yellow isomerization occurs in the B_{12}-enzyme diol dehydrase when a mono-ester derivative of B_{12}-coenzyme is substituted for 5'deoxyadenosylcobalamin. This mono-ester coenzyme analog shows a substrate-dependent red to yellow transition, and the yellow form of the coenzyme analog is about 5% as active as 5'deoxyadenosylcobalamin itself.

Each of the above examples demonstrate that the red (R) to yellow (Y) isomerization reaction for free B_{12} analogs as well as for a B_{12} analog bound to the enzyme diol dehydrase, results in stabilization of the Co-C bond to homolytic cleavage. Recently we have shown that $Pt^{II}Cl_4^{2-}$ forms a one to one complex with the corrin macrocycle and labilizes the Co-C bond to attack by weak electrophiles such as Pt^{IV}, As^{V} and Se^{VI} (Figure 5). No reaction occurs with Pt^{IV}, As^{V} or Se^{VI} alone. Using 270 MHz NMR together with a detailed kinetic and equilibrium study of this reaction the following features emerge:

1. $Pt^{II}Cl_4^{2-}$ forms a one to one "outer sphere" complex with methyl-B_{12}.

2. There is no displacement or interaction with the benzimidazole moiety nor any other region of the molecule which projects below the plane of the corrin macrocycle.

3. The pKa for the displacement of benzimidazole increases

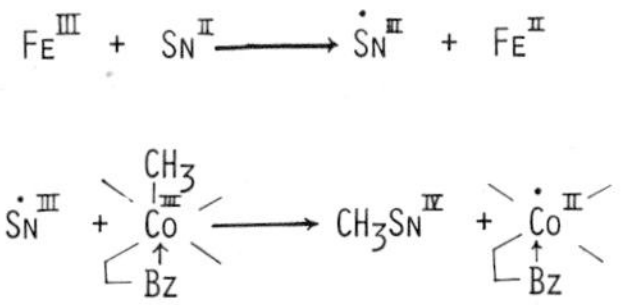

$$Fe^{III} + Sn^{II} \longrightarrow \dot{S}n^{III} + Fe^{II}$$

$$\dot{S}n^{III} + CH_3\text{–}Co^{III}\text{(Bz)} \longrightarrow CH_3Sn^{IV} + \dot{C}o^{II}\text{(Bz)}$$

Figure 3. Reductive Co–C bond cleavage of methyl-B_{12} occurs by a transient Sn^{III} radical which is generated by one equivalent oxidation of Sn^{II}.

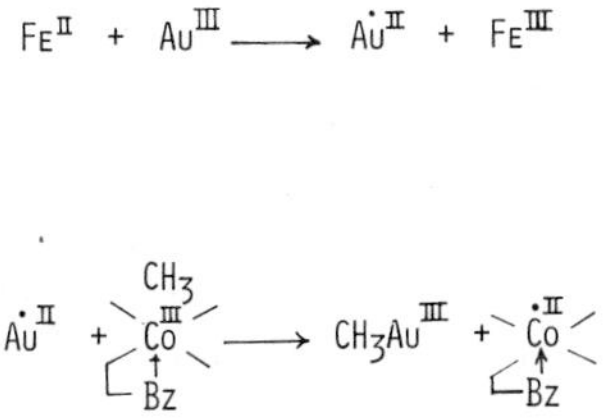

$$Fe^{II} + Au^{III} \longrightarrow \dot{A}u^{II} + Fe^{III}$$

$$\dot{A}u^{II} + CH_3\text{–}Co^{III}\text{(Bz)} \longrightarrow CH_3Au^{III} + \dot{C}o^{II}\text{(Bz)}$$

Figure 4. Cleavage of the Co–C bond requires the generation of Au^{II} by single-electron reduction of Au^{III}.

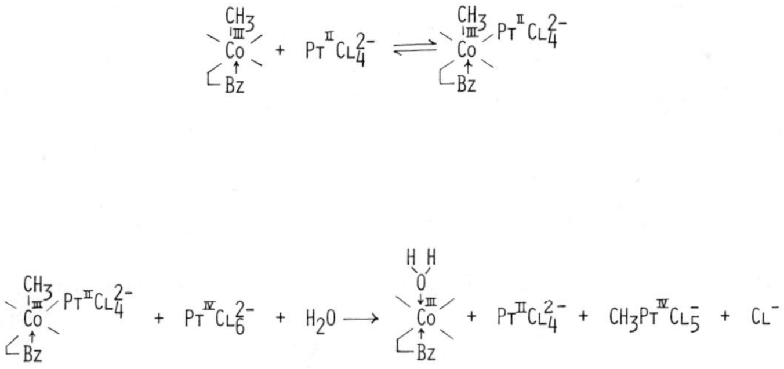

$$CH_3\text{–}Co^{III}\text{(Bz)} + Pt^{II}Cl_4^{2-} \rightleftharpoons CH_3\text{–}Co^{III}\text{(Bz)}\cdot Pt^{II}Cl_4^{2-}$$

$$CH_3\text{–}Co^{III}\text{(Bz)}\cdot Pt^{II}Cl_4^{2-} + Pt^{IV}Cl_6^{2-} + H_2O \longrightarrow H_2O\text{–}Co^{III}\text{(Bz)} + Pt^{II}Cl_4^{2-} + CH_3Pt^{IV}Cl_5^{-} + Cl^{-}$$

Figure 5. $Pt^{II}Cl_4^{2-}$ forms a one-to-one complex with the corrin macrocycle and labilizes the Co–C bond to attack by weak electrolytes such as Pt^{IV}, As^{V}, and Se^{VI}.

by 0.6 units in the $Pt^{II}Cl_4{}^{2-}$-methyl-B_{12} complex.

4. Complex formation is sensitive to both ionic strength and the nature of the anion in solution.

5. A downfield shift of 0.064 ppm is observed for the methyl group σ bonded to the cobalt atom of MeB_{12}-$Pt^{II}Cl_4{}^{2-}$ complex indicating a change in electron density on the methyl-group making it more susceptible to electrophilic attack.

6. The Co-C bond in the $Pt^{II}Cl_4{}^{2-}$-methyl-B_{12} complex is more photolabile than for methyl-B_{12} alone (i.e. the bond is approximately half as stable to light in the complex).

A detailed report on the kinetics and mechanism for methyl-transfer to platinum is in press [6]. The major features of the reaction mechanism are presented in Figure 5. Although we do not know the detailed structure of the $Pt^{II}Cl_4{}^{2-}$-methyl-B_{12} complex, it is clear that activation of the Co-C bond occurs through the formation of this complex. At the present time a detailed 270 MHz NMR study of this interaction is underway with a view to determination of the precise structure of this "activated" methyl-B_{12} species.

Conclusions

The mechanisms for B_{12}-dependent methyl-transfer to a number of metals and metalloids are shown to be determined by the standard reduction potential (E°) for the attacking inorganic salt. The standard reduction potential for molecular oxygen separates those complexes which cleave the Co-C bond by heterolysis from those which cleave this bond by homolysis. In the case of platinum, which has an E° for the Pt^{IV}/Pt^{II} couple close to that for molecular oxygen, we have discovered the formation of an "outer sphere" complex between the Pt^{II} salt and the corrin macrocycle which labilizes the Co-C bond to electrophilic attack by Pt^{IV}. This mechanism has been called a "Redox Switch" and has many similarities to current mechanisms being proposed for labilization of the Co-C bond in the B_{12}-enzymes. We believe that a fundamental study of kinetics and mechanism for B_{12}-dependent methylation of metals and metalloids has not only helped to define environmental conditions for biomethylation reactions, but has also shed some light on the prerequisites required for "activation" of the Co-C bond in B_{12}-enzyme catalyzed reaction.

Acknowledgements

Some of the research supported in this report was subsidized by the U.S. Public Health Service AM 18101, the International Lead Zinc Research Organization and the Northwest Area Foundation.

References

1. Challenger, F. Chem. Rev. (1945) 36, 315.
2. Ridley, W.P., Dizikes, L.J., Cheh, A., and Wood, J.M. Env. Health Perspectives (1977) 19, 43.
3. DeSimone, R.E., Penley, M.W., Charbonneau, L., Smith, S.G., Wood, J.M., Hill, H.A.O., Pratt, J.M., Ridsdale, S., and Williams, R.J.P. Biochim. Biophys. Acta (1973) 304, 851.
4. Schrauzer, G.N., Sibert, J.W., and Windgassen, R.J. J. Amer. Chem. Soc. (1968) 90, 6681.
5. Dizikes, L.J., Ridley, W.P., and Wood, J.M. J. Amer. Chem. Soc. (1978) 100:3, 1010.
6. Fanchiang, Y.-T., Ridley, W.P., and Wood, J.M. J. Amer. Chem. Soc. (1978) 100, 1010.
7. Ridley, W.P., Dizikes, L.J., and Wood, J.M. Science (1977) 197, 329.
8. Wood, J.M. (1978) Lead in the Marine Environment. Proceedings of an International Conference on Lead in the Marine Environment. Rovinj, Yugoslavia. M. Branica, editor.
9. Agnes, G., Hill, H.A.O., Pratt, J.M., Ridsdale, S.C., Kennedy, F.S., and Williams, R.J.P. Biochim. Biophys. Acta (1971) 252, 207.
10. Scovell, W.H. J. Amer. Chem. Soc. (1974) 96, 3451.
11. Fanchiang, Y.-T., Ridley, W.P., and Wood, J.M. Chem. Communs (1978) (submitted) March.
12. Frick, T., Francia, M.D., and Wood, J.M. Biochim. Biophys. Acta (1976) 428, 808.
13. Espenson, J.H., and Seelers, T.D. J. Amer. Chem. Soc. (1974) 96, 94.
14. Agnes, G., Bendle, B., Hill, H.A.O., Williams, F.R., and Williams, R.J.P. Chem. Communs. (1971) 850.
15. Brown, D.G., and Wood, J.M. Structure and Bonding (1972) 11, 47.
16. Hogenkamp, H.P.C., Vergamini, P.J., and Matwiyoff, N.A. J. Chem. Soc. Dalton Trans. (1975) 2628.
17. Cockle, S.A., Hensens, O.D., Hill, H.A.O., and Williams, R.J.P. J. Chem. Soc. Dalton Trans. (1975) 2633.
18. Abeles, R.H. Current Status of the Mechanism of B_{12}-coenzyme. Biological Aspects of Inorganic Chemistry. Wiley Interscience, Ed. D.H. Dolphin 1977) 245.
19. Rich, R.L., and Taube, H. J. Phys. Chem. (1954) 58, 1,6.

Discussion

J. J. ZUCKERMAN (University of Oklahoma): Is there evidence for bioalkylation where the alkyl group is other than methyl?

WOOD: On the question of B_{12}-dependent reactions, ethylcobalamin has been isolated and characterized as a natural product by a group of German workers, but it's there in very small concentrations. It may be possible to get ethyl transfer reactions. If that's the case, I would think that ethylmercury should have been turning up in some biological systems; however, it's only been turning up where people have spilled ethyllead. If you have a chlorine plant next to an alkyllead-producing plant, you get disproportionation in the synthesis of ethylmercury compound.

ZUCKERMAN: I recall that marsh gas has an odor and is probably a mixture of a large number of hydrocarbons. Are these directly incorporated onto arsenic by biological organisms?

WOOD: The synthesis of the arsines and the synthesis of volatile selenium compounds is well known; the synthesis of volatile dimethylmercury is well known. So, volatilization of metal and metalloid alkyls in these biologically active systems obviously occurs. Also, Brinckman's work has shown possible synthesis of methyltins. [Proc. Internat. Con. on Transport of Persistent Chemicals in Aquatic Ecosystems. National Research Council, Ottawa, Canada, 1974, p.II-73.]

ZUCKERMAN: With respect to the alkylation of tin, has it been demonstrated to occur in environmental conditions of pH, temperature, etc?

F. E. BRINCKMAN (National Bureau of Standards): There is a complication here. Salinity is a very large factor, but the results of a change in the microbial population may be more important as Colwell showed [Microbial Ecology (1975), 1, 191]. She made it quite clear that the *Pseudomonas* population, which is the principle actor in this particular case involving the mercury or the tin transformations, was very susceptible to such changes. So I think there is an additional factor of the availability of growth. The growth kinetics of the microorganisms themselves will then of course affect the apparent rate even for the exocellular production of methylcobalamin. It is not yet clear whether the methylcobalamin might be involved in the endocellular or an exocellular *vis-a-vis* these metal ions such as tin or mercury.

ZUCKERMAN: If I recall your paper in *J. Amer. Chem. Soc.* [(1978) 100, 1010] the conditions were rather more severe than biological environmental conditions.

WOOD: The kinetics experiments were run at low pH for the obvious reason of keeping the tin in solution. For the tin tartrate complex, similar kinetics are obtained for reactions at ambient pH and temperature. We chose to do the study under these conditions so that we could do kinetics more readily with tin ion. It is not a simple problem, but if you make the tartrate complex you can do it. There are many complexes in biology that could mimic tartaric acid.

G. E. PARRIS (Food & Drug Administration): With regard to the formation of other alkylmetal compounds, I read an ACS publication entitled "Chemical Carcinogens" [ACS "Advances", No. 173]. In a section regarding naturally occurring carcinogens, it was reported that _E. coli_ produced ethionine and even the adenosyl derivative of ethionine analogous to methionine, which also provides another possible alkylating agent.

WOOD: In that case, you have to look at some nucleophile. So the metalloids would be good candidates for reactions of that kind.

PARRIS: If it is a bimolecular nucleophilic reaction, as the methyl transfer might be in either enzymatic or non-enzymatic cases, the ethyl transfer would suffer considerably, relative to methyl transfer, in terms of rates.

WOOD: When you do these kinetics studies, you find the ethyl always suffers appreciably.

PARRIS: I think there is an implication that cobalamin is preeminent in the methyl transfer reaction _in vivo_, as well as in _in vitro_ studies you have done. Would you comment?

WOOD: Yes, I can put this in perspective. Many microorganisms do in fact produce B_{12}. Almost all of the blue-green algae synthesize large amounts of B_{12}. Methyl-B_{12} has been shown to be a coenzyme in a number of enzyme-catalyzed reactions. Therefore, this coenzyme does turn over, and so you have this constantly regenerated methylating agent just like you have with S-adenosylmethionine. Basically, if you want to look at methylation problems in the environment, you are always chasing the kinetic parameters. The intriguing thing about methylcobalamin is that you are looking at reactions with respectable reaction rates. In fact, the reaction rate for the methylation of mercury is a little better than the turnover number for B_{12}-dependent methionine synthesis, which is a crucial B_{12}-dependent enzyme. This means every time a mercury ion gets in the vicinity of methyl-B_{12} bound in the methionine enzyme, you are more likely to make methylmercury than you are methionine.

PARRIS: With regard to metals other than the electrophilic mercury, or metalloids, do you regard B_{12} as a preeminent methyl donor, or would you then regard S-adenosylmethionine as more likely source of methyl?

WOOD: I think it's very important to do these experiments with isotopes. We do model studies and we ask that people working in the environmental sciences do this, for example, with lead. If you want to find out whether lead is methylated, see where the C_{14} -methyl group comes from in these extremely complex experiments where you have complex media involving mixed bacterial cultures, sludge, etc. It's critically important to do isotope experiments to find out where the methyl group is coming from. If that's done, then you can write a tentative mechanism. If you can write a tentative mechanism, you can test it. What we try to do is to give people a clue about the sort of mechanism that may occur in the environment so that they can go and see whether it does. That has always been our position.

PARRIS: One last comment with regard to the other alkyl groups. In support of the suggestion that transfer from something like S-adenosylmethione to a metalloid would be a nucleophilic reaction. There is one publication in J. Amer. Chem. Soc. [(1976), 98, 3048] concerning the o-alkylation of dihydroxyacetophenone, and in that case the author is using a deuterium isotope. It was suggested that the reaction was probably bimolecular; at least the critical step in the reaction was bimolecular.

WOOD: That's the only example in the literature.

RECEIVED August 22, 1978.

5

Aqueous Chemistry of Organolead and Organothallium Compounds in the Presence of Microorganisms

F. HUBER, U. SCHMIDT, and H. KIRCHMANN

Chemistry Department, University of Dortmund, D 4600 Dortmund 50, Federal Republic of Germany

Organocompounds of lead are less stable than the corresponding compounds of the lighter group IVb elements, following the decreasing strength of the central atom-carbon bond. Their stability is strongly dependent on the nature and also on the number of the organic groups, R, bound to lead. Alkyllead compounds are, in general, distinctly less stable than aryllead compounds, and their stability decreases with decreasing number of R. In aqueous solution, tri- and dialkyllead compounds, R_3PbX and R_2PbX_2 (X = anion), show a more or less marked tendency to decompose, and monoalkyllead compounds, $RPbX_3$ are actually unknown [there is only one report that $Pb(OAc_4)$ (OAc = acetate) reacts with alkylpentafluorosilicates to give (impure) $RPbF_3$ (R = CH_3, C_2H_5, CH_2=CH) (1)]. Tetraalkyllead compounds are only very slightly soluble in water. For a general review see (2).

Alkylthallium compounds, R_2TlX and $RTlX_2$, are in general more stable than the corresponding lead compounds, but only a rather limited number of monoalkylthallium compounds, $RTlX_2$, are known. Trialkylthallium compounds, R_3Tl, hydrolyze to give $R_2Tl^+ + OH^-$ and RH (3).

Decomposition of Organolead Compounds in Water

Me_2PbX_2. We have investigated quantitatively the decomposition of Me_2PbX_2 (Me = CH_3) in aqueous solution and in aqueous salt solutions between 20 and 60°C (4). The reaction proceeds irreversibly according to equation [1]

$$2\ Me_2PbX_2 \rightarrow Me_3PbX + PbX_2 + MeX \qquad [1]$$

and the stoichiometry is not influenced by the type or the concentration of salt added to the solution. (The redistribution of Me_3PbX [see below] is much slower, so it does not appreciably interfere with [1].) The reaction rate, which in all cases corresponds to first order, increases strongly with increasing salt concentration, the anions displaying much stronger influence

0-8412-0461-6/78/47-082-065$05.00/0

than the cations. The anions arranged according to their ability to increase the reaction rate k give a series reflecting their polarizability:

X^-: OAc^- ClO_4^- NO_3^- Cl^- NO_2^- Br^- SCN^- I^-

k_{X^-}/k_{Cl^-}: 0.02 < 0.11 < 0.20 < 1.0 < 2.4 < 130 < (>130) < ($\sim 10^4$)

(Concentrations: 0.036 mol $Me_2PbX_2.L^{-1}$; NaX and KX respectively 2.7 - 0.7 $mol.L^{-1}$). The results indicate that Me_2PbX_2 in a first step gives an intermediate which is bridged by anions X and/or solvent molecules L:

$$2Me_2PbX_2 \xrightarrow[\text{molecules L}]{\text{solvent}} \begin{array}{c} \text{L/X} \quad \text{X/L} \quad \text{X/L} \\ \text{Me-Pb} \quad\quad\quad \text{Pb-Me} \\ \text{Me} \quad \text{X/L} \quad \text{Me} \end{array}$$

In a second rate-determining step, transalkylation is effected when this intermediate redistributes to give Me_3PbX and $MePbX_3$; the latter decomposes immediately and irreversibly to PbX_2 and MeX. Bridging in organolead compounds is not unusual; e.g., Ph_2PbCl_2 (5) or Ph_3PbX (Ph = C_6H_5; X = Cl, Br) (6) are halide-bridged polymers in the solid state; also, emf measurements with the system Ph_2Pb^{2+}/I^- indicated the formation of binuclear complexes, e.g. $Ph_4Pb_2I_2^{2+}$, $Ph_4Pb_2I_3^+$, or $Ph_4Pb_2I_4$ in addition to the usual mononuclear complexes (7).

<u>Me_3PbX</u>. Me_3PbX redistributes in aqueous solution according to [2]

$$2Me_3PbX \underset{b}{\overset{a}{\rightleftarrows}} Me_4Pb + Me_2PbX_2 \qquad [2]$$

Me_2PbX_2 is an educt of [1], but since [2] is reversible, [2b] competes with [1]. The irreversibility of [1], however, finally causes the decomposition of Me_3PbX according to [3] (=[2]+[1])

$$3Me_3PbX \rightarrow 2Me_4Pb + PbX_2 + MeX \qquad [3]$$

The reaction rates of [2a] and of [3] are appreciably smaller than that of [1] so the concentration of Me_2PbX_2 is always small in such solutions. [2b] is much faster than [1]. The rate law is rather complicated, as dissociation and complex equilibria play an important role. The rate of [3] is increased in the same way as mentioned for [1], the dependency on the type of anion, however, being smaller. This can be explained by the fact that the comproportionation [2b] is influenced by the same accelerating factors. Results of experiments on redistribution reactions of methylthallium compounds shall be discussed in a subsequent chapter.

Toxicity of Organolead and Organothallium Compounds

During our redistribution experiments, we became interested in the toxicity of organolead and organothallium compounds toward bacteria. The two topics, apparently having nothing to do with each other, finally brought us to study biomethylation.

Organolead Compounds. Organolead compounds act as biocides (8, 9); in the main, interest in investigation of this property had been directed towards determining minimum concentrations necessary to prevent microbial or fungal growth, and very few data are available on effects of sublethal doses and on the chemical fate of the organolead species. Even less was known regarding the toxicity of organothallium compounds.

For our measurements we used a mixed bacteria population. Culture media were prepared in BOD bottles. Lead compounds and nutrients (peptone, yeast extract) dissolved in dilution water (10, 11) were pipetted into the flasks, simultaneously inoculated with 10 ml of aquarium water (cell density 10^4 - 10^5 cells/mL; from an aerated aquarium, which was repeatedly inoculated with surface water from a freshwater lake), filled with dilution water to the top, and sealed. The bottles stood in the dark at 20°C. Results of the BOD_1 and BOD_5 measurements are listed in table I. In other experiments, dissolved O_2 was analyzed continuously with an oxygen electrode. As growth parameters we determined the

Table I. Dependency of Inhibition of Bacterial Growth after 5 Days (or 24 h*) on Type of Lead Compound Added to Culture (Nutrient: peptone, 5 mg/L; in 24 h experiments 15 mg/L)

Compound	N.I. [mg Pb/L]	T.I. [mg Pb/L]	Compound	N.I. [mg Pb/L]	T.I. [mg Pb/L]
Me_3PbCl	1	(a)	Bu_3PbOAc	0.05*	5*
Me_2PbCl_2	0.5 - 1	8 - 10		0.1	10
Et_3PbOAc	0.05*	1*	$Bu_2Pb(OAc)_2$	0.01	5 - 8
	1	40	Ph_3PbCl	0.01	10
Et_3PbCl	0.01*	10*	Ph_2PbCl_2	0.1	(a)
Et_2PbCl_2	0.01*	10*	$PbCl_2$	1	50

Me = CH_3, Et = C_2H_5, Bu = n-C_4H_9, Ph = C_6H_5, OAc = CH_3COO

N.I. = no inhibition, T.I. = total inhibition

(a) No total inhibition in saturated solution of organolead salt (Solubilities at 20°C: Me_3PbCl ca. 130 g Pb/L, Ph_2PbCl_2 ca. 0.5 g Pb/L) duration of the initial lag phase and the slope of the O_2 consumption rate during the beginning of the log phase, linearly approximated. A typical set of growth curves is shown in Fig. 1; data of growth parameters at various lead and nutrient concentrations are given in table II.

At comparatively low lead concentrations the lag phase is prolonged, while the log phase seems to be unaffected; the cell count of 10^8-10^9 cells/mL after 24 h is equal to Pb-free samples.

At medium lead concentrations, the lag phase is more prolonged; the log phase slope decreases and up to 20% necrotic cells are observed after 24 h. It is important to note that the lag phase sometimes is so prolonged, that short term BOD values would indicate total inhibition. At high lead concentrations no exponential acceleration is observed; cell density rises only to about 10^5 cells/mL (ca. 50% necrotic) and decreases after about 2 days.

The measurements show that Pb^{2+} is less toxic than organolead compounds and that, in general, R_2PbX_2 compounds are more toxic than R_3PbX compounds, though there are exceptions. The effect of Me_3PbCl is astonishingly small, compared with that of other compounds. Tetraorganoleads are more toxic than R_3PbX or R_2PbX_2; the corresponding effects on bacterial growth were observed at concentrations one or two orders of magnitude lower (in the case of Pb^{2+}, at concentrations one order of magnitude higher). The rule that the toxicity of organolead compounds increases with increasing chain lengths of R (8) proved not to be true under all conditions; apparently nutrient and also organolead concentrations are of appreciable influence (see table II).

Organothallium Compounds. The results of the measurements of the toxicity of Tl compounds showed astonishing differences compared to the results with Pb compounds. Tl^+ is more toxic than the R_2TlX compounds investigated; the smallest concentration at which inhibition was observed is 0.01 mg Tl^+/L, total inhibition was found at 1000 mg Tl^+/L. As Fig. 2 shows, the inhibition of bacterial growth in various cases was greater at lower R_2TlX concentrations, and there also is no straightforward relationship of chain length of R and toxicity.

Biomethylation of Pb^{2+} and Organolead Compounds.

Considering the facts that organolead compounds redistribute in solution and that different species show different toxicities, it was convient to study the influence of redistribution on the toxicity and vice versa to control the stoichiometry of the redistribution reactions in a nonsterile solution.

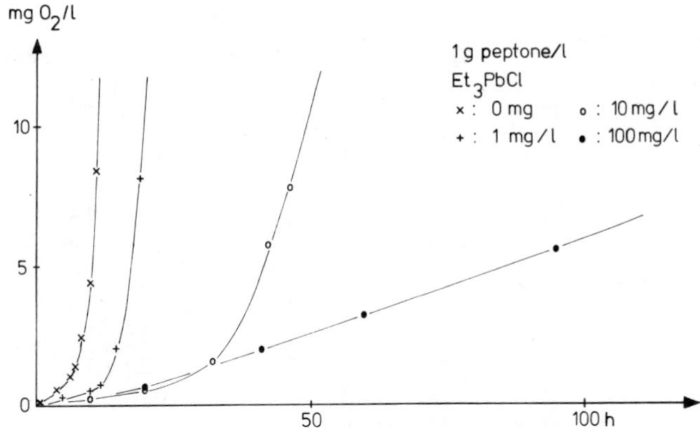

Figure 1. Typical growth curves at various Et_3PbCl concentrations showing O_2 consumption as a function of time

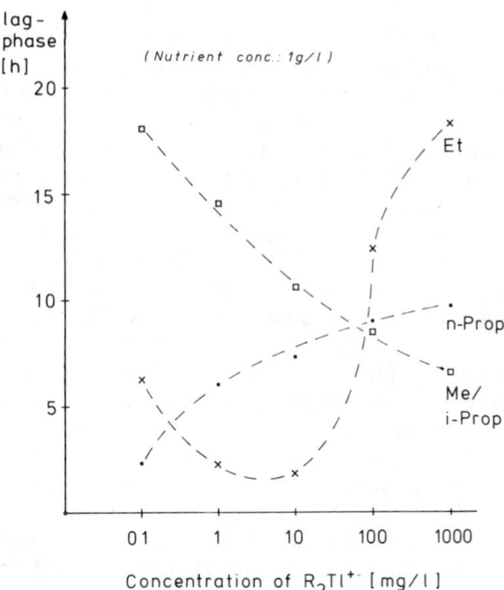

Figure 2. Inhibition of bacterial growth by diorganothallium compounds

TABLE II. Inhibition of Bacterial Growth by Organolead Compounds at Various Lead and Nutrient Concentrations

Concentr.		Me_3PbCl		Me_2PbCl_2		Et_3PbCl		Et_2PbCl_2	
Pb[a)]	Nu[b)]	lag	log	lag	log	lag	log	lag	log
0	0.2	7	2.9	7	2.9	6	1.3	6	1.3
0	1	5	3.7	5	3.7	6.5	2.4	6.5	2.4
0.1	0.2	7	3.4	20	2.4	10	0.5	18	1.4
0.1	1			10	2.2	6.5	2.4	8	0.84
1	0.2	12	2.9	45	2.7	13	0.36	30	0.23
1	1			23	2.6	12	1.1	15	0.64
10	0.2					25	0.25	50	0.14
10	1					30	0.42	25	0.37
100	0.2	80	0.8	95	0.2		0.03		0.05
100	1	30	0.8	95	0.4		0.06		0.05

lag = period of lag phase (h); log = slope of log phase (mg O_2/L·h)

a) Concentration of lead compound (mg Pb/L)
b) Concentration of nutrient (g/L)

In no case was there a noticeable effect of redistribution on the growth curves, and with rather low cell densities no remarkable influence on redistribution was observed. But in continually aerated cultures with higher cell densities, e.g., corresponding to those in the activated sludge of a sewage plant, a very appreciable increase in the redistribution rate was observed. This was accompanied by oxidative degradation of 50 - 60% of the added methyllead species to Pb^{2+}. This could be interpreted as a detoxification by the bacteria, as Pb^{2+} is less toxic than the added Me_3PbX or Me_2PbX_2. Under discontinuous conditions (in BOD flasks) the oxidative degradation amounted to about 15 - 20%.

Biomethylation of Pb^{2+}. When we followed the redistribution of Me_3PbX in anaerobic cultures (bacteria from the surface of a natural lake, grown under N_2, or from the anaerobic sediment of a small pond), we also observed a rate increase, but less Pb^{2+} and more Me_4Pb than were expected from the stoichiometry of equation [3].

Inferring from the deficit of Pb^{2+} and the extra amount of Me_4Pb that Pb^{2+} might have been methylated by bacteria, we added Pb^{2+} [as $Pb(OAc)_2$] to anaerobic cultures in gas wash bottles and cultivated these cultures under N_2 in the dark at 30°C. After 7 to 14 days volatile products were flushed with N_2 into a 0.2 N methanolic I_2 scrubber solution. Ten mL $KI \cdot I_2$ solution were added to ensure quantitative transformation of organolead species to Pb^{2+}.The latter was determined photometrically after reduction of I_2 with Na_2SO_3 in ammoniacal buffer solution and after complexation with PAR [4-(2-pyridylazo)-resorcinol] (12, 13). A blank and also a sterile solution containing Pb^{2+} or methyllead compounds showed no Pb content in the methanolic solution after the same treatment. We therefore concluded that Me_4Pb was the volatile species produced in the biomethylation of Pb^{2+} by bacteria (14). We could further prove the identity of Me_4Pb in the head-space gas above the cultures by GC analysis. The production rate of Me_4Pb was about 2.5 µg Pb/d.

The biomethylation of Pb^{2+} proceeded reproducibly, provided a) the Pb^{2+} concentration was controlled (taking into consideration the results of the toxicity measurements), b) the concentration of sulfur compounds in the solution was not too high (otherwise H_2S produced by the anaerobes precipitates PbS, leaving only a concentration of Pb^{2+} which is too small for the generation of detectable amounts of Me_4Pb), c) the inoculum was not more than 6 - 7 weeks old. Optimum results were obtained with glucose and urea or amino acids as nutrients (supply of sulfur is maintained by SO_4^{2-} in the dilution water) and at concentrations of 1 - 10 µg Pb^{2+} /mL. Only slightly smaller production rates were usually observed using concentrations of ca. 100 µg Pb^{2+} /mL [ca. 15 mg $Pb(OAc)_2$/100 mL]. It was favorable to add $CaCO_3$ to avoid greater decrease of pH.

Biomethylation of Me_3PbX. On addition of Me_3PbX to the anaerobic cultures, the Me_4Pb production was much higher than from cultures containing Pb^{2+}, and also higher than from the redistribution of Me_3PbX in sterile solutions. This indicated a high proportion of Me_4Pb production by chemical redistribution. After we had obtained these results, Wong, Chau and Luxon reported (15) that they had detected Me_4Pb above the sediment of a lake, and that addition of Me_3PbOAc and, in some cases, of $Pb(NO_3)_2$ or $PbCl_2$, increased Me_4Pb production; pure species of bacterial isolates, however, were not able to produce Me_4Pb from PbX_2. Another paper presented a controversial explanation denying alkylation of Pb^{2+} by microorganisms (16): Me_4Pb and Et_4Pb (Et = C_2H_5) should be products of chemical alkylation of Me_3PbOAc and Et_3PbOAc, respectively, in an anaerobic sediment system. As a possible mechanism, initial formation of $(R_3Pb)_2S$ followed by decomposition to give R_4Pb as one product was proposed.

Our results explain the observations unequivocally (see figure 3); in an anaerobic bacterial culture, Pb^{2+} is methylated to give Me_4Pb (reaction [4] in Fig. 3). Since alkyllead compounds redistribute in aqueous solution according to [1] and [2] to give Me_4Pb and Pb^{2+}, in a bacterial culture both chemical and biological production of Me_4Pb occurs. In media containing high concentrations of sulfur compounds (which was the case in the work of ref. [16]), redistribution of R_3PbX proceeds rather fast, as sulfide present has a high polarizability; Pb^{2+}, though produced simultaneously in appreciable amounts, is precipitated as PbS. Therefore, the amount of Me_4Pb produced by biomethylation only can be rather low in such systems, and the Me_4Pb production essentially is caused by chemical redistribution.

A rough estimate of the ratio of chemically and biologically produced Me_4Pb from our cultures (100 mL solution, containing 200 mg glucose, 10 mg urea; 56 mg Me_3PbCl were added after 1 week of incubation) is possible by comparing concentrations of Me_4Pb, Me_3PbX and PbX_2, which were a) analyzed and b) calculated from equations [2] and [3]: after 7 days we found 5.3 μmol Me_4Pb; in the solution 0.3 μmol Me_2Pb^{2+} were analyzed (13), which according to [2] correspond to 0.3 μmol Me_4Pb produced chemically; we found 2.1 μmol Pb^{2+}, which according to [3] correspond to an additional amount of 4.2 μmol Me_4Pb produced chemically. (The sediment contained no detectable amount of Pb.) The difference between the analyzed and the calculated amount of Me_4Pb is 0.8 μmol. This amount, however, must still be corrected in two respects. Any Pb^{2+} transformed biologically to Me_4Pb was produced chemically together with Me_4Pb; on the other hand, one has to make allowance for a certain amount of Me_4Pb produced by immediate biomethylation of Me_3PbX (see below). We assume that these two quantities are about similar. One therefore can estimate that not more than 0.8 μmol Me_4Pb have been produced biologically after 7 days, i.e., not more than about 16% of the total amount of Me_4Pb found. From four additional experiments we obtained similar estimates of 15, 16, 17, and 19% for the portion of Me_4Pb which was produced biologically. The production rate of Me_4Pb from sterile nutrient solutions (100 mL solution, containing 200 mg glucose, 10 mg urea, amino acids; 0.1 and 10 mg Pb respectively, added as Me_3PbCl or Me_2PbCl_2) after 7 days was 2.9 and 3.4 μg Pb/d (Me_3PbCl) and 3.1 and 3.3 μg Pb/d (Me_2PbCl_2). In analogous experiments similar rates were observed; in autoclaved cultures maximum rates of 6 - 8 μg Pb/d have been measured. The production of Me_4Pb from cultures containing Me_3PbCl (56 mg Me_3PbCl/100 mL = 40 mg Pb/100 mL) and Me_2PbCl_2 (60 mg Me_2PbCl_2/100 mL), respectively, varied between 85 and 157 μg/d and between 45 and 124 μg/d.

Biomethylation of Et_3PbX. To check whether R_3PbX is also biomethylated by anaerobes, we added Et_3PbCl (up to 100 mg Pb/L) to a 4 L culture and observed a maxmium production rate of 500 μg Pb/d. As nutrient, a solution of 0.1 g NH_4Cl, 0.052 g $K_2HPO_4 \cdot 3H_2O$, 0.1 g $MgCl_2 \cdot 6H_2O$, 1 g EtOH in dilution water (10, 11) with added amino acids was used. The gas above the culture solution was slowly passed (using a slow N_2 stream) through petroleum ether to absorb the tetraalkyllead compounds. The solution, which was separated by GC on a Chromosorb column with Apiezon (15%), contained besides solvent Et_4Pb (83%), Et_3MePb (13%) and Me_4Pb (3 - 4%; total amount of $R_{4-n}R'_nPb$ = 100%). Practically, neither Et_2Me_2Pb nor $EtMe_3Pb$ were found. (The composition of the products from other experiments was different.) To exclude the possibility that any compounds in the cultures catalyze the redistribution of mixtures of Et_4Pb and Me_4Pb, 0.5 ml of these compounds were placed in sterile nutrient solutions and autoclaved cultures; after 2 weeks no change in the composition of the R_4Pb mixture had occurred. According to these results which are summarized in figure 4, Et_3PbX is biomethylated directly to Et_3MePb. Et_4Pb is produced chemically by redistribution of Et_3PbX; Et_2PbX_2, the other redistribution product which could not be detected (13) in the filtered solution, apparently redistributes (according to [1])so fast, that no appreciable amount is available for biomethylation. Pb^{2+}, one of the final products of redistribution, is biomethylated to give Me_4Pb. Considering the different modes of production (Et_4Pb chemically, Et_3MePb and Me_4Pb biologically) the percentage of the different tetraalkyllead species in the mixture roughly corresponds to the estimate of chemically and biologically produced Me_4Pb from cultures containing Me_3PbX.

Biomethylation of Tl^+. An intriguing problem raised by the discovery of biomethylation of Pb^{2+} is the increase of the formal oxidation number of Pb^{2+} during its biological transformation to Me_4Pb. For this and other obvious reasons we sought to find out whether Tl^+, isoelectronic with Pb^{2+}, is also subject to biomethylation in anaerobic mixed bacterial cultures, during which process it too increases its formal oxidation number. In the literature there were no reports on biomethylation of Tl^+. Model experiments showed that methylcobalamin is not demethylated by Tl(I) (17), yet is demethylated by Tl(III) (17, 18); from spectral titrations it was concluded that $MeTl^{2+}$ is produced (19).

The experimental conditions we used to biomethylate Tl^+ essentially corresponded to those we had applied during biomethylation of Pb^{2+}. In 250 mL gas wash bottles we incubated three types of solutions which had been inoculated with 5 g anaerobic sediment. The solutions were prepared from 100 ml dilution water (10, 11) and contained 1 g peptone (solution A) or 0.1 g NH_4Cl, 0.052 g $K_2HPO_4 \cdot 3H_2O$, 0.01 g $MgCl_2 \cdot 6H_2O$ and either 1.0 g $Ca(OAc)_2$ (solution B) or 1.0 g C_2H_5OH (solution C). 1.9 g solid $CaCO_3$ were added to neutralize acid from metabolic processes. Starting at the third day of incubation, 100 mg TlOAc were added during 7 days. After 2-3 weeks, Me_2Tl^+ could be detected in samples as well as Tl^+; that Me_2Tl^+ was the only methylated thallium species found is understandable in view of the normal behavior of alkylthallium compounds: $MeTlX_2$ compounds in general are unstable or tend to decompose in aqueous solution (20, 21); Me_3Tl decomposes instantaneously in water to form stable Me_2Tl^+ and methane (3).

To determine the amount of Me_2Tl^+ produced by biomethylation

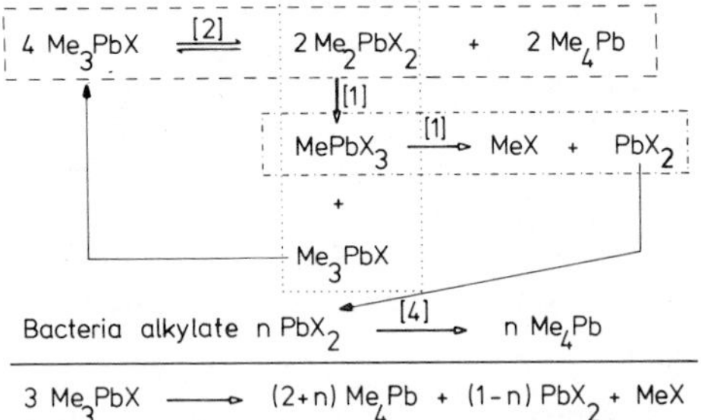

Figure 3. Sources of Me_4Pb in an anaerobic bacterial culture

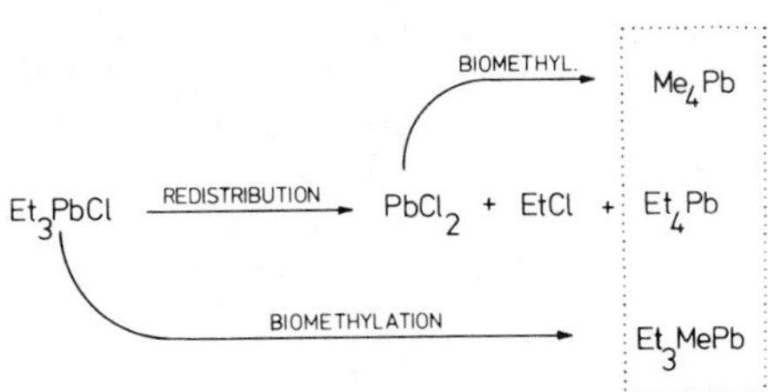

Figure 4. Biomethylation of Et_3PbCl

of Tl^+, 10 mL samples after filtering through a diaphragm were mixed with EDTA to mask Tl^+. Then 1-(2-pyridylazo)-2-naphthol (PAN) was added at pH 10-12 to complex Me_2Tl^+, and after extraction with $CHCl_3$ the Me_2Tl-PAN concentration was measured photometrically at 570 nm. [Masking of Tl^+ is necessary as Tl^+ forms a complex with PAN (22) which also absorbs at 570 nm.]

The results showed that in the cultures of solutions B and C after 21 days, 10 µg and 36 µg Me_2Tl^+/mL, respectively, had been formed. In solution A only about 1 - 2 µg Me_2Tl^+/mL were detected after 21 days; a low yield of methylated species from cultures containing peptone as nutrient was also observed in biomethylation experiments with Pb^{2+}.

The observation that Tl^+ is biomethylated to Me_2Tl^+ under increase of the oxidation number shows that "oxidative methylation" is not unique with Pb^{2+}. In this context one has to recall that an increase in oxidation number occurs also during biomethylation of As compounds (23).

Mechanistic Considerations

The overall process of the *in vivo* methylation of metal ions is certainly complex. Hence, use of results of mechanistic *in vitro* experiments to postulate general *in vivo* mechanisms should be done only with great care. Moreover, well-founded knowledge is still too scarce to allow one of the various pathways which are conceivable for "oxidative methylation" to be favored.

One might assume that methylation involves methylcobalamin which conveys Me^- to an electrophilic thallium or lead species; a principal question is then whether oxidation occurs first followed by methylation, or vice versa; the two steps could also be simultaneous. It is hard to see which oxidizing agent could overcome the high oxidation potential of Tl^+ and Pb^{2+} (24), and therefore it seems reasonable to think of methylation as the first step, particularly since $MePb^+$ was recently reported as a product of the reaction of a dimethylcobalt complex and Pb^{2+} (25). In this case the primary methylation product of Pb^{2+} and Tl^+ could disproportionate to give Me_2Pb^{2+} or Me_2Tl^+ and the element. Since we could not detect elementary Pb or Tl in the samples, we conclude that dimethyl species are not formed by disproportionation of intermediate $MePb^+$ or MeTl. We therefore would prefer to assume that Tl^+ and Pb^{2+}, present in appropriate complexed form at specific sites of the cell, are simultaneously oxidized and methylated to MeTl(III) and MePb(IV) moieties which are stabilized by complexation. The importance and effectiveness of complexation for stabilizing unstable monoorganolead compounds is known: Organolead trihalides, $RPbX_3$ ($X \neq F$), are unknown; however, it is possible to prepare rather stable complex derivatives $M'[PhPbX_4]$ and

$M''_2[PhPbX_5]$ (M' = Ph_4P, Ph_4As; M'' = Me_4N; X = Cl, Br) (26).

Another pathway might start with an electrophilic attack by CH_3^+, which could easily solve the problem of oxidation. In this case the reaction of the "oxidizing agent" CH_3^+ with the metal ion (in a specifically complexed form, mainly with negative ligands) would lead formally to $MeTl^{2+}$ or $MePb^{3+}$. These could disproportionate, or, being strong electrophiles, could be further methylated at a specific CH_3^--conveying methylating site of the cell system.

Behavior of $MeTlX_2$ in Solution. In order not to end this paper with speculation, some reactions in aqueous solution relevant to considerations on methyl transfer in biological systems by redistribution will be described.

Decomposition of $MeTlX_2$. Monomethylthallium compounds are unstable in aqueous solution. The decomposition of $MeTl(OAc)_2$ follows a first order rate law (X = OAc) (21) and produces, according to the reaction sequence [5],

$$\begin{array}{l} MeTl(OAc)_2 \rightarrow TlOAc + CH_3COOCH_3 \\ CH_3COOCH_3 + H_2O \rightleftarrows CH_3COOH + CH_3OH \end{array} \qquad [5]$$

TlOAc, CH_3COOCH_3, and, as products of hydrolysis of the latter, CH_3COOH and CH_3OH (21, 27). Based on kinetic and conductometric data and by comparison with data calculated from derived rate equations, a bimolecular S_N2 mechanism with a rate determining step of the attack of OAc^- at the Tl-bonded Me-group of the very electrophilic $MeTlOAc^+$ could be established (21). Other nucleophilic agents also attack $MeTl(OAc)_2$ and can thereby be methylated: pyridine and other N-bases react with $MeTl(OAc)_2$ to N-methylated compounds according to [6]:

$$MeTl(OAc)_2 + N\!\!<\; \rightleftarrows \; Me\overset{+}{N}\!\!< \; + OAc^- + TlOAc \qquad [6]$$

Since the nucleophilic character of these N-bases is higher than that of OAc^-, CH_3COOCH_3 is formed only in minor amounts (28, 29).

In methanolic solution $MeTl(OAc)_2$ decomposes over a period of several weeks to TlOAc and CH_3COOCH_3 (28). If thioanisole is added, the decomposition is complete in 2-3 days. This "catalytic effect" is caused by a two-step reaction (29). At first, $MeTl(OAc)_2$ methylates thioanisole according to [7]:

$$MeSPh + MeTlOAc^+ \rightleftarrows [Me_2SPh]^+ + TlOAc \qquad [7]$$

and in the second step, OAc^- is methylated by S-methylthioanisole according to [8]:

$$[Me_2SPh]^+ + OAc^- \rightleftarrows MeSPh + CH_3COOCH_3 \qquad [8]$$

This behavior of $MeTl(OAc)_2$, which is shown for other monomethylthallium compounds (<u>28</u>, <u>30</u>), also helps to explain why experiments to prepare $MePbX_3$ were doomed to failure: $MePb^{3+}$, being still more electrophilic than $MeTl^{2+}$, methylates most effectively solvent molecules or/and anions and is decomposed during this reaction to Pb^{2+}; this reactivity is highly promoted by the "inert pair effect".

<u>Disproportionation of $MeTlX_2$.</u> Disproportionation reactions of monoalkylthallium compounds have not yet been reported in the literature. In solutions of $MeTl(OAc)_2$ in CD_3COCD_3/CH_3OH (5:1; composition chosen to optimize conditions for NMR measurements), the equilibrium concentrations of the products of disproportionation according to equation [9]:

$$2\ MeTl(OAc)_2 \rightleftarrows Me_2Tl(OAc) + Tl(OAc)_3 \qquad [9]$$

are too small to be detected. However, if one adds a reducing agent to remove $Tl(OAc)_3$, one gets quantitative transformation of $MeTl(OAc)_2$ to $Me_2Tl(OAc)$ (<u>30</u>) according to [9]. The reaction of $Tl(OAc)_3$ and $(MeO)_3P$ proceeds according to [10]:

$$Tl(OAc)_3 + (MeO)_3P + CH_3OH \rightarrow TlOAc + (MeO)_3PO + HOAc + CH_3COOCH_3 \qquad [10]$$

The reaction products have been identified by NMR spectroscopy and by GC. Equations [9] and [10] add to give the overall reaction [11]:

$$2\ MeTl(OAc)_2 + (MeO)_3P + CH_3OH \rightarrow Me_2Tl(OAc) + TlOAc + (MeO)_3PO + HOAc + CH_3COOCH_3 \qquad [11]$$

Reaction [10] is instantaneous, while [11] is complete only after some minutes. We therefore conclude that the exchange of the methyl groups in the redistribution reaction [9] is rate determining.

Regarding the behavior of $MeTl(OAc)_2$ in solutions, and presupposing that $MeTl^{2+}$ is an intermediate of biomethylation of Tl^+, one arrives at an important conclusion concerning the open question as to which pathways might be of significance. There are in principle two possibilities for the transformation of $MeTl^{2+}$ to Me_2Tl^+, one chemical and one biological (the vertical and the horizontal branch respectively of the following scheme) in which [12] and [13] formally symbolize the complexation of the monomethylthallium species in solution:

$$Tl^+ \xrightarrow{+Me^+} MeTl^{2+} \qquad Tl^+ \xrightarrow[-2\,e^-]{+Me^-} MeTl^{2+}$$

$$MeTl^{2+} \xrightarrow{+Me^-} Me_2Tl^+$$

$$MeTl^{2+} \underset{[12]}{\overset{+X^-}{\rightleftharpoons}} MeTlX^+$$

$$MeTlX^+ \underset{[13]}{\overset{+X^-}{\rightleftharpoons}} MeTlX_2 \qquad MeTlX^+ \overset{[5]}{\rightleftharpoons} MeX + TlX$$

$$MeTlX_2 \overset{[9]}{\rightleftharpoons} Me_2TlX + TlX_3$$

If one assumes that methylations of $MeTl^{2+}$ and of the isoelectronic Hg^{2+} (<u>31</u>) proceed at a similar rate, the chemical pathway for the formation of Me_2Tl^+ (redistribution according to [9]) has no appreciable chance of competing with the biotransfer of Me^- (via methylcobalamin), since the rate of [9] is comparatively small, even if TlX_3 is removed extremely rapidly from the equilibrium. An eventual competition of the decomposition [5] (being faster than the non-catalyzed reaction [9]) and biomethylation (presumably much faster than [5]) can be neglected just as well. So, according to these considerations for the transformation of $MeTl^{2+}$ to Me_2Tl^+ in bacterial cultures, biological pathways should be favored over chemical ones.

Analogous considerations apply for the discussion of different pathways of the biomethylation of Pb^{2+}. Since $MePb^{3+}$ is still more electrophilic than $MeTl^{2+}$, the rate of its methylation of Me^- might be still higher, as might the rate of its decomposition. One can therefore expect that the biomethylation of Pb^{2+} is also, on the whole, a stepwise biological methylation involving no chemical steps via disproportionation reactions.

<u>Significance of Biomethylation of Tl^+ and Pb^{2+}</u>

Me_2Tl^+ compounds are less toxic to bacteria than Tl^+, so biomethylation of Tl^+ appears as a detoxification of the bacterial environment (but only in a relative sense, since Tl remains in the solution). This aspect, however, must not of necessity be the determining reason for the occurrence of methylation, as biomethylation of Pb^{2+} leads to Me_4Pb, which is much more toxic to bacterial cultures than Pb^{2+}. However, the solubility of Me_4Pb in water is extremely low, and Pb is therefore finally removed from the solution.

Concerning conclusions on the ecological significance of the biomethylation of Tl^{+} and Pb^{2+}, one should remember that biomethylation only has been observed in an anaerobic medium, and that results on this reaction and those on toxicity of Tl and Pb compounds were obtained from laboratory experiments and should be transferred to natural conditions only with great care and not without specific experimental examination. Nevertheless, one should be alert to (and investigate) the possibility that Pb^{2+} and Tl^{+}, in anaerobic regions of metal-contaminated natural waters, are methylated and are transported as organometallic compounds to other regions of the natural system, there showing the different behavior of organometallic species or being reconverted to Pb^{2+} and Tl^{+}.

Acknowledgement

The support of this work by Deutsche Forschungsgemeinschaft and Herbert-Quandt-Stiftung is greatfully acknowledged.

Literature Cited

1. Shapiro, H., Frey, F.W., "The Organic Compounds of Lead", Wiley-Interscience, New York, 1968.
2. Müller, R., Reichel, S., Dathe, C., Inorg. Nucl. Chem. Letters (1967) 3, 125.
3. Hart, C.R., Ingold, C.K., J. Chem. Soc. (1964) 4372
4. Haupt, H.-J., Huber, F., Gmehling, J., Z. anorg. allg. Chem. (1972) 390, 31.
5. Mammi, M., Busetti, V., DelPra, A., Inorg. Chim. Acta (1967) 1, 419.
6. Preut, H., Huber, F., Z. anorg. allg. Chem. (1977) 435, 234.
7. Stafford, S., Haupt, H.-J., Huber, F., Inorg. Chim. Acta (1974) 11, 207.
8. Van der Kerk, G.J.M., Biodeterior. Mater., Proc. Internat. Biodeterior. Symp. (1971, publ. 1972) 1.
9. Lorenz, J., Biodeterior. Mater., Proc. Internat. Biodeterior. Symp. (1971, publ. 1972) 443.
10. Standard Methods for the Examination of Water and Wastewater, 13th ed., 489 (edit. by APHA, AWWA, WPCF, Washington, 1971)
11. Deutsche Einheitsverfahren zur Wasseruntersuchung, H 5, 6; L 12 (edit. by GDCh, Fachgruppe Wasserchemie, Verlag Chemie, Weinheim, 1972)
12. Dagnall, R.M., West, T.S., Young, P., Talanta (1965) 12, 583.
13. Schmidt, U., Huber, F., Anal. Chim. Acta (1978) 98, 147.
14. Schmidt, U., Huber, F., Nature (1976) 259, 157.
15. Wong, P.T.S., Chau, Y.K., Luxon, P.L., Nature (1975) 253, 263.

16. Jarvie, A.W.P., Markall, R.M., Potter, H.R., Nature (1975) 255, 217.
17. Agnes, G., Bendle, S., Hill, H.A.O., Williams, F.R., Williams, R.J.P., Chem. Comm. (1971) 850.
18. Agnes, G., Hill, H.A.O., Pratt, J.M., Ridsdale, S.C., Kennedy, F.S., Williams, R.J.P., Biochim. Biophys. Acta (1971) 252, 207.
19. Abley, P., Dockal, E.R., Halpern, J., J. Amer. Chem. Soc. (1973) 95, 3166.
20. Lee, A.G. "The Chemistry of Thallium", Elsevier, Amsterdam, London, New York, 1971.
21. Pohl, U., Huber, F., J. Organometal. Chem. (1976) 116, 141.
22. Cheng, K.L., Bray, R.H., Anal. Chem. (1955) 27, 782.
23. Challenger, F., Chem. Rev. (1945) 36, 315.
24. Ridley, W.P., Dizikes, L.J., Wood, J.M., Science (1977) 197, 329.
25. Witman, M.W., Weber, J.H., Inorg. Chem. (1976) 15, 2375.
26. Lindemann, H., Huber, F., Z. anorg. allg. Chem. (1972) 394, 101.
27. Kurosawa, H., Okawara, R., J. Organometal. Chem. (1967) 10, 211.
28. Pohl, U., Huber, F., J. Organometal. Chem. (1977) 135, 301.
29. Knips, U., Huber, F., unpublished results.
30. Pohl, U., Huber, F., unpublished results.
31. DeSimone, R.E., Penley, M.W., Charbonneau, L., Smith, S.G., Wood, J.M., Hill, H.A.O., Pratt, J.M., Ridsdale, S., Williams, R.J.P., Biochim. Biophys. Acta (1973) 304, 851.

Discussion

J. H. WEBER (University of New Hampshire): Concerning some of the products you got in your reactions, particularly with the microbiological alkylation of lead, if you alkylated lead(II) you could have an unstable species of dimethyllead(II). This is not a well-known species and is considered a transient species which decomposes to tetramethyllead and lead metal. Have you thought of this possibility and looked for lead metal in your reactions?

HUBER: Yes, we did and we did not find lead. Similarly, we did not find metallic thallium.

J. M. WOOD (University of Minnesota): Do you assume 10 days is a good time for steady state when you determine how much is biological and how much is chemical disproportionation?

HUBER: It is a time which gives reproducible results. Experiments over a longer period were similar, so we chose the shorter time.

WOOD: The microbes were in the stationary phase, and they were just sitting there when you had these different chemical species in solution at the time you did the analyses? [HUBER: Yes]. Now, I think it is critically important to start isotope experiments with C-14 labels, probably C-14-labelled methionine which should get into the cells. This will establish what sort of methyl transfer is occurring in these systems. I have a difficulty in trying to rationalize a mechanism without knowing what the biological methyl donor is in the system. Do you plan to do some isotope experiments and find out?

HUBER: Yes, we plan to do so. We are studying the thallium compound. Currently we are using methylcobaloxime and methylcobalamin to see if methylation occurs with addition of an oxidant for the Pb^{2+} and Sn^{2+}. We have positive results with Sn^{2+} and negative results with Pb^{2+}.

WOOD: When you add thallium(I) to a complex system, have you any idea what the oxidation state is of the active thallium species? We can rationalize methyl transfer of thallium(III). In fact the reaction goes quite well. Thallium(I) is unusual; if you look at the standard reduction potential idea for thallium(III) to thallium(I), it is fairly high. If, for example, methyl B_{12} were involved in your reaction conditions, I'm sure that conditions are so extreme that the oxidizing agent would certainly break the cobalt-carbon bond anyway. Therefore, there is a real difficulty with a CH_3^- suggestion.

HUBER: If we start with thallium(I) we cannot postulate that we have an oxidation in the first place followed by methylation, because the oxidation would give Tl(III), and Tl(III) would oxidize the methylcobalt compound. Therefore, we have some kind of simultaneous methylation and oxidation, where two electrons are taken away and some kind of CH_3^- results to give $MeTl^{2+}$. We have attempted to rationalize this possibility in a way which is consistent with our experimental results, as shown in the Scheme on page 12.

RECEIVED August 22, 1978.

6

Bioorganotin Chemistry: Stereo- and Situselectivity in the Monooxygenase Enzyme Reactions of Cyclohexyltin Compounds

RICHARD H. FISH, JOHN E. CASIDA, and ELLA C. KIMMEL

Pesticide Chemistry and Toxicology Laboratory, College of Natural Resources, Wellman Hall, University of California, Berkeley, CA 94720

The in vitro reactions of organotin compounds with monooxygenase enzymes utilizing rat liver microsomes were previously studied with tributyltin derivatives (1a,b). The results from that study confirmed carbon-hydroxylation as the primary biochemical reaction occurring with these compounds. Furthermore, the tin-carbon sigma electrons were implicated in the possible stabilization of carbon free radicals generated on the α and β carbon atoms to the tin atom. Additionally, by using several criteria, we established that the metabolism of these tributyltin derivatives involved the interesting and biologically important cytochrome P-450 dependent monooxygenase enzyme system (1a).

Recent studies on this system have concluded that all the available evidence points to a heme-iron-monooxygen complex which converts carbon-hydrogen bonds to carbon-hydroxyl bonds (2), Figure 1. The reaction has been shown to be highly stereospecific

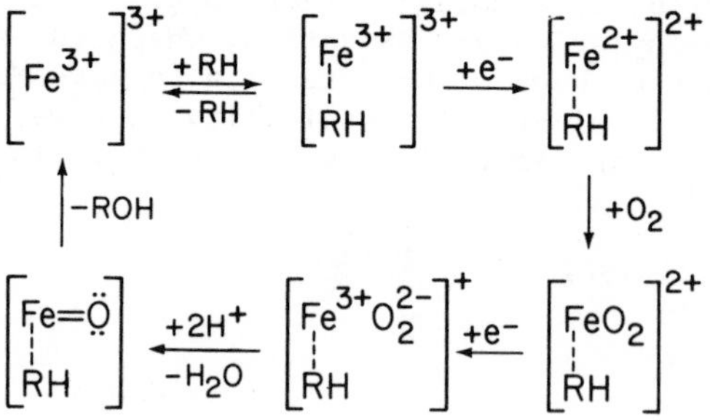

$$[Fe^{3+}]^{3+} \underset{-RH}{\overset{+RH}{\rightleftharpoons}} [Fe^{3+}\cdots RH]^{3+} \xrightarrow{+e^-} [Fe^{2+}\cdots RH]^{2+} \xrightarrow{+O_2} [FeO_2\cdots RH]^{2+} \xrightarrow{+e^-} [Fe^{3+}O_2^{2-}\cdots RH]^{+} \xrightarrow[-H_2O]{+2H^+} [Fe{=}\ddot{O}\cdots RH] \xrightarrow{-ROH} [Fe^{3+}]^{3+}$$

Figure 1. Mechanism of Cytochrome P-450 enzyme hydroxylation reaction

(3); however, surprisingly few investigations have been concerned with the cyclohexyl ring system (4). Our work in this area was logically extended to cyclohexyltin compounds for several reasons. Firstly, these compounds are used as agricultural miticides on

0-8412-0461-6/78/47-082-082$05.00/0

food crops and a study of their monooxygenase enzyme reactions would be important for biological and toxicological reasons. Secondly, we wanted to ascertain the stereo as well as situ-selectivities of these reactions with appropriate cyclohexyltin model compounds, since this aspect, as far as we could establish with the cytochrome P-450 monooxygenase enzyme system, has not been elucidated in a definitive manner.

We decided initially (5) to study cyclohexyltriphenyltin, **1**, because we found in an earlier investigation (1b) that the triphenyltin derivatives were not hydroxylated under *in vitro* reaction conditions. More importantly, the synthesis of potential metabolites, which with the cyclohexyl system involves the *cis*- and *trans*-hydroxycyclohexyltin isomers in the 2,3, and 4 positions (triphenyltin being position 1), would be more convenient using the triphenyltin group as a synthetic handle (6). Additionally, the extension to cyclohexyldiphenyltin acetate, **9**, would only involve electrophilic cleavage of a phenyl group in order to prepare potential metabolites for this model substrate (7).

A discussion of the stereo and situselectivity involved in the P-450 monooxygenase reactions of **1** and **9** and the consequence of their conformation at the active site will be presented.

Results

The preparation of [1-^{14}C]cyclohexyltriphenyltin, **1**, was readily accomplished by the reaction of [1-^{14}C]cyclohexylmagnesium bromide with triphenyltin chloride in 33% yield with a specific activity of 1.26 mCi/mmole (Eq 1). The use of [^{14}C]-

$$\text{C}_6\text{H}_{11}\text{Br}^{*} \xrightarrow[\text{2) } (\text{C}_6\text{H}_5)_3\text{SnCl}]{\text{1) Mg/THF}} \text{C}_6\text{H}_{11}^{*}\text{Sn}(\text{C}_6\text{H}_5)_3 \;\; \mathbf{1} \qquad (1)$$

labelled organotin substrates is mandatory in these metabolism studies, since the amount of metabolism is generally to the extent of 10% of the starting radiolabelled substrate. Thus, compound **1** (0.05 μmole) was incubated with our source of cytochrome P-450, rat liver microsomes (1a) (10.6 mg protein), for 1 hr at 37° in 2.0 ml phosphate buffer (pH 7.4) containing NADPH (2 μmole) the essential cofactor. After chloroform extraction, we utilized thin layer chromatography (TLC) to separate the metabolites and liquid scintillation counting to quantify them. They were identified by a combination of TLC cochromatography, preparative TLC in conjunction with 360 MHz ^{1}H FT nmr spectroscopy and by specific degradation reactions (Eq 2). The

$$\text{(2)}$$

1 → (P450-Fe=O, NADPH, pH 7.4, 1 h) 2 (85.6%) + 3 (6.5%) + 4 (3.0%) + 5 (1.6%) + 6 (1.9%) + 7 (1.4%)

percentages in Eq 2 represent normalized values of identified metabolites and account for 8% of starting substrate, 1. The remainder was 1 (70%) and unidentified materials (22%).

One important aspect of this type of work is the ability to synthesize potential metabolites and to understand their subsequent chemistry. This facilitates their ultimate identification and allows the use of cochromatography as one criterion for this purpose. All the metabolites, 2-7, were separated from one another using neutral TLC solvent systems (5) and then cochromatographed with synthetic standards (6,7). Furthermore, compound 2 was purified by preparative TLC and a 360 MHz ^{1}H FT nmr spectrum was obtained (13,600 acquisitions, $CDCl_3$, TMS) confirming that the major metabolite (85.6%) was trans-4-hydroxycyclohexyltriphenyltin, 2. The nmr spectrum showed the axial methine proton on the carbon (C4) bearing the hydroxyl group as a 9 line multiplet (3.58 ppm, J_{ax}-J_{ax} = 11.1 Hz; J_{ax}-J_{eq} = 4.0 Hz) consistent with a spectrum of the authentic compound 2 (6,7). We were able to detect the corresponding cis isomer of 2 by this nmr technique; however, none was observed in the nmr spectrum of metabolite 2.

The cis- and trans-3-hydroxy metabolites, 3 and 4, were identified only by TLC cochromatography with authentic compounds, because of the low amounts produced in the biological oxidation reaction. In this regard, we are confident of their assigned structures, since both isomers are readily separable by this technique.

Metabolite 5 was identified by TLC cochromatography and by a specific degradation reaction. The metabolite mixture was acidified with glacial acetic acid and TLC analysis showed the disappearance of metabolite 5. Experiments with authentic 5 revealed that this trans-2-hydroxy compound undergoes a facile 1,2-deoxystannylation reaction giving cyclohexene, triphenyltin acetate and water. In agreement with this, we assume that metabolite 5 reacted similarly under acidic conditions (Eq 3).

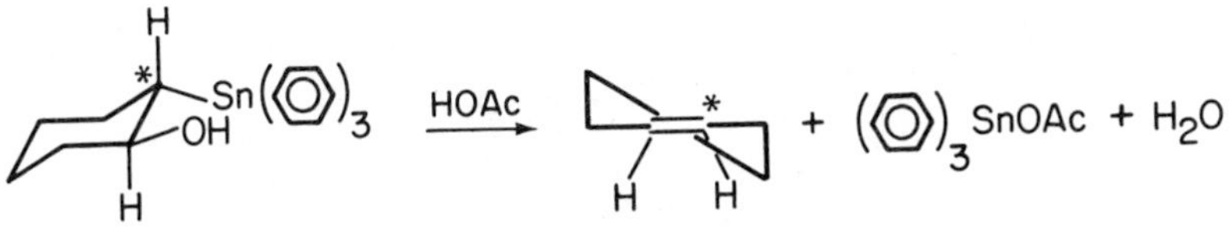

$$\text{trans-2-hydroxycyclohexyl-}Sn(C_6H_5)_3 \xrightarrow{HOAc} \text{cyclohexene} + (C_6H_5)_3SnOAc + H_2O \quad (3)$$

The trans metabolite, 5, but not the corresponding cis isomer of 5, can form cyclohexene, since a trans diaxial conformation (I) is needed for reaction to take place. The

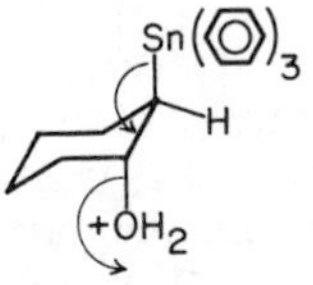

I

corresponding cis-methyl ether, a model for the cis-alcohol we were not able to synthesize, did not react even after three days with glacial acetic acid (Eq 4).

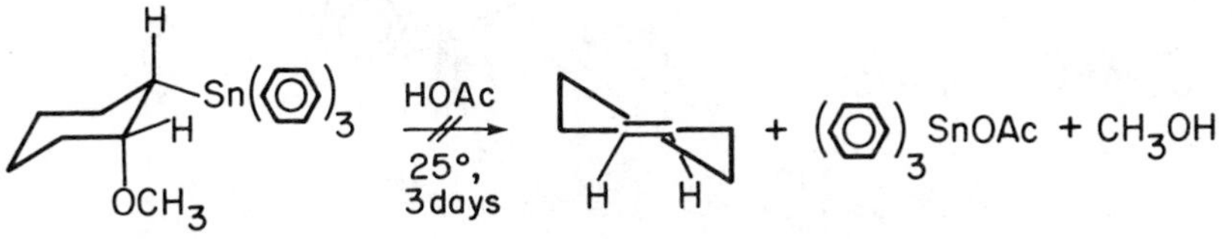

$$\text{cis-2-methoxycyclohexyl-}Sn(C_6H_5)_3 \underset{25°,\ 3\,days}{\overset{HOAc}{\not\longrightarrow}} \text{cyclohexene} + (C_6H_5)_3SnOAc + CH_3OH \quad (4)$$

These results are also consistent with those found with the corresponding cis- and trans-2-hydroxycyclohexyltrimethylsilicon derivatives under weakly acid conditions (8). Accordingly, any cis-2-hydroxy metabolite that might form would have been detected and was not. We also found that the ketones 6 and 7 from the alcohols 2 and 3, 4 were produced, and verified this result by TLC cochromatography, but again, we are confident of their assigned structures using this technique.

Compound 8, 1-hydroxycyclohexyltriphenyltin, a metabolite that might also be formed, could not be detected because of its probable low concentration and experimental difficulties associated with quantifying it. We found that 1-hydroxyalkyltin derivatives undergo electrophilic cleavage reactions of the tin-carbon bond bearing the hydroxyl group with hydrochloric acid to give the alcohol and the corresponding dealkylated tin derivative (1a). Consequently, reaction of 8 if present with hydrochloric acid would provide [1-^{14}C]cyclohexanol and triphenyltin chloride (Eq 5). In control experiments with [1-^{14}C]cyclohexanol, determined as its phenylcarbamate derivative, we could detect levels of this derivative down to 2% but not lower. Similar experiments with the reaction mixture gave no detectable phenylcarbamate of [1-^{14}C]cyclohexanol and thus we rationalized

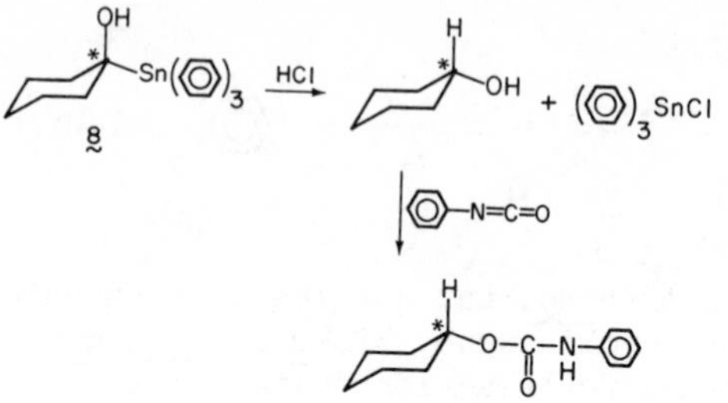

(5)

that < 2% or none of 8 was formed.

The study of [1-^{14}C]cyclohexyldiphenyltin acetate, 9, in those monooxygenase enzyme reactions was accomplished by converting compounds 1-5 to their corresponding cyclohexyldiphenyltin bromides. For example, 1 reacted with bromine in isopropanol/chloroform to give an excellent yield of [1-^{14}C]cyclohexyldiphenyltin bromide (Eq 6). Preparative TLC using diisopropyl ether/acetic acid (49:1) converted the bromide to the acetate, 9, as analyzed by 90 MHz ^{1}H nmr spectroscopy. Reaction of 9 with rat liver microsomes, as with 1, gave the following results after acidification of the reaction mixture and extraction with chloroform (Eq 7). The percentages in Eq 7 represent

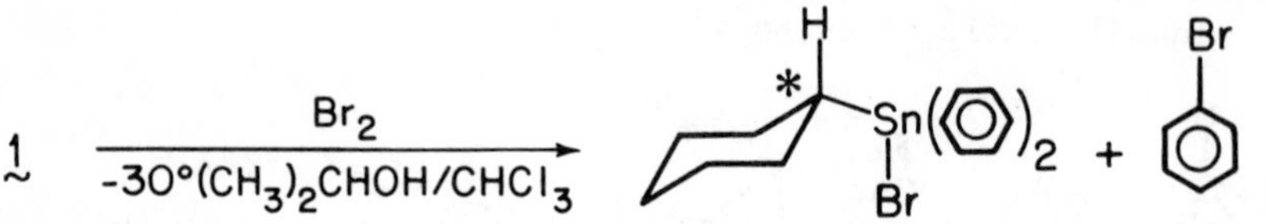

(6)

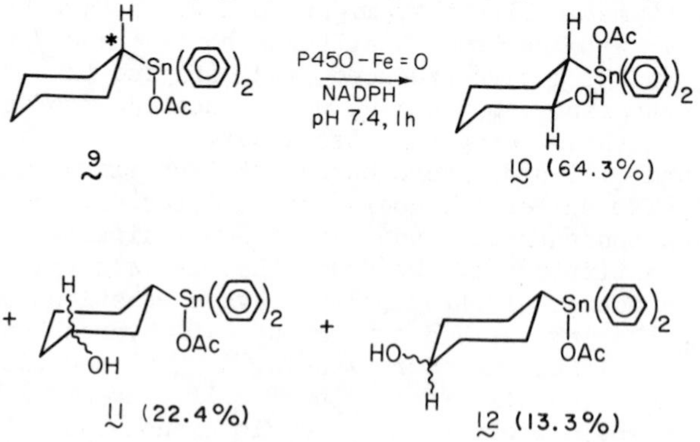

(7)

normalized values of identified metabolites which account for 10% of starting compound, 9. The remaining materials were 9 (82%) as well as 8% unidentified compounds.

The corresponding ketones from 11 and 12, although available, were not positively identified as metabolites because of interfering compounds and inconsistent cochromatography results, respectively, upon TLC analysis. We also analyzed for [1-^{14}C]-cyclohexanol, as previously described for 1-hydroxycyclohexyltriphenyltin, 8, but were not able to detect this compound as its phenylcarbamate due to experimental difficulties. Thus, 1-hydroxycyclohexyldiphenyltin acetate was not determined using this method.

Compounds 11 and 12 were identified by TLC cochromatography, but unfortunately, separation of both the cis- and trans-3- or -4-hydroxyl isomers was not successful by TLC and therefore no assignments could be made.

The trans stereochemistry of 10 was ascertained, as with 5, by reaction with glacial acetic acid. Since the TLC analysis of compounds 10-12 was performed in acidic solvents, metabolite 10 could not be directly analyzed as was done with metabolite 5. Consequently, trapping of the [1-^{14}C]cyclohexene was essential for quantification of 10. This was accomplished by reaction of the [1-^{14}C]cyclohexene with mercuric acetate in methanol to give the oxymercuration product, 13, which could be recrystallized to constant specific activity and quantified (Eq 8).

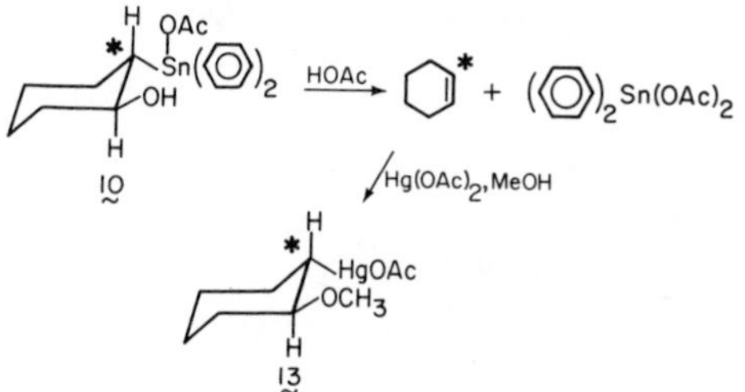

10 —HOAc→ cyclohexene* + $(C_6H_5)_2Sn(OAc)_2$; cyclohexene* —$Hg(OAc)_2$, MeOH→ 13 (8)

Discussion

Cytochrome P-450 monooxygenase enzyme reactions have been widely studied; however, the use of monosubstituted cyclohexyl derivatives as substrates has received only limited attention.

One such in vitro study (4) concerned methylcyclohexane, 14, and it was shown that P-450 enzyme hydroxylation gave the

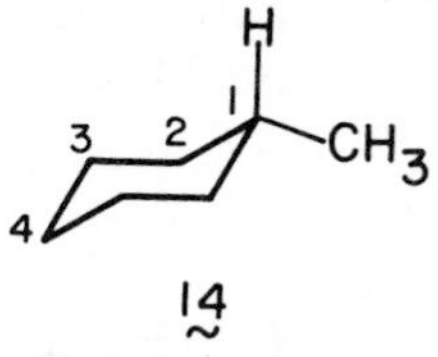

14

following situselectivity on a per hydrogen basis, i.e., $C_4 : C_3 : C_2 : C_1$ of 5 : 5 : 1 : 11. No stereochemical assignments were made, unfortunately, thus, this important aspect cannot be compared to our results. In contrast to the observed situselectivity for 14, compound 1 presents a dramatically different result. Thus, the situselectivity for 1 on a per hydrogen basis for $C_4 : C_3 : C_2 : C_1$ is 109 : 7 : 1 : 0.

This striking situselectivity for 1 is also complemented by a high degree of stereoselectivity for predominantly equatorial hydroxylated metabolites, with compounds 2, 3 and 5 representing 95% of the metabolites formed and giving an equatorial/axial ratio of 59.

While steric effects could be invoked to explain the situselectivity differences between 1 and 14, it should be noted that the carbon-tin bond length is 2.18 Å as compared to a carbon-carbon bond length of 1.54 Å and this should alleviate the steric bulk of the triphenyltin group. What could possibly account for the situselectivity is the highly specific manner by which 1 must bind to the membrane in proximity to the P-450-Fe=O complex. By viewing Drieding models of 1, it was realized that the triphenyltin group could provide a template for the possible positioning of the cyclohexyl group towards the Fe=O complex as shown in Figure 2.

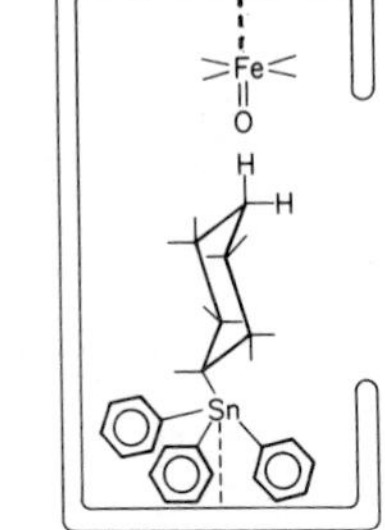

Figure 2. One possible conformation of 1 at the active enzyme site

Interestingly, removal of one phenyl group, i.e., compound 9, provides a vastly different situselectivity, with the ratio per hydrogen for $C_4 : C_3 : C_2$ being 1.2 : 1 : 2.9. Evidently, the conformation of 9 with respect to binding in proximity to the Fe=O complex is different than 1. It is highly conceivable that with the triorganotin derivative a tetrahedral configuration rather than a trigonal-bipyramidal configuration is more important. Since 9 has a replaceable acetate group, binding may be depicted as in Figure 3. This would make hydrogens on the 2,

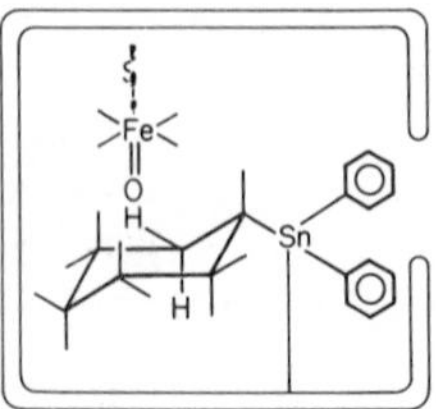

Figure 3. One possible conformation of 9 at the active enzyme site

and 3 carbon atoms of 9 more accessible than for comparable hydrogens in 1 (Figure 2).

We wished to learn more about the binding differences between compounds 1 and 9. Visible spectrophotometry has been utilized to determine the binding characteristics of a wide variety of compounds in proximity to oxidized (Fe^{3+}) P-450 (9). We thought similar spectra might show the usual optical differences associated with type I and II binding. Figure 4 shows the optical difference spectra for compounds 1 and 9 as well as for triphenyltin and tributyltin chlorides at ~ 1.5×10^{-5}M with rat liver microsomes (0.43 nmole P-450/ml phosphate buffer, pH 7.4). Compounds 1 and 9 showed troughs at 386-389 nm and a peak at 418 nm. This was somewhat similar to triphenyltin chloride, 385 and 414 nm, but opposite to tributyltin chloride which had a peak at 410 nm and a trough at 425 nm. The former compounds are shifted type II spectra (trough at 390-410 nm and peak at 424-430 mm), while the latter gave a shifted type I curve (peak at 385-390 and trough at 416-420 nm). These dramatic differences between the aryl- and alkyltin compounds might provide one explanation as to why the triphenyltin group does not get hydroxylated, i.e., the binding places this group away from the Fe=O complex (Figure 2). The spectral features of 1 and 9, while slightly different, cannot be definitively interpreted as evidence for the possible binding changes we envision for both substrates (Figures 2 and 3).

The stereoselectivity provides at each carbon atom the thermodynamically most stable equatorial hydroxylated isomer. In our previous studies with the tributyltin derivatives, we postulated from the observed situselective hydroxylation pattern and other criteria that a free radical mechanism was operative in these biological oxidation reactions (1a). The present results with compounds 1 and 9 clearly demonstrate that this free radical process can be highly stereospecific (Eq 9).

H abstraction → OH trapping (9)

In regards to this specificity, a recently reported result (10) using a purified P-450 enzyme preparation from rabbit liver microsomes with exo-D_4-norbornane reveals a pronounced stereoselectivity for exo hydrogen abstraction (k exo/k endo = 7)

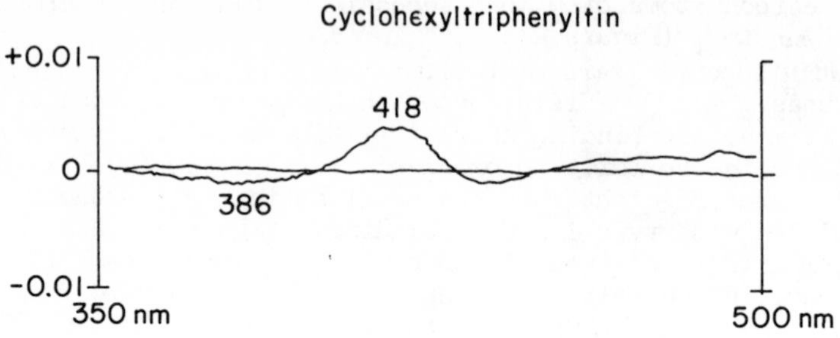

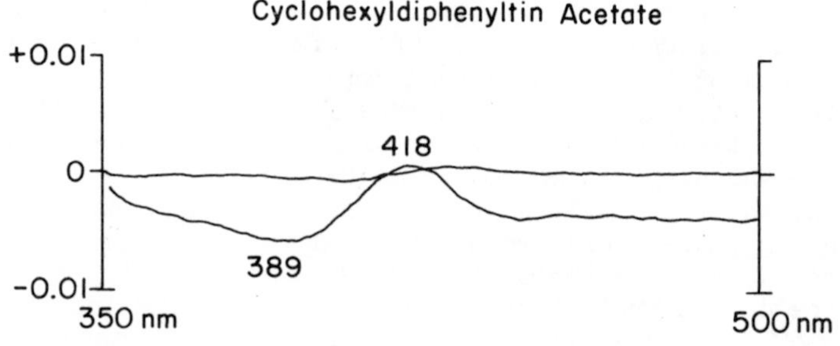

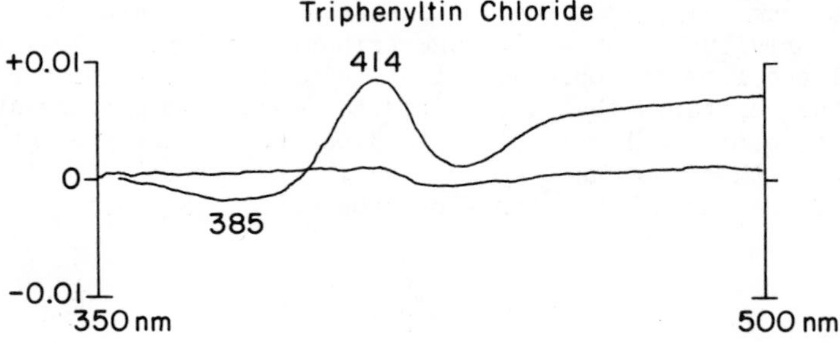

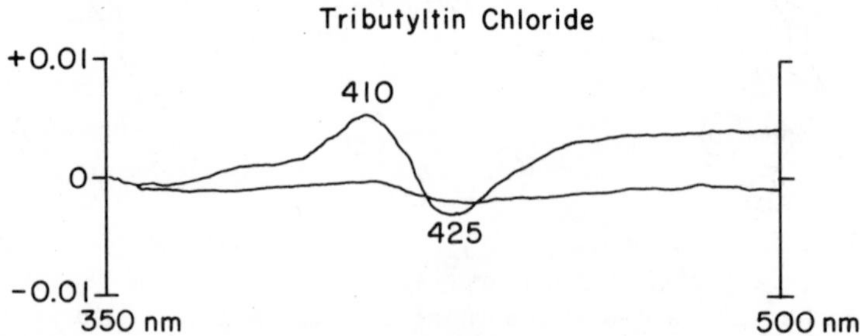

Figure 4. Optical difference spectra with P-450 (Fe^{3+}) oxidized enzyme

and a large kinetic isotope effect (k_H/k_D = 11). Groves *et al.* (10) interpreted these results in terms of a homolytic C-H bond scission giving a long lived free radical capable of epimerizing, i.e., *endo* hydrogen abstraction to give *exo* alcohol (25% of *exo*-norborneol had retained all deuterium).

The important question in our case is whether the Fe=O complex homolytically removes equatorial hydrogens in a totally stereospecific reaction (as implied in Figure 2) or whether some of the equatorial products come from axial hydrogen abstraction and can only be ascertained with specific deuterium labelling. Our results and that of Groves *et al.* (10) make the free radical hydroxylation with the P-450 Fe=O complex a viable pathway and our results show that this can occur with high stereoselectivity.

In conclusion, the reactions of cyclohexyltin compounds, **1** and **2**, with a P-450 monooxygenase enzyme system provide hydroxycyclohexyl metabolites in a stereo and situspecific manner and indicate that the conformational aspects of the substrate, upon binding near the Fe=O complex, can control these biotransformations.

Abstract

The reactions of a cytochrome P-450 monooxygenase enzyme system from rat liver microsomes with cyclohexyltriphenyltin and cyclohexyldiphenyltin acetate were studied. The stereo and situselectivity with cyclohexyltriphenyltin showed a strong preference for equatorial hydroxylation occurring predominately at the 4-position (*trans*-4-hydroxycyclohexyltriphenyltin). Thus, the conformation of the cyclohexyl group in proximity to the P-450 ferryl oxygen complex seems to be an important factor in determining the situselectivity. Furthermore, by removing one phenyl group, i.e., cyclohexyldiphenyltin acetate, the pattern of carbon-hydroxylation is more equally distributed on the 2, 3 and 4 carbon positions. The stereochemistry in the case of this latter system was only verified for the *trans*-2-hydroxycyclohexyldiphenyltin acetate metabolite. The results with cyclohexyldiphenyltin acetate further substantiate the conformational factor as critical for the observed situselectivity.

Acknowledgments

We wish to thank Dr. W. W. Conover of the Stanford Magnetic Resonance Laboratory for obtaining the 360 MHz ^1H nmr spectrum of metabolite **2**. SMRL is supported by grants NSF GR 23633 and NIH RR 00711. The work reported herein was supported by the National Institute of Environmental Health Sciences (Grant P01 ES00049).

References

1. a. R. H. Fish, E. C. Kimmel, J. E. Casida, J. Organometal. Chem. (1976) 118, 41.
   b. E. C. Kimmel, R. H. Fish, J. E. Casida, J. Agr. Food Chem. (1977) 25, 1.
2. J. H. Dawson, J. R. Trudell, G. Barth, R. E. Linden, E. Bunnenberg, C. Djerassi, R. Chiang, L. P. Hager, J. Amer. Chem. Soc. (1976) 98, 3707 and references therein.
3. K. R. Hanson, I. A. Rose, Accounts Chem. Res. (1975) 8, 1.
4. U. Frommer, V. Ullrich, H. Staudinger, Hoppe-Seyler's Z. Physiol. Chem. (1970) 351, 903.
5. R. H. Fish, J. E. Casida, E. C. Kimmel, Tetrahedron Lett. (1977) 3515.
6. R. H. Fish, B. M. Broline, J. Organometal. Chem. (1977) 136, C41.
7. R. H. Fish, B. M. Broline, Ibid. (1978) in press.
8. M. De Jesus, O. Rosario, G. L. Larson, Ibid. (1977) 132, 301.
9. A. P. Kulkarni, R. B. Mailman, E. Hodgson, J. Agr. Food Chem. (1975) 23, 177 and references therein.
10. J. T. Groves, G. A. McClusky, R. E. White, M. J. Coon, Biochem. Biophys. Res. Comm. (1978) 81, 154.

Discussion

J. J. ZUCKERMAN (University of Oklahoma): Could there be another pathway with the cleavage of the cyclohexane group from tin, and would you know anything about the steps along that route?

FISH: You mean without hydroxylation?

ZUCKERMAN: With or without, but with separation from the tin?

FISH: I think that you need hydroxylation in order to separate the cyclohexyl group; either 1-hydroxylation (which we didn't see) that would result in the loss of tin or the cyclohexyl group, and/or formation of the 2-hydroxy compound which is also labile under protic conditions.

F. HUBER (University of Dortmund): I would like to comment on the mechanism you proposed. I feel a bit uneasy that you propose a coordination number of five (a trigonal bipyramid) when you have the hydroxylation of the triphenyltin compound, and you propose only a tetrahedral coordination when you have a diphenyltin acetate. It's normal in the coordination chemistry of these organotin compounds that there is no great tendency for pentacoordination with tetraorganotin compounds, and the tendency for coordination increases when you reduce the number of organic ligands at the tin center.

FISH: You mean the mechanism of the binding (cf. Figures 2 and 3). I think tetraorganotin compounds can, in fact, form trigonal bipyramidal-type intermediates in solution or on a protein surface. You do believe that the triorganotin compounds can bind the way I depicted?

HUBER: Yes, they certainly can but they show much greater tendency to have a trigonal bipyramidal configuration than the other ones.

ZUCKERMAN: We have isolated a complex of a tetraalkyltin compound with a bromide ion. I think Dr. Fish has this intermediate as a step in a kinetic pathway, and maybe it wouldn't have to last very long to do its job.

HUBER: It's the same problem we have when we study reactions of tetraorgano-tin or -lead compounds. We have to assume some kind of coordination for a short time.

A. J. CANTY (University of Tasmania): You mentioned that the phenyl groups do not get hydroxylated and this has also been shown for phenylmercury compounds in rat liver. Is this a general phenomenon for aryl organometallics?

FISH: I'm not sure. I don't think a wide spectrum of phenylmetal compounds has been studied. One of the problems is the binding; i.e., indiscriminate binding at other sites. It also could be the mechanism of hydroxylation.

J. M. WOOD (University of Minnesota): The hydroxylation of aromatic compounds by P450 in the liver is well known, i.e., hydroxylation of the steroids. It could be that liver is a bad choice with these metal complexes. I'm sure you could find some microbial communities that will produce P450 and hydroxylate aromatic rings attached to metal ions.

FISH: It may happen in vivo, where we see triphenyltin being metabolized to diphenyl or monophenyl, but it's not apparent what the mechanism is if we can't do it in vitro.

RECEIVED August 22, 1978.

7

Anaerobic and Aerobic Alkylation of Arsenic

BARRY C. McBRIDE, HEATHER MERILEES, WILLIAM R. CULLEN, and WENDY PICKETT

Departments of Microbiology and Chemistry, University of British Columbia, Vancouver, British Columbia, V6T 1W5 Canada

Arsenic and its derivatives are important as herbicides and are used in, or are by-products of, many industrial processes. Large quantities of this potentially toxic compound are mobilized and placed in new environments where they accumulate in concentrations which may exceed the normal arsenic burden of these ecosystems. The microflora in these environments possess the metabolic machinery to transform these compounds into gaseous arsines. Challenger (1) described the formation of trimethylarsine by fungi. Cox and Alexander (2) have shown that soil organisms will produce trimethylarsine and McBride and Wolfe (3) reported that the anaerobic methane bacteria synthesized dimethylarsine. A consequence of this microbial activity is the modification of arsenic to new compounds possessing different chemical, physical, and biological properties.

Methanogenic Bacteria

The methanogenic bacteria are a unique group of microorganisms which produce methane as their principal metabolic end product. They are found in large numbers in anaerobic ecosystems when organic matter is decomposing. As a group they are morphologically diverse, embracing coccal, bacillary and spiral forms. They are extremely sensitive to O_2, a factor which has contributed to our limited understanding of their biochemical activities. A restricted number of substrates can be reduced to methane these include: CO_2, formate, acetate, and methanol. Hydrogen is the preferred source of electrons. The pertinent characteristics of the methane bacteria are summarized in Table I. There are several (4-8) reviews of these organisms which deal with their isolation, characterization and biochemistry.

The methane bacteria function as the terminal members of the

0-8412-0461-6/78/47-082-094$05.50/0

Table I. Characteristics of the Methanogenic Bacteria

Organisms	Habitat	Morphology	Substrates
Methanobacterium ruminantium	rumen sludge	coccus to short rods	H_2+CO_2 formate
Methanobacterium strain M.oH	mud sludge	irregularly curved rods	H_2+CO_2
Methanobacterium formicicum	mud, sludge	irregularly curved rods	H_2+CO_2
Methanobacterium mobilie	rumen	short rod, motile	H_2+CO_2 formate
Methanosarcina barkerii	mud, sludge	sarcina	H_2+CO_2 methanol, acetate
Methanococcus vannielii	mud	motile coccus	H_2+CO_2 formate
Methanospirilium hungatii	mud,sludge	spirillum	H_2+CO_2 formate
Methanobacterium thermoautotrophicum	mud, sludge hot springs	irregularly curved rod	H_2+CO_2 formate

anaerobic food chain, scrubbing the environment of potentially toxic components and releasing a non-toxic end product which eventually escapes into oxygenated environments.

A generalized scheme (6) illustrating the essential steps required to synthesize a molecule of methane is summarized in Fig. 1. The scheme accounts for all the known methane precursors and indicates the reactions required to form a completely reduced carbon molecule. Compound X is a carrier which may represent one or more molecular species. Our understanding of the steps in methane biosynthesis is sketchy and is almost entirely limited to the terminal methyl transfer reactions. No intermediates between CO_2 and $X\text{-}CH_3$, have been identified. The carrier molecule X has been postulated to be required because no free reduced C-1 intermediates have been identified.

Progress has been made in understanding the mechanism of the methyl transfer reactions. Blaylock and Stadtman (9) demonstrated that methyl cobalamin would serve as a substrate for methane biosynthesis in cell extracts of Methanosarcina barkerii. Subsequently Wolin et al (10) found that methylcobalamin was a substrate for methane biosynthesis in Methanobacterium omelianskii. Blaylock was able to resolve the Methanosarcina (11) system into a number of components, one of which was a corrinoid containing protein. Wood (12) was successful in isolating a cobalamin containing protein from M. omelianskii. The protein stimulated CH_4 production when added to cell extracts (13) of the organism. These studies suggested that B_{12} might be the X factor involved in methyl transfer. This seemed particularly feasible in light of the role of $CH_3\text{-}B_{12}$ in the CH_3 transfer reactions leading to methionine biosynthesis.

Unfortunately, Wood's culture was found to consist of two organisms; a methanogen (Methanobacterium strain M.oH) which reduced CO_2 to CH_4 and a second organism which supplied electrons for methanogenesis by oxidizing ethanol to acetate and hydrogen (14). Results from investigations with the mixed culture must be viewed with caution because it is impossible to identify with certainty the genetic origin of the B_{12} protein which stimulated methanogenesis.

Biochemical studies have been complicated by the lability of the extracts and until a few years ago it was not possible to fractionate the cells or resolve them for specific components of the methane synthesizing system. As a consequence it was necessary to use stimulation of CH_4 biosynthesis as the assay for constituents which had been fractionated from other extracts. In this type of assay the reaction mixture was complete and would support a limited amount of CH_4 biosynthesis. Fractionated cell extract was then added to the reaction mixture and if it stimulated CH_4 production it was implicated in the terminal methyl transfer reactions. A difficulty with this type of assay in a complex multistep reaction is that one can only measure stimulation of a rate limiting reaction. In methane biosynthesis, stimulation could result from increased electron transport, activation of ATP, or transfer of methyl groups. Thus the implication of a corrinoid protein in CH_4 biosynthesis by *M. omelianskii* was based on rather tenuous evidence, but certainly the best evidence that could be developed at that time.

The isolation of the pure culture of *M.oH* facilitated the study of methanogenesis; cell extracts were not as labile and catalyzed CH_4 formation at an increased rate. In addition these extracts reduced CO_2 to CH_4 (7).

Attempts to isolate the corrinoid protein from *M.oH* have been unsuccessful (15). In fact no B_{12} containing enzymes appear to be present in this organism. The red protein isolated by Wood was shown to be synthesized by the non-methanogenic contaminant (15) and thus could not be a part of the *in vivo* methane synthesizing system. The inability to find a corrinoid protein cast some doubt on the importance of CH_3-B_{12} in terminal methyl transfer reactions.

The question was answered with the discovery of the methyl donor, CH_3-CoM. CoM is a low molecular weight, heat stable cofactor found in all methane bacteria that have been examined (16). Chemically CoM is 2,2'-dithiodiethane sulfonic acid (17). It can be reduced and methylated chemically or biologically to form 2-(methylthio) ethanesulfonic acid. In the presence of cell extract, ATP, and H_2,CH_3-CoM is reductively cleaved to yield CH_4 and CoM (Fig. 2). Extracts can be resolved for CoM by anaerobic dialysis and will synthesize CH_4 only if they are supplied with CoM. It seems reasonable to assume that B_{12} is not involved in CH_4 biosynthesis in *MoH*, and that this reaction is catalyzed by CoM. It is not clear at this time whether *Methanosarcina* depends on B_{12}, as this organism has been shown to possess CoM in addition to a corrinoid protein. Possibly this organism possesses 2 methane synthesizing pathways, one dependent and one independent of B_{12}.

Alkylation of Metals

M.oH has been implicated as an organism which will alkylate mercury (18) and arsenic (3). Alkylation of metals by *M.oH* could theoretically occur at any biosynthetic step which generated a C-1 group. Because the methanogenic bacteria process large numbers of C-1 units it is not unreasonable to assume that the CH_4 biosynthetic pathway could be important in these reactions, and that CoM would be a molecule of interest. It would be unwise to forget that methyl groups must be generated for the synthesis of methionine and that intermediates in this reaction could be involved in metal alkylation. Two methionine biosynthetic pathways operate in bacteria. One system involves transfer of a methyl group from N^5-methyl-tetrahydrofolic acid (N^5, CH_3THFA) to B_{12} and then to homocysteine. The methylation of homocysteine requires catalytic amounts of S-adenosyl methionine (SAM). The second system requires N^5, CH_3-THFA tri-glutamate and SAM. The inability to find a corrinoid protein in M.oH suggests that the latter mechanism may operate in these organisms.

Arsine Biosynthesis

Cell extracts of M.oH produce a strong garlic odor when they are incubated with arsenate. The following section of this paper will deal with the biochemical studies which led to the identification of the garlic smelling compound and to a scheme for its biosynthesis. This will be followed by a study of arsenic transformation in natural anaerobic systems.

Alkylarsine Assays. The reaction vessel and conditions used for the biosynthesis of alkylarsines were similar to those described for methane biosynthesis (19). To trap alkylarsines the reaction vessel was connected by polyethylene tubing to a glass tube which contained 2 mL of 2M HNO_3. The trapping tube was placed in an ethanol-ice bath. A slow stream of H_2 was passed into the reaction flask and the volatile alkylarsines were swept into the nitric acid, where they were condensed and trapped by oxidation to non-volatile acids (20). The contents of the trap were assayed for arsenic by atomic absorption spectrometry and for ^{14}C by counting in Bray's scintillation fluid.

The isotope, ^{74}As was used to follow the formation of volatile alkylarsine. Extracts were incubated with [^{74}As]-Na_2HAsO_4 and the volatile methylated ^{74}As was measured. Advantage was taken of the property of alkylarsines to react with the red-rubber serum stoppers which were used to seal the reaction flasks. At the appropriate times the reaction mixture was inactivated by heating on a steam cone. The flasks were then incubated for an additional 20 min to ensure that all the volatile dimethylarsine was adsorbed onto the rubber stopper. The serum stopper was removed from the flask, rinsed in water, cut in half, and placed in a scintillation vial together with 15 mL of Bray's scintillation fluid. This mixture was either counted directly, as the rubber stopper did not quench the high-energy β particles emitted by ^{74}As, or the stopper was removed after the ^{74}As had been leached from the rubber (1 hr.). The rubber stopper technique was found to be an efficient and specific means of separating [^{14}C] dimethylarsine from [^{14}C]CH_4.

Requirements for Biosynthesis. Conditions for the biosynthesis of alkylarsine derivatives are described in Table II. In these experiments cell extracts were incubated under anaerobic conditions with [^{74}As]Na_2HAsO_4, and the formation of volatile ^{74}As compounds was measured. A methyl donor, H_2, ATP, and arsenate were required for the reaction; with the exception of arsenate these same components are required in the CH_4-synthesizing system. A total of 34 μg of arsenic was found in an alkylarsine trap linked to a reaction flask which contained arsenate and cell extract, providing evidence that a volatile alkylarsine was being synthesized. Volatile alkylarsine compounds were not detected in a similar control reaction mixture which contained arsenate and boiled extract.

Substrates. CH_3-B_{12} and CO_2 were shown to be methyl donors for alkylarsine synthesis. Arsenate, arsenite and methylarsonic acid were reduced to dimethylarsine in the presence of a C-1 donor. Cacodylic acid was reduced in the absence of a C-1 donor. Methylarsonic acid was found in extracts incubated with AsO_4^{3-} and a C-1 donor.

Whole Cells. The ability of whole cells to synthesize alkylarsine is shown in Figure 3. In this experiment a sample of an actively growing culture was removed anaerobically from a 12-L fermentor and was incubated anaerobically under a H_2:CO_2 atmosphere.

Analysis of the Alkylated Arsine. The structure of the alkylarsine formed in cell extracts was indicated by two experimental procedures. In one procedure Na_2HAsO_4 was incubated with [^{14}C]methylcobalamin, and the trapped alkylarsine was analyzed for arsenic and ^{14}C. The results of two such experiments are presented in Table III. Dissolved [^{14}C]CH_4 was removed by bubbling CH_4 through the trapping solution after it had been disconnected from the reaction flask. The number of methyl groups in the trap was calculated from the specific activity of [^{14}C]CH_3-B_{12}. This value was divided by the amount of arsenic to obtain a methyl group to arsenic ratio. The ratios obtained (1.8:1 and 1.9:1) indicate that the compound is dimethylarsine.

These results were substantiated by a double-labeling experiment in which [^{74}As]Na_2HAsO_4 and [^{14}C]methylcobalamin were substrates. The results of two such experiments are shown in Table IV. Alkylarsine was separated from contaminating [^{14}C]CH_4 by trapping in a rubber serum stopper as described previously. The specific activity of the substrates was used to calculate the micromoles of arsenic as well as methyl groups. The ratios of methyl groups to arsenic (2.1:1 and 1.9:1) suggest that the garlic-smelling compound is dimethylarsine.

A pathway for dimethylarsine synthesis has been proposed (3) Fig. 4. Arsenate is first reduced to arsenite which is then methylated to form methylarsonic acid. This compound was found in reaction mixtures when extracts were incubated with either arsenate or arsenite and CH_3-B_{12}. Methylarsonic acid is reduced and methylated to form dimethylarsinic acid. The acid is then reduced to form dimethylarsine. All intermediates have been shown to be converted to dimethylarsine by cell extracts. This pathway is based on studies in which CH_3-B_{12} was used as methyl

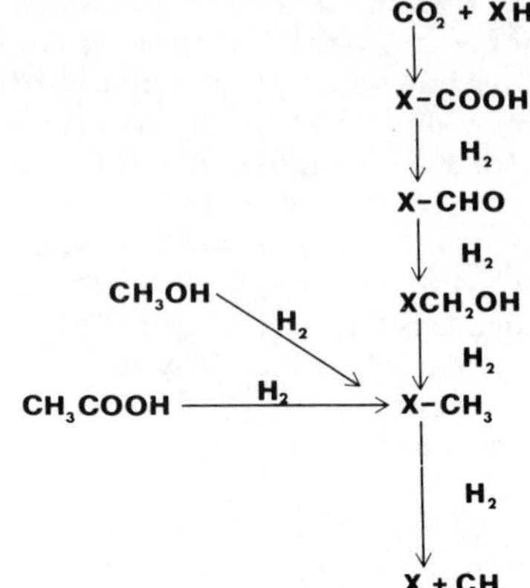

Figure 1. Possible pathway for methane biosynthesis. C-2 of acetate is preferentially reduced to methane.

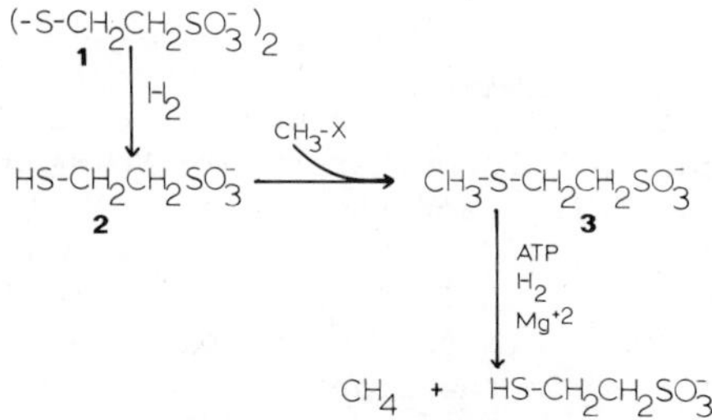

Figure 2. Role of CoM in methane biosynthesis. (1) 2,2′-dithiodiethane sulfonic acid; (2) 2-mercaptoethane sulfonic acid; (3) 2-(methylthio)ethane sulfonic acid.

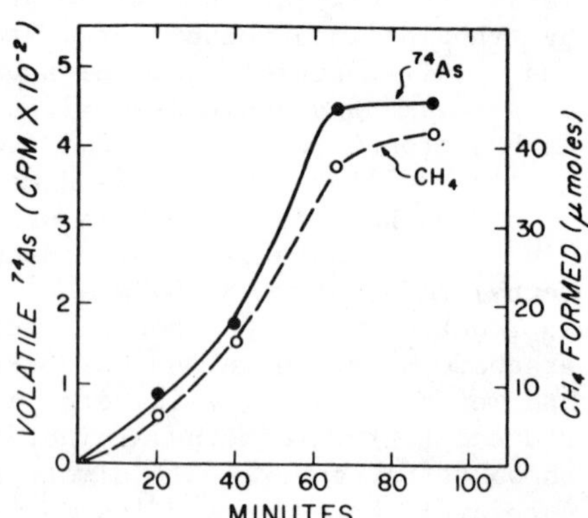

Figure 3. Synthesis of alkylarsine by whole cells of Methanobacterium strain M.oH. *Each reaction mixture contained whole cells, ^{74}As–Na_2HAsO_4, and H_2–CO_2 (80:20).*

Table III. Identification of Dimethylarsine.

Each reaction flask contained: potassium phosphate buffer, adenosine triphosphate, $[^{14}C]CH_3$-B_{12}, Na_2HAsO_4, crude cell extract; the arsines were trapped in 2 N HNO_3 and assayed for arsenic by atomic absoprtion spectrometry and for radioactivity by scintillation counting in Bray's solution. Calculated from specific activity of $[^{14}C]CH_3$-B_{12} added. Determined by atomic absorption spectrometry.

Flask	Omissions	Volatile ^{74}As (cpm)
1	None	13,400
2	$-CH_3$-B_{12}	100
3	$-H_2$	0
4	-ATP	3,600
5	-Cell extract	0
6	None (reaction mixture boiled)	0

Table II. Requirements for Dimethylarsine Synthesis by Cell-Extracts of Methanobacterium Strain M.O.H.

Complete reaction mixture contained; TES buffer, adenosine triphosphate, CH_3-B_{12}, crude cell extract, $[^{74}As]Na_2HAsO_4$, H_2, air replaced H_2 in flask 3; Flask 6 was heated for 5 min. in a boiling-water bath prior to incubation.

Expt	CH_3 Groups (μmole)	As (μmole)	Ratio CH_3:As
1	0.058	0.033	1.90
2	0.096	0.053	1.81
Control	0	0	

donor; it is postulated that methylation by the product of CO_2 reduction follows the same scheme.

Arsenic transformation in anaerobic ecosystems.

The studies on pure cultures of M.oH proved that the organism could make dimethylarsine. It then became of interest to determine if the methanogenic bacteria would synthesize the molecule in a natural ecosystem. Sewage sludge from an anaerobic digester was chosen as the methanogenic source because it is an area of intense biological activity, populated by large numbers of methanogenic bacteria actively synthesizing methane.

Sludge was collected fresh daily from an anaerobic digester; exposure to oxygen was minimized. The sludge was transferred to 18 mm test tubes and amended with $[^{74}As]Na_2HAsO_4$. The tube was then gassed with O_2 free H_2/CO_2 (80/20 v/v) and sealed with a rubber stopper. Strips of rubber suspended at various points in the sludge were used as arsenic traps. The traps were removed at designated intervals, washed thoroughly to remove unreacted ^{74}As, and then placed in Bray's scintillation fluid and counted immediately.

The transformation of arsenate to a form which will react with a rubber stopper is described in Fig. 5. As can be seen, transformation is rapid and complete within 8 hours. There was no transformation to a rubber reactive derivative when the sludge was heated to 90° for 15 min. prior to adding the arsenate. The sensitivity of the system to heat suggests that transformation is a biological process requiring living cells and is not likely to be a reaction mediated by compounds present in the sewage. The reaction was dose dependent; addition of increasing levels of arsenate resulted in a corresponding increase in the amount of transformed arsenic.

It was not possible to identify the arsenic compound in the rubber stoppers using the methodology employed in pure culture studies because the levels of carbon precursors could not be accurately established. Attempts were made to isolate the products by anaerobically extracting the sewage mixture at various points during the reaction. Rubber stoppers were left out of the reaction in the hope that the transformed metal would remain free and thus be available for chemical analysis. None of the extraction procedures produced detectable levels of alkylated arsenic. It seemed possible that the transformed arsenic, which is believed to be an arsine, was reacting with some other constituent in the sewage sample and was further transformed before it was extracted. To test this hypothesis an experiment was designed in which traps were placed in the reaction mixture at various times after addition of the arsenate and then all the traps were removed at the same time. The control experiment in which stoppers were in the reaction mixture when arsenate was added and were removed at various times show the typical result of increasing incorporation of the transformed arsenic (Fig. 6).

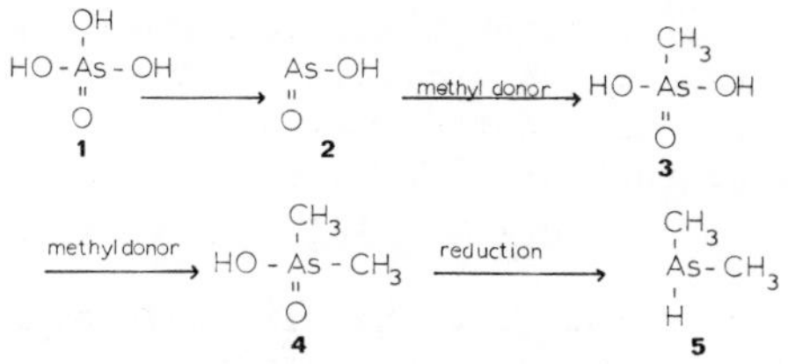

$$\underset{\mathbf{1}}{HO{-}As(=O)(OH){-}OH} \longrightarrow \underset{\mathbf{2}}{As(=O){-}OH} \xrightarrow{\text{methyl donor}} \underset{\mathbf{3}}{HO{-}As(=O)(CH_3){-}OH}$$

$$\xrightarrow{\text{methyl donor}} \underset{\mathbf{4}}{HO{-}As(=O)(CH_3){-}CH_3} \xrightarrow{\text{reduction}} \underset{\mathbf{5}}{H{-}As(CH_3){-}CH_3}$$

Figure 4. Proposed pathway for the biosynthesis of dimethyl arsine. (1) arsenate; (2) arsenite; (3) methyl arsonic; (4) cacodylic acid; (5) dimethyl arsine.

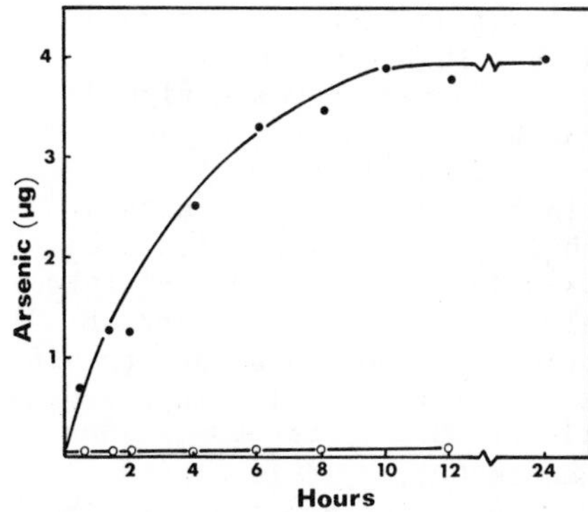

Figure 5. Transformation of arsenic in sludge. Sewage sludge was mixed with ^{74}As–Na_2HAsO_4 *and incubated anaerobically. Rubber traps were removed at the specified intervals, washed, and counted. Experimental, ●; sludge was boiled for 15 min prior to adding* ^{74}As–Na_2HAsO_4, ○.

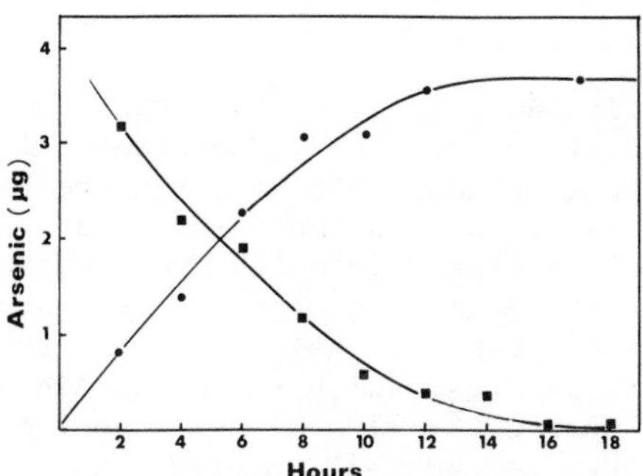

Figure 6. Lability of transformed arsenic. Rubber traps added at zero time and removed at times indicated, ●; Rubber traps added at times indicated and removed at 18 hr, ■.

In the experimental situation, there is a decrease in the number of counts in the traps which correlates with the time when the traps were placed in the reaction mixture. It can be concluded from this experiment that arsenic is transformed to a labile product which rapidly loses its ability to bind to rubber stoppers.

This experiment points out the difficulties of defining metal transformation in natural systems and indicates that the final product in an ecosystem may represent the culmination of a number of biological activities occurring in a specific sequence. It is important to know the nature of the intermediate products because they may have significantly different physical, chemical or biological properties than either the starting material or the final product. A knowledge of the intermediates may help explain how a metal will move through the food chain.

Nature of the Transformed Product. Assuming that the compound synthesized was dimethylarsine and that this compound was alkylated by methanogenic bacteria, an experiment was designed which took advantage of the fact that C-2 of acetate is the principle precursor of CH_4 in sewage sludge (21).

Reactions were set up as described previously but in this case ^{14}C-acetate was added as a methyl donor. The results are shown in Table V. Label from 2-^{14}C-acetate is incorporated into the traps in significant amounts only when there is concurrent addition of arsenate. There is little endogenous arsenate to act as a receptor for the methyl groups. Acetate labelled in C-1 did not label the arsenate as efficiently presumably because it was diluted by equilibration with endogenous CO_2. It is a good control showing that acetate itself is not incorporated into the traps or into arsenic. While this data can in no way be construed as proof that the transformed arsenic is an alkyl arsine we feel that it is at least highly suggestive that this is the case.

Role of Methanogenic Bacteria. The fact that the methyl carbon of acetate is readily incorporated into arsenic in the sludge mixture suggests that methanogenic bacteria are involved in the biotransformation reaction. Further proof that this is the case was provided by an experiment in which $CHCl_3$ was used to selectively poison the methane bacteria. Methanogenic bacteria are particularly susceptible to halogenated hydrocarbons (22) at levels which do not interfere with the metabolic activity of other organisms. When $CHCl_3$ (10^{-6}M) is incubated with sewage in the presence of arsenate and 2-^{14}C-acetate, CH_4 biosynthesis is inhibited as is the biosynthesis of arsine. The correlation between CH_4 biosynthesis and formation of ^{14}C-labelled arsenic was noted in other experiments in which the sludge had decreased methanogenic activity.

Arsenic Transformation in Other Anaerobic Ecosystems. Samples from a number of anaerobic ecosystems were incubated with arsenate to determine if they would produce a rubber reactive derivative. As shown in Table VI the ecosystems which produced CH_4 also transformed arsenate. Particular note should be made of rumen fluid which is particularly effective at transforming arsenate. This has important implications for animals which feed on arsenic treated foods and substantiates the suggestion of Peoples (23) that arsenic compounds are transformed in the rumen.

In summary we can say that the cell extracts of methanogenic bacteria will convert arsenate to dimethylarsine. An arsine is produced by whole cells of this organism but the nature of the arsenic derivative has not been determined. The methanogenic bacteria appear to play an important role in the biotransformation of arsenate in anaerobic ecosystems where methane is synthesized. The nature of the arsenic derivative is not known but it is probably an alkyl arsine which reacts rapidly with unidentified constituents in the environment.

Aerobic Methylation of Arsenic

Challenger and his coworkers (1) showed that trimethylarsine was the volatile garlic-smelling product, "Gosio gas", obtained when the mould *Scopulariopsis brevicaulis* growing on bread crumbs is exposed to arsenite. They also found that *S. brevicaulis* methylates, arsenate, methylarsonate, and cacodylate to trimethylarsine, and that other alkylarsonic and dialkylarsinic acids are methylated to the appropriate methylarsine, e.g. $(C_2H_5)(C_3H_7)AsO(OH)$ yields $(C_2H_5)(C_3H_7)As(CH_3)$. Challenger also established that aerobic alkylation by other microorganisms is substrate dependent. Thus *Penicillium notatum* does not methylate arsenate. In an interesting example of the same phenomenon the alkylation of benzene arsonic acid to $(C_6H_5)As(CH_3)_2$ by *Candida humicola* proceeds with ease (24) yet the same substrate is unaffected by *S. brevicaulis* (1). This result is particularly pertinent in view of the wide spread use of aromatic arsonic acids as food additives for chickens, turkeys, and swine (25) and the report of garlic like odors above soils treated with these compounds (26,27).

C. humicola was first investigated as an arsenic methylating agent by Cox and Alexander (2,25) who found that it acted on the same inorganic and methyl arsenical substrates as *S. brevicaulis* (1). More recent work shows that this organism also methylates longer chain aliphatic arsonic acids such as $C_4H_9AsO(OH)_2$. This result is seen in experiment 14, Table VII (28). For these experiments *C. humicola* was grown in liquid culture and the arsine evolved from the added arsenical was usually trapped and examined by VPC/mass spec. or by direct injection into a mass spectrometer. In a number of experiments L-methionine-methyl-d_3, $CD_3SCH_2CH_2$-$CH(NH_2)COOH$, was added and the results show considerable incorporation of the CD_3 label in the evolved arsine. Similar

Table IV. Identification of Dimethylarsine.

Each reaction flask contained: potassium phosphate buffer, adenosine triphosphate, crude cell extract, $[^{74}]Na_2HAsO_4$, $[^{14}C]CH_3-B_{12}$, H_2, atmosphere. Arsine was trapped in the serum stopper, serum stoppers were washed, placed in Bray's scintillation fluid for 1 hr. and then removed. The scintillation fluid was counted simultaneously on two channels in a Mark III scintillation counter (Nuclear-Chicago).

Expt	Compound	Volatile cpm	mole of CH_3 Groups or As	Ratio $^{14}C:^{74}As$
1	^{74}As	28.0×10^3	0.0036	1.9
	^{14}C	6.7×10^3	0.0071	
2	^{74}As	16.7×10^3	0.0020	2.1
	^{14}C	4.26×10^3	0.0044	

Table V. Incorporation of ^{14}C from ^{14}C-acetate into Arsenic.

Sewage sludge was incubated anaerobically. The rubber traps were removed after 8 hrs. washed and counted.

	^{14}C (CPM x 10^4)	
	Heated	Unheated
2-^{14}C-Acetate	0	210
2-^{14}C-Acetate + Arsenate	0	7800
1-^{14}C-Acetate	0	75
1-^{14}C-Acetate + Arsenate	0	500

Table VI. Transformation of Arsenic in Anaerobic Ecosystems.

Samples from anaerobic ecosystems were placed in a $H_2:CO_2$(80:20 v/v) atmosphere and mixed with $Na_2{}^{74}AsO_4$; a rubber trap was placed for CH_4 and the trap for ^{74}As. Control assays were performed on samples which had been heated to 80° for 30 min. prior to the addition of the isotope and the trap.

	As (CPM)		CH_4 (nmoles)	
	Heated	Unheated	Heated	Unheated
sewage	70	7900	0	7000
marine mud	120	200	0	-7
rumen	900	4500	0	1200
compost	100	1524	0	5

Table VII. Volatile Arsines Produced by Microorganisms. Samples were collected during the fourth and fifth days after inoculation. [a] *VPC/mass spec determination.* [b] *Substrate added two days after inoculation.* [c] *Ethionine.* [d] $C_4H_9As(CH_3)_2$. [e] $C_4H_9As(CH_3)(CD_3)$. [f] $C_4H_9As(CD_3)_2$. [g] *Collection was made over a 7-day period.*

Expt. number		Substrates (mM): Arsenic compound	Substrates (mM): L-$CD_3SCH_2CH_2CH$-(NH_2)COOH	Deuterated species (%): $(CH_3)_3As$	$CD_3As(CH_3)_2$	$(CD_3)_2AsCH_3$	$(CD_3)_3As$
Candida humicola							
1[a]	As_2O_3	5	2	0.6	3.2	13.3	83
2[a]	$CH_3AsO_3Na_2$	5	2	0.7	7.8	84.7	6.8
3[a]	$CD_3AsO_3Na_2$	5	2	<1	2.7	10.1	87
4	$CD_3AsO_3Na_2$	5	-	8	92	-	-
5	$(CH_3)_2AsO_2H$	5	5	12.6	80.5	3.6	3.2
6	$(CH_3)_2AsO_2H$	5	0.5	57	39	2.6	1.4
7	$(CH_3)_2AsO_2H$	1	1	28.6	71.4	-	-
8	$(CH_3)_2AsO_2H$	1	0.1	59.3	40.7	-	-
9[a]	$(CH_3)_2AsO_2H$	5	5	12.1	82.5	2.4	2.7
10[b]	$(CH_3)_2AsO_2H$	5	2.5	11.6	85.3	2.1	1.0
11[b]	$(CH_3)_2AsO_2H$	5	1.0	20	78	0.6	1.4
12[b]	$(CH_3)_2AsO_2H$	5	0.5	24.5	74	0.7	0.7
13[b]	$(CH_3)_2AsO_2H$	0.5	1.0	-	- no arsines detected		
14	n-$C_4H_9AsO_3Na_2$	5	2.5	6.06[d]	16.1[e]	77.9[f]	-
15	$(CH_3)_2AsO_2H$	5	2.5[c]	-	- no arsines detected		
Scopulariopsis brevicaulis							
16[b]	$CD_3AsO_3Na_2$	2.5	-	4.3	95.7	-	-
17	$CH_3AsO_3Na_2$	2.5	1.25	56.5	25.7	10.3	7.5
18	$(CH_3)_2AsO_2H$	2.5	2.5	33.9	41.3	24	0.8
19	n-$C_4H_9AsO_3Na_2$	2.5	1.25		54[e]	45.9[f]	-
20	$(CH_3)_2AsO_2H$	2.5	2.0[c]	-	no arsines detected		
Gliocladium roseum							
21[g]	$CH_3AsO_3Na_2$	5	2.5	73	23	5.2	-
22	n-$C_4H_9AsO_3Na_2$	5	2.5	-	no arsines detected		

incorporation are found for S. brevicaulis and for Gliocladium roseum (28).

The Mechanism of the Aerobic Methylation of Arsenic

The results in Table VII strongly indicate that methionine, or S-adenosylmethionine, SAM, is the source of the methyl group in the biological alkylation of arsenic. They support the following oxidation-reduction pathway involving carbonium ions originally suggested by Challenger (1).

$$As^{V}O_4^{3-} \xrightarrow[-O^{2-}]{2e} \ddot{A}s^{III}O_3^{3-} \xrightarrow{CH_3^+} CH_3As^{V}O_3^{2-}$$

$$\xrightarrow[-O^{2-}]{2e} CH_3\ddot{A}s^{III}O_2^{2-} \xrightarrow{CH_3^+} (CH_3)_2As^{V}O_2^{-} \xrightarrow[-O^{2-}]{2e} (CH_3)_2\ddot{A}s^{III}O^{-}$$

$$\xrightarrow{CH_3^+} (CH_3)_3As^{V}O \xrightarrow[-O^{2-}]{2e} (CH_3)\ddot{A}s^{III}$$

This scheme was proposed on the basis of the ability of S. brevicaulis to produce trimethylarsine from all the arsenic(V) precursors and to produce labeled trimethylarsine from arsenite when ^{14}C labeled D,L-methionine, $^{14}CH_3SCH_2CH_2CH(NH_2)COOH$, was added to the culture medium.

The mechanism is essentially the same as that of Fig. 4 except that the methyl donor is identified as the chemically reasonably (29) methyl carbonium ion and that the reaction proceeds further, before the final reduction, in the aerobic system.

Challenger reports that crude cell extracts of S. brevicaulis do not produce trimethylarsine when exposed to arsenicals. This and other experiments (1) indicated that biological methylation by this organism is confined to the mould cell, and does not take place in the medium. In this connection some recent work on cell extracts of C. humicola are of considerable interest (30). It is found that when the extract is incubated with ^{74}As labeled arsenate, SAM as methyl source, and NADPH as reducing agent, and the resulting supernatant liquid applied to a Dowex 1 ion exchange column, several distinct arsenic containing fractions can be eluted as shown in Fig. 7. These can be identified following electrophoresis and autoradiography as being largely arsenite, cacodylic acid, methylarsonic acid, and arsenate in fractions 1 to 4 respectively. Experiments of this sort can be refined further and reveal the presence of other arsenic containing fractions including trimethylarsine oxide. Thus, providing SAM is added, all the arsenic(V) intermediate in Challenger's scheme can be identified in cell extracts of an organism which is known to produce trimethylarsine.

The result of experiment 15, Table VII is particularly interesting since, when ethionine is added in place of methionine, no arsines are produced. The absence of ethylarsines argues against a purely chemical transfer of an alkyl group from sulphur to arsenic, and the absence of trimethylarsine argues strongly for a methionine based enzyme involved synthetic path since ethionine is a well known antagonist to methionine (31). Along the same lines it has been found (2,25) that phosphates inhibits the formation of trimethylarsine from arsenate, arsenite, and methylarsonate, but not from cacodylate, by growing cultures of *C. humicola*. The same organism can be preconditioned by cacodylate to produce trimethylarsine at a greater than usual rate from both arsenate and cacodylate (32). This is seen in Fig. 8. On the other hand, using the same organism, preconditioning with arsenate results in a dramatic reduction in trimethylarsine production from cacodylate while it only slightly stimulates production from arsenate.

Model Studies

Each step in Challenger's mechanism is chemically reasonable. Indeed the alkylation steps are related to the well known Meyer reaction (20,32), which uses methyl iodide or dimethylsulphate as the source of CH_3^+. Since SAM is a sulphonium compound (34) other simpler analogues have been studied with respect to their ability to methylate arsenic (35). At pH 12 and 80°C $(CH_3)_3P^+PF_6^-$ can be used for all the steps in Challenger's scheme which involve CH_3^+ (these reactions can be monitored by NMR techniques since the reactant and products have distinct chemical shifts) and each arsenic(V) compound in the scheme can be reduced to the appropriate arsenic(III) derivative using SO_2. Thus the whole sequence can be easily duplicated.

Methyl methionine is often invoked as a model for SAM (34) and this compound slowly but incompletely, methylates $CH_3AsO_2^{2-}$ at 25°C and at the more realistic pH of 5.8. However a methyl sulphonium derivative of CH_3-CoM under the same conditions failed to transfer its methyl group.

Arsenic Cycle in the Environment

Studies with aerobic and anaerobic organisms have shown that the former produce trimethylarsine whereas the latter produce the more chemically reactive dimethylarsine. Taking these observations into account, it is possible to postulate that there is an arsenic cycle in nature, a cycle that relies upon both biological and abiotic reactions. Such a scheme is outlined in Fig. 9. It is important not to view the reactions as being confined to sediment, water, or air but rather to ecosystems which are either aerobic or anaerobic because it is the availability of oxygen which will determine the nature of the microbial flora and will influence the fate and subsequent movement of the arsine. Arsenate, arsenite, and methylarsonate react in a similar manner

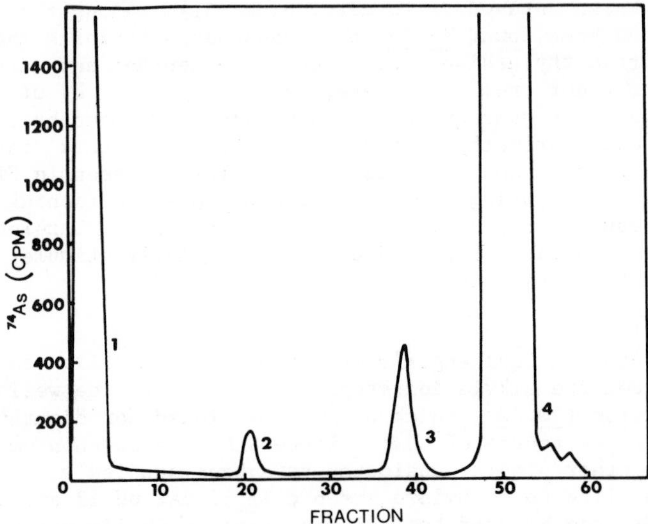

Figure 7. Ion-exchange separation of C. humicola *extracts incubated with* ^{74}As–Na_2HAsO_4*, SAM, and NADPH. Peaks 1–4 are arsenite, cacodylic acid, methylarsonic acid, and arsenate, respectively.*

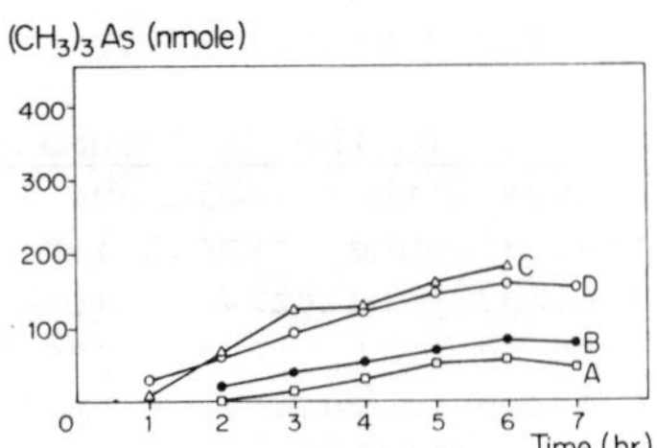

Figure 8. Amount of $(CH_3)_3As$ *in the headspace above growing cultures of* C. humicola *preconditioned in cacodylate (C and D) and not preconditioned (A and B). In Curves A and C the arsenical substrate is arsenate, in B and D it is cacodylate.*

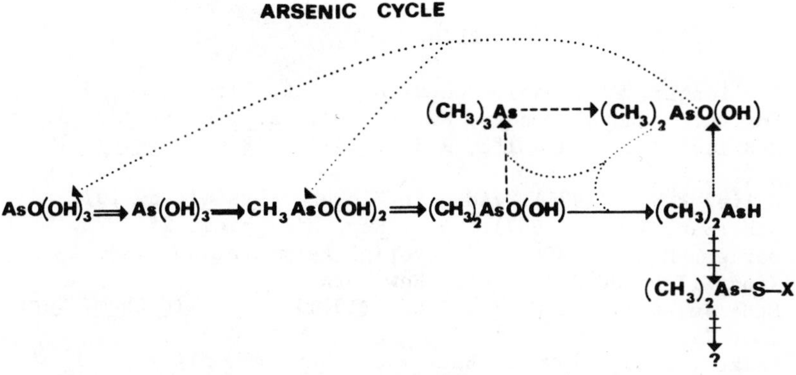

Figure 9. Biological arsenic cycle. (=) aerobic or anaerobic; (· · ·) aerobic biotic or abiotic; (——) anaerobic; (– – –) aerobic; (++) these reactions are probably abiotic.

in both aerobic and anerobic ecosystem. Cacodylate is an important branch point; aerobes reduce and methylate this compound to form trimethylarsine, anerobes reduce it to dimethylarsine. The trimethylarsine is eventually oxidized to cacodylate which can be incorporated directly into the cycle or further modified to methylarsonate or arsenate by microbial activity. Dimethylarsine can be oxidized to cacodylate but it may react with other chemical constituents in its environment and thus initiate an entirely new set of reactions. The latter is a reasonable hypothesis when one considers the reactivity of the arsine and the environment in which it is produced.

Acknowledgements

This paper is a contribution from the Bioinorganic Chemistry Group supported in part by operating and negotiated development grants from N.R.C. and M.R.C., Canada.

BIBLIOGRAPHY

1. Challenger, F. (1945). Chem. Rev., 36, 315.
2. Cox, D.P. and Alexander, M. (1973). Applied Micro. 25, 408.
3. McBride, B.C. and Wolfe, R.S. (1971). Biochemistry, 10, 4312.
4. Wolfe, R.S. (1971). Adv. in Microb. Physiol. 6, 107.
5. Stadtman, T.C. (1967). Ann. Rev. Microbiol. 21, 121.
6. Barker, H.A. (1956). "Bacterial Fermentations", P. 1 John Wiley and Sons, Inc., New York.
7. McBride, B.C. and Wolfe, R.S. (1971). Adv. in Chem. Ser. 105, 11.
8. Zeikus, J.G. (1977). Bacteriol. Rev. 41, 514.
9. Blaylock, B.A. and Stadtman, T.C. (1966). Biochem. Biophys. Res. Commun. 11, 34.
10. Wolin, M.J., Wolin, E.A. and Wolfe, R.S. (1963). Biochem. Biophys. Res. Commun. 12, 465.
11. Blaylock, B.A. (1968). Archs. Biochem. Biophys. 124, 314.
12. Wood, J.M. and Wolfe, R.S. (1966). Biochemistry 5, 3598.
13. Roberton, A.M. and Wolfe, R.S. (1969). Biochim. Biophys. Acta. 192, 420.
14. Bryant, M.P., Wohn, E.A., Wolin, M.J. and Wolfe, R.S. (1967). Arch. Mikrobiol. 59, 20.
15. McBride, B.C. (1970). Dissertation. University of Illinois, Urbana, Illinois.
16. McBride, B.C. and Wolfe, R.S. (1971). Biochemistry 10, 2317.
17. Taylor, C.P. and Wolfe, R.S. (1974). J. Biol. Chem. 249, 4879.
18. Wood, J.M., Kennedy, F.S. and Rosen, C.G. (1968). Nature 220, 173.
19. Wolin, E.A., Wolin, M.J. and Wolfe, R.S. (1963). J. Biol. Chem. 238, 2882.

20. Raiziss, G.W. and Gavron, J.L. (1923). Organic arsenical compounds, New York, N.Y. The Chemical Catalog Co., pp. 38-49.
21. Smith, P.H. and Mah, R.A. (1966). Appl. Micro. 14, 368.
22. Bauchop, T. (1967). J. Bact. 94, 171.
23. Peoples, S.A. (1975). Arsenical Pesticides A.C.S. Symposium Sèries #7, p. 1.
24. Avelino, N., Cullen, W.R., McBride, B.C. Unpublished results.
25. Woolson, E.A., Ed. (1975). Arsenical Pesticides A.C.S. Symposium Series #7, Washington, D.C.
26. Woolson, E.A. and Kearney, P.C. (1973). Environ. Sci. Technol. 7, 47.
27. Isensee, A.R., Kearney, P.C., Woolson, E.A., Jones, G.E. and Williams, V.P. (1973). Environ. Sci. Technol. 7, 841.
28. Cullen, W.R., Froese, C.L., Lui, A., McBride, B.C., Patmore, D.J. and Reimer, M. (1977). J. Organometal. Chem. 139, 61.
29. Zingaro, R.A. and Irgolic, K.J. (1975). Science 187, 7651.
30. Cullen, W.R., McBride, B.C. and Pickett, A.W. Unpublished results.
31. Simmonds, S., Keller, E.B., Chandler, J.P. and duVigneaud, V. (1950). J. Biol. Chem. 183, 191.
32. Cullen, W.R., McBride, B.C. and Reimer, M. Bull. Environ. Contam. Toxicol., in press.
33. Quick, A.J. and Adams, R. (1922). J. Amer. Chem. Soc. 44, 805.
34. Salvatore, F. Borek, E., Zappia, V., Williams-Ashman, H.G., Schlenk, F., Eds. (1977). The Biochemistry of Adenosylmethionine, Columbia University Press, New York, 1977.
35. Chopra, A.K., Cullen, W.R., Dolphin, D. Unpublished results.

Discussion

J. M. WOOD (University of Minnesota): I want to ask this question on behalf of Professor Challenger, because in his review article he alludes to the possible role of coenzyme M in methylation of arsenic by Methanobacteria. He has a speculative paragraph in which he suggests that the dimethylsulphonium derivative would be a very good candidate for methylation of arsenic. The methyl coenzyme M itself is a bad candidate. Have you any idea whether his suggestion is a good one?

McBRIDE: We haven't tried any other compound on methyl coenzyme M.

W. R. CULLEN (University of British Columbia): In model systems, that methyl transfer does take place from the methyl coenzyme M to arsenic(III). Any sulphonium compound with a methyl transfers to arsenic.

G. E. PARRIS (Food and Drug Administration): Do you have kinetic data for this transfer that you are talking about?

CULLEN: Yes, we have attempted to measure some kinetics on this. It's a very difficult system to work on.

M. O. ANDREAE (Scripps Institute of Oceanography): Do you have evidence for the formation of larger compounds; would you see them if they were formed?

McBRIDE: By larger compounds do you mean complexes?

ANDREAE: Yes, arseno-betaine or something like that.

McBRIDE: We don't have any good evidence, but recently we ran a Sephadex column, a gel filtration column, and found a radioactive fraction that indicates a high molecular weight.

ANDREAE: Did you do a mass balance that will tell you how much of the label can be accounted for by arsenate and other compounds?

McBRIDE: Yes, we can account for 100%.

ANDREAE: In your experiments with the marine mud, were you using a rubber trap to assay for the arsine or was that determined by head space analysis?

McBRIDE: No, there was a trap.

F. E. BRINCKMAN (National Bureau of Standards): I'm intrigued by this apparent complexation on the Dowex column. Is it engendered by the bioactivity? That is, might there be a third component that leads to complexation of these arsenic species? [McBRIDE: Yes] Typically, under the conditions you use for elution, these are an ionic species from the pKa values. In our laboratory, for HPLC, we use both weak and strong anion and cation columns and can make satisfactory separations of demethylated metabolites from soil bacteria involving the methylarsenical pesticides. We see no evidence of this kind of complexation. We are concerned about higher molecular weight eluants because of your work and that of Professor Irgolic concerning the betaine or other possible anionic species with a large molecule pendant.

McBRIDE: It seems to be real, and it seems to be associated with a biologically active extract. If you inactivate the extract you don't see these. You can use labels (C-14 cacodyl, C-14 methylarsonic acid, As-74 arsenate) and you do not see these complexes form. Everything elutes at the appropriate point, but when you have a biologically active system, then you see this change to what we think is a complex formation. It's very dramatic when you look at C-14 cacodyl as a substrate rather than as arsenate. Almost 99% of the material moves into a fraction which elutes with water.

BRINCKMAN: This is a very clear demonstration that caution should be exercised in speciating these trace materials, metabolites particularly, when using methodology like Braman's technique of reductive volatilization to form hydrides. You may lose a rather critical molecule which is important for coupling several ionic species and which may be indicative of the mechanistic pathway.

K. J. IRGOLIC (Texas A & M University): You allude to the possible transformation of trimethylarsine to dimethylarsinic acid in some of your systems. In test tube reactions under aerobic conditions, it's a relatively slow reaction, but you break down the compound, lose one methyl group, and end up with dialkylarsinics from trialkylarsines.

ANDREAE: Your marine mud produces hardly any methane. Where did you get that mud? Some marine muds produce methane, some do not.

McBRIDE: We didn't find any marine mud that was producing a lot of methane although obviously they exist.

ANDREAE: Was your mud sample from a relatively shallow level in the core?

McBRIDE: No, we went fairly deep. There was a lot of sulfur in these muds, whether that was influencing what was going on, I don't know.

RECEIVED August 22, 1978.

8

Arsenic Uptake and Metabolism by the Alga *Tetraselmis Chui*

N. R. BOTTINO—Department of Biochemistry and Biophysics, Texas A & M University, College Station, TX 77843

E. R. COX—Department of Biology, Texas A & M University, College Station, TX 77843

K. J. IRGOLIC, S. MAEDA[1], W. J. McSHANE, R. A. STOCKTON, and R. A. ZINGARO—Department of Chemistry, Texas A & M University, College Station, TX 77843

Not too many years ago there was no great public or even scientific concern about substances present in very low concentrations in the air, water or soil. Knowledge about their presence was not at hand. Now, because of the efforts of many analytical chemists a number of these trace elements and their compounds have been identified and quantified. We know now that a whole spectrum of compounds is present at trace level concentrations in the environment. We also know that many of them are, or may be potentially harmful.

Arsenic is one of the elements which is widely distributed in the earth's environment. Arsenic has a complicated chemistry. It forms a great variety of organic and inorganic compounds of variable potency with regard to their effects on living organisms. Arsenic compounds are amenable to chemical transformation through interactions with biological systems. It has been known for almost one hundred years, that molds convert the arsenic present in green paint pigments into a volatile arsenic compound which was identified as trimethylarsine (1). The biological methylation of arsenic compounds producing methylarsines is now a well-established fact (2, 3, 4, 5). Methylcobalamin has been implicated as the methyl donor (2), but recently, Cullen and co-workers (5a) have made the important discovery that when L-methionine-methyl-d_3 is added to certain microbial cultures, the CD_3 label is incorporated into the evolved arsenic. This is good evidence that S-adenosylmethionine is involved in the biological methylation process.

Lunde (6) found arsenic compounds in the lipids of marine and limnetic organisms. Acid hydrolysis of the arsenic-containing lipid fractions produced water soluble, organic arsenic compounds. The isolation of arsenobetaine from rock lobsters was reported by Edmonds *et al.* (7) in 1977. These findings prove that transformations more complex than simple methylation do occur to

[1] Present Address: Department of Applied Chemistry, Faculty of Engineering, Kagoshima University, Kagoshima 890, Japan.

0-8412-0461-6/78/47-082-116$05.00/0

arsenic compounds in biological systems.

Arsenic Incorporation and Algal Growth. Three years ago an investigation on the metabolism of arsenate by marine algae was initiated at Texas A&M University. The goal of the project was and still is the isolation, identification and characterization of organic arsenic compounds formed by the algae from arsenate under carefully controlled conditions. Preliminary experiments with a variety of marine algae led to the use of *Tetraselmis chui*, a green flagellate (Chlorophyta), as a convenient experimental organism. This alga grows well, takes up arsenic efficiently and is rather insensitive toward mechanical shock which accompanies the harvesting operations. *T. chui* was found to tolerate well exposure to 10 ppm As(arsenate) when grown in artificial sea water, and in fact, at the beginning of the stationary phase the growth medium contained more than one million cells per ml (8). The arsenic uptake by *T. chui* depended on the light intensity. At a light intensity of 12,500 lux the arsenic content of the cells increased first to a pronounced maximum within several days and then decreased rapidly to a very low level. At 7000 lux the arsenic level in the cells remained at half the concentration reached at 12,500 lux. The detailed causes for this behavior remain unknown, but it can be postulated that since arsenate absorption is an endergonic process (9) it might compete with cell growth for the available photosynthetic energy.

Large-Scale Algal Growth and Arsenic-Protein Interactions. The characterization of the arsenic-containing compounds present in *T. chui* required the isolation of 300-500 mg amounts of these compounds for further purification and analysis. This could not be done with the test-tube cultures used in the preceding experiments. Consequently, an alga growing facility was constructed to raise *T. chui* under controlled conditions including protection from other organisms such as bacteria. The facility consists of a converted cold storage room, which houses four 1500-liter tanks. This culture room is air-conditioned, well insulated and is kept under positive air pressure. Ultraviolet lights are strategically placed to minimize bacterial contamination. The tanks are illuminated by banks of fluorescent lights. Ballasts are mounted outside the culture room to reduce the heat load on the air conditioner. The growth medium is mixed in a control room. The tanks can be filled, inoculated, aerated, sampled and harvested without entering the culture room, through an appropriate system of pipes and valves. Each large tank produces approximately one liter of tightly packed algae within 20 days.

Figure 1 illustrates the typical type of experiments run with the large-scale growing facility. When ^{74}As-labeled disodium hydrogen arsenate was added to the growth medium at the same time as the inoculum, the algae reproduced rapidly following

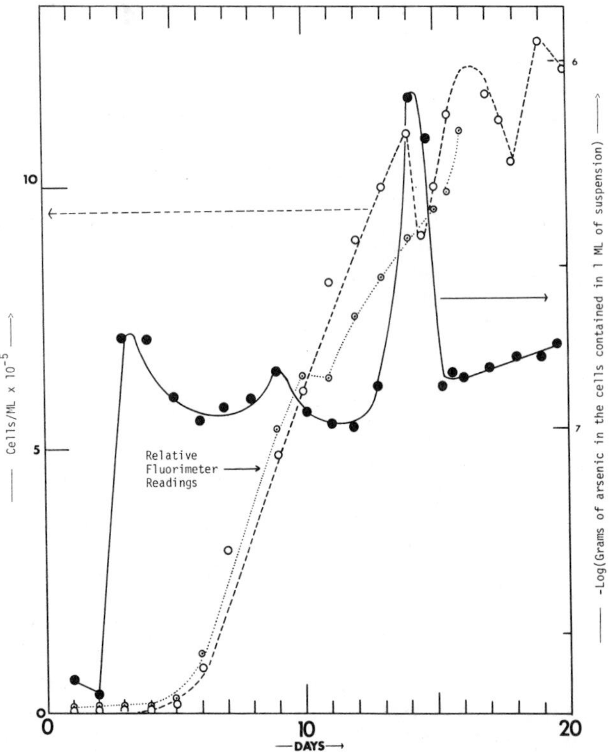

Figure 1. Large-scale growth and arsenic uptake by T. chui *in instant ocean medium containing 10 ppm As (arsenate) added at the time of inoculation*

an initial lag period, in artificial sea water containing 10 ppm arsenate. In separate experiments, the arsenate was added to the growing cultures during their exponential phase or stationary phase. In the experiment depicted in Figure 1 the arsenate was added at zero time, i.e., at the time the medium was inoculated. The curves show peculiar, but recurrent fluctuations resulting from changes in the rate of incorporation and efflux of arsenic into and out of the cells. The two phenomena seem to reach an equilibrium at the time the algal cells reach the stationary phase of growth (about day 14). Other experiments (not shown) indicated that intake and efflux of ^{74}As equilibrated at approximately the same level shown in Figure 1, whether the ^{74}arsenate was added at the time of inoculation, during the exponential growth phase, or during the stationary phase.

The algal cells were also grown in the presence of arsenate, then collected at a time when there were at least one million cells per ml and the cell population did not change appreciably for two or three days. The cells were then centrifuged and washed repeatedly with an As-free medium to remove any arsenate adhering to the cell surfaces. The algal cells were then homogenized in chloroform/methanol (2:1 v/v) and extracted repeatedly with this solvent mixture until all the green pigments had been removed and the residue assumed a grayish-white appearance. Approximately one half of the arsenic was in the organic extract; the other half remained in the residue. The arsenic distribution between these two phases varied somewhat from batch to batch.

When the residue was treated with 1 M HCl and the resulting solution was distilled, all the arsenic remained in the distillate as arsenite as indicated by polarography and by reduction to arsine. Hot water extracted considerable amounts of arsenic from the residue. When four ml of methanol were added to one ml of aqueous extract the precipitate contained all the arsenic. When increasing amounts of Na_2HAsO_4 were added to fixed volumes of the water extract and the resulting solutions were mixed with methanol, all the arsenate precipitated quantitatively when 500 μg of As or less were added. Arsenic added in excess of 1 mg remained in the supernatant. The arsenic content of the precipitate reached a maximum value of approximately 1 mg As (Figure 2), in the precipitate obtained from one ml of the aqueous extract. Most of the algal growth experiments were carried out with arsenate tagged with the gamma-emitting ^{74}As isotope. When the precipitations with methanol after addition of non-radioactive Na_2HAsO_4 were repeated, but only the ^{74}As-content of the precipitates and the supernatants determined, the results summarized in Figure 3 were obtained. Exchange of non-radioactive arsenate for the radioactive arsenic compounds occurred only after approximately 500 μg As(arsenate) had been added to one ml of aqueous extract. Both the precipitate formed upon methanol addition, and the supernatant gave positive ninhydrin and biuret tests indicating the presence of

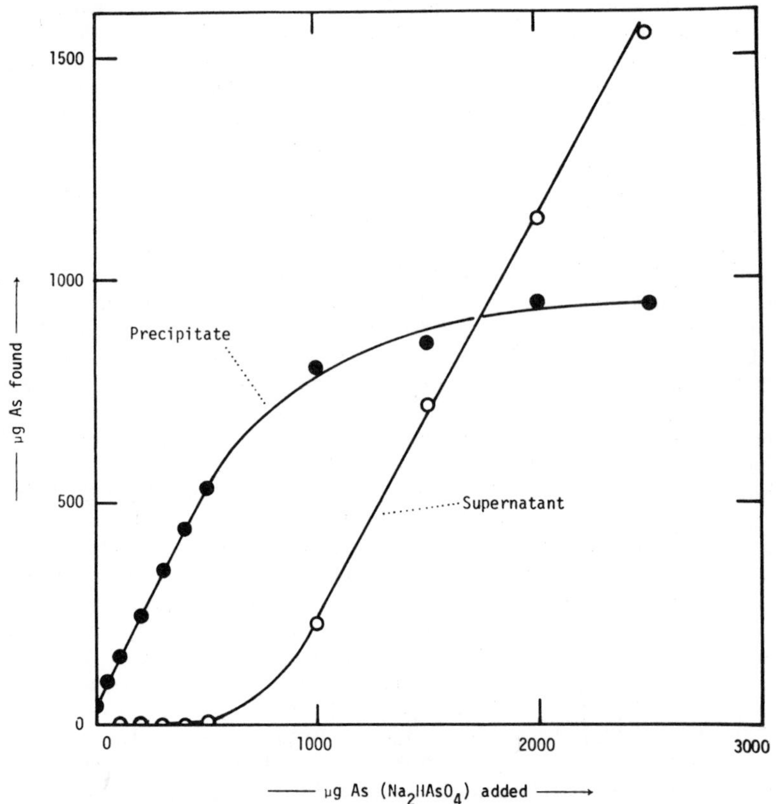

Figure 2. The arsenic content of the precipitates and supernatants obtained upon addition of arsenate and then methanol to the aqueous arsenic-containing extracts of the residue from the $CHCl_3/CH_3OH$ extraction of T. chui

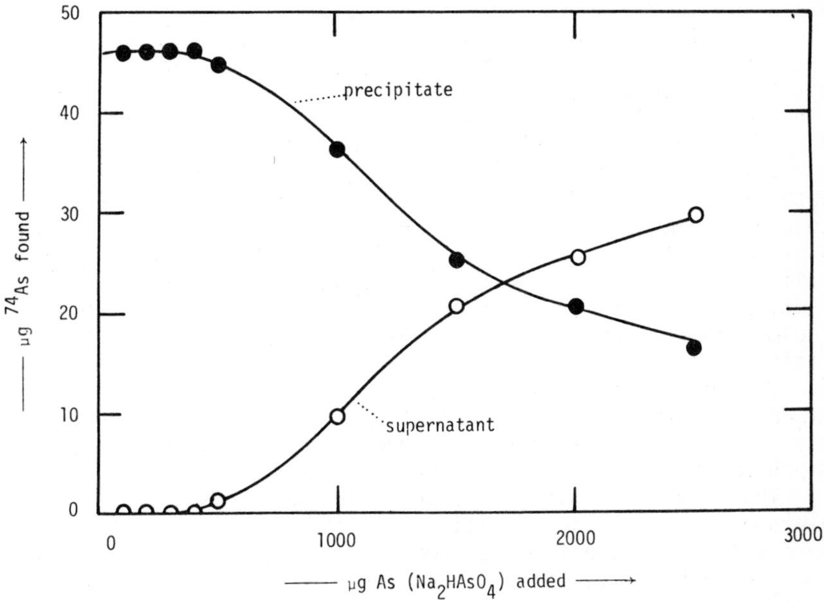

Figure 3. The distribution of ^{74}As activity between the precipitates and supernatants obtained upon addition of arsenate and then methanol to the aqueous arsenic-containing extracts of the residue from the $CHCl_3/CH_3OH$ extraction of T. chui

amino groups and protein. These observations are consistent with the working hypothesis that arsenate forms a complex with a protein located in the algal cell membrane prior to its chemical transformation by the biochemical apparatus of the cell. The separation of this complex by chromatographic techniques is in progress.

Isolation and Characterization of As-Containing Lipids. The combined organic solutions obtained by homogenizing the algal cells in chloroform/methanol were dark green. The removal of these pigments from the arsenic compounds was difficult. The separation scheme which led to the isolation of an arsenic-containing lipid fraction free of green compounds is summarized in Figure 4. Addition of water to the extract produced a chloroform layer which contained all of the arsenic. Several precipitations with acetone removed the phospholipids and all the arsenic from the chloroform phase (10). Gel filtration chromatography on Sephadex LH-20 of the green precipitate produced green, arsenic-containing bands, which were then chromatographed on DEAE Cellulose employing a sequence of solvents ranging from chloroform, chloroform/methanol, (9:1 v/v) chloroform/methanol/acetic acid (3:1:1 v/v/v) to acetic acid. The brownish-green arsenic fractions were further purified by preparative high pressure liquid chromatography on Silica gel with chloroform/methanol/acetic acid/water (17:3.5:2:1 v/v/v/v). A slightly brown oil was obtained. Analytical HPLC of this oil on Silica gel using a Hitachi Zeeman Graphite Furnace Atomic Absorption Spectrometer as an arsenic-specific detector showed that at least two arsenic compounds were present (Figure 5).

Thin-layer chromatography on Silica gel H using a mixture of chloroform methanol acetic acid water (50:25:8:4, v/v) (11) as the developing solvent showed two major arsenic-containing compounds with R_f values of 0.41 (compound A) and 0.61 (compound B) and one major component without arsenic (compound C) with an R_f of 0.85. Fraction A was the largest, followed by fractions B and C. Compound A co-chromatographed with phosphatidyl choline, compound B with phosphatidyl serine and compound C with phosphatidyl ethanolamine and monogalactosyl diglyceride. A standard of sulfoquinovosyl diglyceride was not available at the time these experiments were run, but its R_f in the solvent system used would have been considerably lower than 0.41. Further experiments are in process which involve the use of other solvent systems, appropriate standards and color reactions for the recognition of specific chemical groups. The fatty acid compositions of compounds A, B and C were determined by gas-liquid chromatography and are shown in Table I.

The most remarkable characteristic of these data is the extremely low level of C_{20} to C_{22} highly unsaturated fatty acids. Such acids have been found at much higher concentrations in the phospholipids of several algae (12). It is possible that the

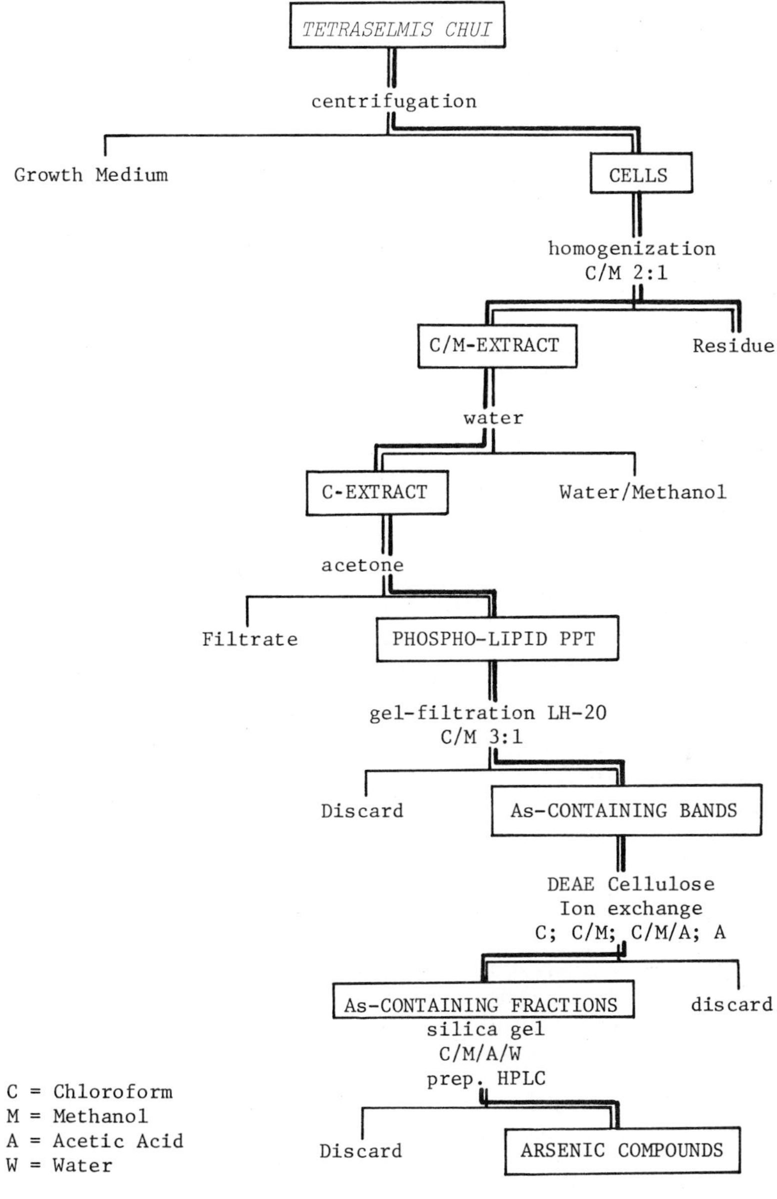

Figure 4. Scheme for the chromatographic isolation of an arsenic-containing phospholipid fraction

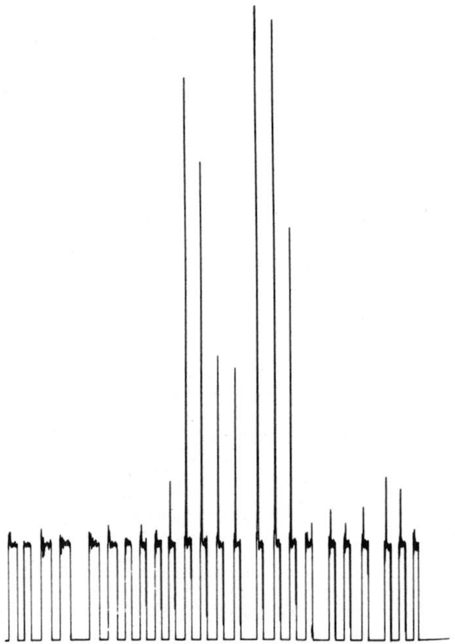

Figure 5. Chromatographic separation and detection of two arsenic compounds in a phospholipid fraction isolation from T. chui. *(HPLC Microporosil (25 cm); $CHCl_3/CH_3$-$OH/CH_3COOH/H_2O$, 17:3.5:2:1; flow rate, 0.5 mL min^{-1}/1-min fractions; Hitachi–Zeeman AA Model 170–70, 193.6 nm.)*

Table I. Major Fatty Acids of Polar Lipids[a] from *Tetraselmis chui*

Fatty Acid[b]	Weigh Percent			
	Unfrac-tionated	TLC fractions[c]		
		A[d]	B[d]	C[d]
14:0	3.6	5.5	4.3	5.2
16:0	20.4	27.8	15.7	8.3
18:0	3.8	18.3	10.5	7.4
24:0	2.6	1.4	0.9	28.5
13:1	12.8	10.4	11.3	29.6
16:1 w7[e]	--	1.7	8.3	1.2
18:1 w9[e]	17.5	11.9	21.2	3.3
20:1 w9	10.2	2.8	2.0	0.1
18:2	3.9	5.7	17.0	2.3
20:4	0.4	3.1	1.8	1.8
UNKNOWN[f]	17.4	5.4	--	--

a/ Only fatty acids present at a level of 2% or more are included.
b/ Chain length: number of double bonds. w = position of first double bond counting from the methyl end.
c/ TLC = thin-layer chromatography with a mixture of chloroform: methanol:acetic acid:water (50:25:8:4, v/v) as developing solvent.
d/ R_f of Fractions: A = 0.41; B = 0.61; C = 0.85.
e/ It may contain other isomers.
f/ Retention time relative to 18:1 = 0.60, on siliconized diethylene glycol succinate polyester at 170°C.

prolonged extraction procedure might have caused their loss.

Elemental analyses of the arsenic-containing lipids after chromatography on Silica gel (Figure 4) showed arsenic present at a level of 0.5 percent and confirmed the presence of C, H, P and N in ratios characteristic of those in phosphatidyl choline. Pure phosphatidyl arsenocholine would have an arsenic content of 8.8 percent. HPLC separation and enzymatic and chemical hydrolysis experiments are in progress to isolate the arsenic-containing components from the algal lipid fraction.

In 1976 at the International Conference on Environmental Arsenic at Fort Lauderdale, we suggested (13) that arsenic could replace nitrogen in choline. The arsenocholine *1* could then become part of a lipid *2* bonded to the phosphate group of a phosphatidyl residue.

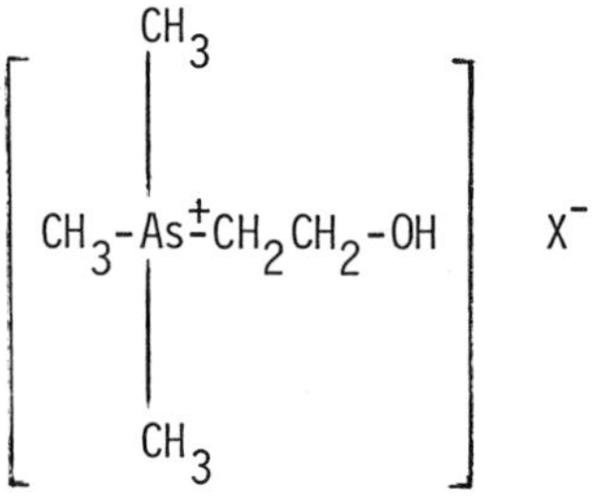

$$\left[CH_3-\overset{CH_3}{\underset{CH_3}{As^{+}}}-CH_2CH_2-OH \right] X^-$$

1

$$CH_3-\overset{CH_3}{\underset{CH_3}{As^{+}}}-CH_2COO^-$$

3

$$\begin{array}{l} CH_2-O-\overset{O}{\overset{\|}{C}}-R \\ | \\ CH-O-\overset{O}{\overset{\|}{C}}-R \\ | \\ CH_2-O-\overset{O}{\overset{\uparrow}{\underset{O^-}{P}}}-O-CH_2CH_2-\overset{+}{As}(CH_3)_3 \end{array}$$

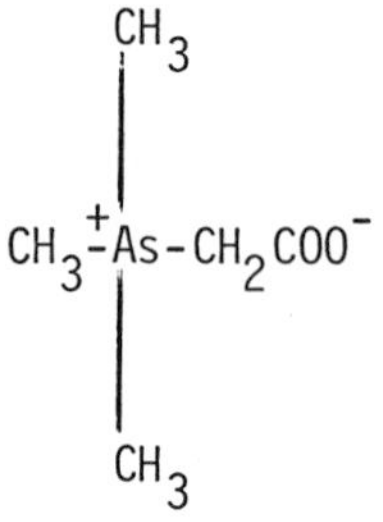

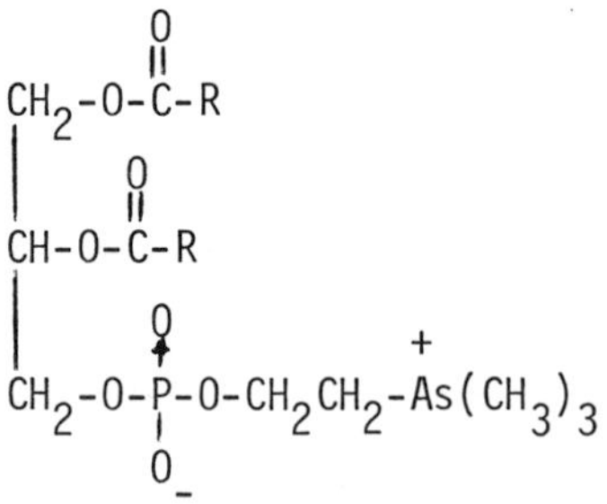

2

The results obtained thus far with the algal lipids are consistent with this hypothesis. The isolation by Edmonds et al., of arsenobetaine *3* from rock lobsters (7) shows that an arsenic compound very similar to arsenocholine exists in nature. It has been observed (14, 15) that arsenic-containing compounds from various organisms would yield volatile arsines upon reduction with sodium borohydride, but only after the sample had been digested with 2N sodium hydroxide. We subjected synthetic samples of arsenobetaine and arsenocholine first to sodium hydroxide digestion and then to sodium borohydride reduction to determine if methylarsines were generated. Arsenocholine, under these conditions, was found to produce, at best, only tiny traces of dimethylarsine or trimethylarsine, but arsenobetaine did readily. The arsenic-containing lipids isolated from *T. chui* do form dimethylarsine or trimethylarsine under these conditions. Because betaine is not a common constituent of lipids, it is unlikely that arsenic is incorporated into lipids in the form of arsenobetaine. Arsenocholine might, however, be capable of being converted to arsenobetaine. The question as to whether such a conversion is significant in organisms, can be answered after the arsenic-containing lipids have been identified.

The chemical structure of the arsenic compounds formed by *T. chui* from arsenate should soon be known. The starting material and the end products of this biological conversion will then have been identified. Future work will have to be carried out to elucidate the biochemical pathway of this transformation. Investigations of this nature will finally lead to the development of metabolic charts for arsenic and other trace elements similar to those now available for lipids, carbohydrates, nucleic acids and proteins. Until such trace element metabolic pathways have been worked out, the details of the interactions of trace elements with living organisms cannot be understood. Basic research that would eventually achieve this important goal has been initiated in several laboratories.

Acknowledgement: These investigations were supported by the National Institute of Environmental Health Sciences (Grant No. 5 R01 ES01125), by the Robert A. Welch Foundation of Houston, Texas, and by the Texas Agricultural Experiment Station.

Literature Cited

1. Challenger, F., Chem. Rev., (1945) 36, 315.
2. McBride, B. C., and Wolfe, R. S., Biochemistry, (1971) 10, 4312.
3. Schrauzer, G. N., Seck, J. A., Holland, R. J., Beckham, T. M., Rubin, E. M., and Sibert, J. W., Bioinorg. Chem., (1972) 2, 93.
4. Ridley, W. P., Dizikes, L., Chek, A., and Wood, J. M., Environ. Health Perspect., (1977) 19, 43.
5. Wood, J. M., Science, (1974) 183, 1049.

5a. Cullen, W. R., Froese, C. L., Lui, A., McBride, B. C., Patmore, D. J., and Reimer, M., J. Organometal. Chem., (1977) 139, 61.
6. Lunde, G., Environ. Health Perspect., (1977) 19, 47.
7. Edmonds, J. S., Francesconi, K. A., Cannon, J. R., Raston, C. L., Skelton, B. W., and White, A. H., Tetrahedron Lett., (1977) (18) 1543.
8. Bottino, N. R., Newman, R. D., Cox, E. R., Stockton, R., Hoban, M., Zingaro, R. A. and Irgolic, K. J., J. Exp. Mar. Biol. Ecol., in press.
9. Blasco, F., Physiol. Veg., 13 (1975) 185.
10. Kates, M., "Techniques of Lipidology: Isolation, Analysis and Identification of Lipids". American Elsevier Publishing Co., New York, N. Y., 1972; p. 393.
11. Skipski, V. P., Peterson, R. F., and Barclay, M., Biochem. J., (1964) 90, 374.
12. Nichols, B. W., "Comparative Lipid Biochemistry of Photosynthetic Organisms", in Harborne, J. B., ed., "Phytochemical Phylogeny". Academic Press, London, 1970, p. 105.
13. Irgolic, K. J., Woolson, E. A., Stockton, R. A., Newman, R. D., Bottino, N. R., Zingaro, R. A., Kearney, P. C., Pyles, R. A., Maeda, S., McShane, W. J. and Cox, E. R., Environ. Health Perspect., (1977) 19, 61.
14. Crecelius, E. A., Environ. Health Perspect., (1977) 19, 147.
15. Edmonds, J. S., and Francesconi, K. A., Nature, (1977) 265, 436.

Discussion

Y. K. CHAU (Canada Centre for Inland Water Research): How do you separate the methylation due to bacteria which generate the methylarsenic compounds from concentration of arsenicals by algae?

IRGOLIC: First of all, we try to keep the bacteria out of the algae. We checked whether any trimethylarsine is formed in our algal cultures. We couldn't find any with As-74 tracers. These cultures are not completely bacteria free, but we believe most of the transformation is done within the algal cell and not by bacteria.

M. O. ANDREAE (Scripps Institute of Oceanography): A word of confirmation. We tried to see if there is a contribution by bacteria by comparing cultures and monocultures of the same organism. There was no measurable difference, so I think that bacteria do not contribute significantly in the kind of systems that you use. Did you make any attempt to identify the substance that was released by your algal system after 5 days or whenever that big peak occurred?

IRGOLIC: No, but there is one experiment which I didn't describe. We took these algae, grew them in 10 parts per million arsenate, and then determined their arsenic level. We then resuspended the algae in arsenate-free medium and found that between 50% and 75% of the arsenic comes right out. We would like to know what form the arsenic is; it could simply be arsenate. The arsenate has to go through the cell wall, perhaps to form an arsenate complex there. The cell does the transformation; it incorporates the arsenic perhaps into the lipids. If you disturb the cell, the arsenate can migrate right back out. We intend to do this; we have now the sensitivity to determinate this arsenate by polarography.

ANDREAE: Did you try to deacylate the fraction in your lipid-soluble extract and see if it could be identified by electrophoresis or chromatography?

IRGOLIC: We tried some phospholipases with the arsenic activities partitioned between the organic phase and the aqueous phase. We have not yet interpreted the results.

ANDREAE: We tried some deacylation of algal lipids, and used the water extract. Coelectrophoresis with arsenocholine or arsenobetaine was unsuccessful.

G. E. PARRIS (Food and Drug Administration): Is there any comment or suggestion regarding formation of a bond between arsenic and the two-carbon unit in betaine or arsenobetaine or arsenocholine? How does it occur?

IRGOLIC: Somehow we have to go from arsenate to whatever that organic compound is. If it's arsenocholine or arsenobetaine, that is similar. But if there are many steps in between, we really can't talk intelligently about what is happening unless we understand that metabolic pathway. There is some indication that organoarsenic compounds produced by transformations in marine organisms, e.g., shrimp or crabs, do not seem to be toxic when ingested by man, as shown by Dr. Crecelius ["Methods and Standards for Environmental Measurement", W. H. Kirchhoff, ed., Proc. Eighth Materials Res. Symp., NBS Spec. Publ. 464, Washington, D.C., 1977, p. 495].

W. R. CULLEN (University of British Columbia): A compound isolated from Atlantic fish by Environment Canada was repeatedly purified, and we got back an analysis recently; it had no arsenic in it. Basically, it was betaine or something very like betaine. We suggest that somehow there is probably an arsenate that gets associated with the betaine in some way and can be lost through purification.

IRGOLIC: It was interesting to hear Dr. McBride mention the association which then breaks up. We might see something like this in some of the cell walls (extract residues). Unfortunately, I think we are in the dark about these associations.

RECEIVED August 22, 1978.

9

The Chemistry of Organometallic Cations in Aqueous Media

R. STUART TOBIAS

Department of Chemistry, Purdue University, West Lafayette, IN 47907

A number of organometallic cations that are at least moderately stable in aqueous solution are known, and among these are the following species: Group IVB, R_3Ge^+, R_3Sn^+, R_2Sn^{2+}, R_3Pb^+, R_2Pb^{2+}; Group IIIB, R_2Ga^+, R_2In^+, R_2Tl^+; Group IIB, RHg^+; Group IB, R_2Au^+; Group VIIIA, R_3Pt^+. Several of the methyl derivatives, R = CH_3, can be produced by the action of methanogenic bacteria or by reaction of methylcobalamin with inorganic compounds of the appropriate metal, but data on methylation under environmental conditions are mainly limited to mercury(II). Most of the information available on the aqueous chemistry pertains to the methyl species since these have the highest solubility. Characteristically these ions are sigma-bonded carbanion complexes of the metal in its maximum oxidation state. In 1966, the aqueous chemistry of organometallic cations was reviewed (1), but interest in environmental effects of several of these species in recent years has stimulated a good deal of new work.

Although the reactions of metal ions in biological systems are exceedingly complex, a knowledge of the hydrolysis constants and stability constants with a limited number of ligand types will permit a number of predictions about the ionic binding and transport. [In this discussion, unless otherwise indicated, hydrolysis will refer to proton transfer from a coordinated water molecule rather than M-C or M-X bond cleavage in the presence of water.] Thermodynamics will play a large part in governing the reactions of these organometallic species. Without exception, the methyl derivatives are highly labile to substitution at the metal center. In certain cases with bulky alkyl groups, reactions may proceed more slowly because of steric effects. Heterogeneous reactions, e.g. dissolution, are slower with large alkyl groups. Because of their hydrophobic nature, they restrict attack at the metal center.

0-8412-0461-6/78/47-082-130$05.00/0

This review will concentrate on the interaction of these organometallic cations with the solvent water, and attention will be focused on those ions of environmental interest, namely the derivatives of mercury(II), tin(IV), and lead(IV). The interaction with water is a strong one, it is not normally encountered in the organometallic chemistry of these elements, and it has a major influence on reactions at the metal center. Some of the more acidic species exhibit amphoteric behavior, i.e. the organometallic oxides dissolve in acid to give cationic and in base to give anionic species. Strong solvent interactions are responsible for the absence from the list above of organometallic cations of certain perfectly stable organometallic moieties, e.g. $R_2Ge(IV)$ and RSn(IV). These interact so strongly with water that they normally are encountered only as the hydrated oxides. An even more extreme example would be cacodylic acid, $(CH_3)_2AsO(OH)$, which exists in solution only in the neutral or anionic form, $(CH_3)_2AsO_2^-$.

Until about a decade ago, it generally was felt that these metal-carbon bonds were rather weak and that their stability was largely kinetic and not thermodynamic (2). Recent synthetic work on metal alkyls indicates that the metal-carbon sigma bonds of even highly reactive metal alkyls can be quite strong. For example, the mean bond dissociation energy of the Ta-C bond in $Ta(CH_3)_5$ has been determined to be 62 ± 2 kcal mol^{-1} (3). The observation of biological methylation of Hg(II) (4,5) showed that the synthesis of a least some metal-carbon bonds was possible even in aqueous media. Nevertheless, in spite of the strength of the metal-carbon bonds, cleavage reactions involving water generally will be thermodynamically favorable, because both metal-oxygen and carbon-hydrogen bonds are formed.

Reactions Involving Cleavage of the Metal-Carbon Bonds

Very little information is available on the pathways by which the metal-carbon bonds are cleaved in aqueous systems. For example, most of these cations are strongly resistant to concentrated aqueous acids. For analysis, $CH_3Hg(II)$ compounds have been decomposed by heating with concentrated nitric acid in a bomb at 150° for an hour (6). Preparative data and mechanistic studies with nonaqueous media, some of which are described elsewhere in this volume (7), can give some general guidelines to factors affecting stability. From studies on protonolysis of dialkylmercury(II) compounds, it has been found that the rate of the reaction depends upon the electron density in the Hg-C bond being cleaved and the polarizability of the metal center (8). With, for example, the very stable linear dialkyltin(IV) cations, protonolysis should be slow because the R'Sn(IV) residue is of low polarizability.

$$R-Sn^{IV}(X)-R' + H-X \longrightarrow R\text{-}H + -Sn(X)-R' \qquad [1]$$

In addition, there are no strong electronic effects releasing electron density in the transition state as occurs with certain transition metal compounds that exhibit strong trans effects. For example, strong acids, HX, cleave one alkyl group rapidly from R_3AuL (9), but the product R_2AuXL is stable so long as X is a reasonable good ligand. The high reactivity of the trialkylgold(III) compounds is due to high trans effect of alkyl groups which facilitates electron release to the proton in the transition state in attack on R_3AuL.

Strong trans effect ligand

$$R_3Au(X) + H-X \longrightarrow R-H + R_2Au-X \qquad [2]$$

Most of the decomposition reactions that have been studied in organic solvents are dissociative and involve as the first step the loss of one or more of the ligands besides the carbanion from the first coordination sphere.

$$R_nML_m \longrightarrow R_nML_{(m-1)} + L$$
$$R_nML_{(m-1)} \longrightarrow \text{Decomposition products} \qquad [3]$$

An example is the protonolysis of the R_3AuL compounds discussed above which involves dissociation of L in the first step.

An alternative mechanism that has been suggested for protonolysis involves an oxidative addition of HX, i.e. a prior protonation of an electron-rich metal center. This is not a reasonable path for organo-mercury(II),-tin(IV), or -lead(IV) compounds, because there is no stable oxidation state two units higher. Such a mechanism may operate in the reaction of alkylplatinum(II) compounds with HCl (10).

$$CH_3Pt^{II}ClL_2 + HCl \longrightarrow CH_3(H)Pt^{IV}Cl_2L_2 \qquad [4]$$
$$\downarrow$$
$$Pt^{II}Cl_2L_2 + CH_4$$

Another type of reaction that is observed in the decomposition of alkyl metal complexes is reductive elimination, [5].

$$R_nML_m \longrightarrow R_nML_{(m-1)} + L$$
$$R_nML_{(m-1)} \longrightarrow R_2 + R_{(n-2)}ML_{(m-1)} \qquad [5]$$

This intramolecular mechanism requires a relatively stable

oxidation state two units lower than that characteristic of the alkyl complex. For example $R_2AuL_2^+$ cations decompose rapidly to R_2 + AuL_2^+ when L is a relatively weak donor (11). In aqueous solution, $[(CH_3)_2Au(OH_2)_2]^+$ decomposes at 25° in hours with the production of colloidal gold and ethane. The low thermodynamic stability of tin(II) relative to tin(IV) makes this decomposition path energetically unfavorable for alkyltin(IV) compounds and accounts in part for their high stability. It is possible that coupling occurs with dialkyllead(IV) solutions which are much less stable than their tin(IV) counterparts. The lead(II) ion is one of the observed decomposition products.

Most of the organometallic ions are quite resistant to nucleophilic attack. In strongly alkaline solution, dimethyltin(IV) has been observed to decompose according to [6] (12).

$$3(CH_3)_2Sn^{2+} \xrightarrow[H_2O]{12OH^-} 3(CH_3)_2Sn(OH)_4^{2-} \xrightarrow{\text{Slow}} 2(CH_3)_3SnOH + Sn(OH)_6^{2-} + 4OH^- \quad [6]$$

Reactions in Which the Metal-Carbon Bonds Remain Intact

A. Proton Transfer from Coordinated Solvent: Hydrolysis. The aqueous chemistry of many of these species is dominated by the hydrolysis reactions, i.e. by proton transfer from coordinated water molecules, and by subsequent condensation reactions of the hydroxo complexes to form clusters with several organometallic cations.

The hydrolysis reactions of the RHg^+, R_3Sn^+, and R_2Sn^{2+} ions were studied in some detail over ten years ago, and this and other early work was reviewed in 1966. (1) The hydrolyses of the alkylmercury(II), -tin(IV), and-lead(IV) species are the only reactions of organometallic cations included in the recent monograph on the hydrolysis of cations by Baes and Mesmer (13). These reactions are somewhat simpler than is characteristic of strongly hydrolyzing monatomic metal species, because the alkyl groups effectively block coordination sites at the metal center. Consequently, the characteristic polycondensation reactions tend to give smaller clusters with fewer metal atoms.

The alkylmercury(II) ions are unusually strong acids for univalent cations, comparable to Hg^{2+}, reaction [7].

$$R\text{-}Hg^+_{(aq)} + H_2O_{(l)} \rightleftharpoons RHgOH_{(aq)} + H_3O^+_{(aq)} \quad [7]$$

Equilibrium constants for R = CH_3, C_2H_5, n-C_3H_7, and n-C_4H_9 are collected in Table I. These ions have $pK_a \simeq 5$ and are acids comparable in strength to the carboxylic acids. As a consequence, at pH 7 the ratio $[CH_3HgOH]:[CH_3Hg^+]$ very much favors the hydroxide, and the cation is a relatively unimportant species. At pH values in the vicinity of the pK_a, condensed species also are important, reaction [8]. The fraction of the

$$CH_3Hg^+{}_{(aq)} + CH_3HgOH \rightleftharpoons H_3CHg(OH)HgCH_3{}_{(aq)}^+ \quad \log K = 2.37 \quad [8] (14)$$

organometal in the polycondensed species depends upon both pH and total organometal concentration. Very concentrated solutions appear to contain small amounts of $(CH_3Hg)_3O^+$, but this is not significant below 0.2 M (15). In general polynuclear complexes are unimportant below total metal concentrations of ca. 10^{-4} M, so they probably play little part in the reactions of mercurials in the environment. The distribution of $CH_3Hg(II)$ species as a function of pH is illustrated in Figure 1.

Table I. Hydrolysis Constants for Univalent Organometallic Cations at 25°[a]

Species	log $*K_{pq}$	Reference
CH_3HgOH	-4.59^b	(14)
$(CH_3Hg)_2OH^+$	$CH_3Hg^+ + CH_3HgOH \rightleftharpoons (CH_3Hg)_2OH^+$ log K = 2.37^b	(14)
$(CH_3Hg)_3O^+$	$CH_3HgOH + (CH_3Hg)_2OH^+ \rightleftharpoons (CH_3Hg)_3O^+ + H_2O$ log K = 0.3-0.7	(15)
C_2H_5HgOH	-4.98	(44)
$n\text{-}C_3H_7HgOH$	-5.12	(44)
$n\text{-}C_4H_9HgOH$	-5.17	(44)
$(CH_3)_3SnOH$	-6.60	(16)
$(C_2H_5)_3SnOH$	-6.81	(16)
$(CH_3)_3PbOH$	-9.1	(17)

a-The symbols used throughout for the equilibrium constants are those of the Chemical Society Tables of Stability Constants (37); b - 20°.

Most of the other univalent organometallic ions are rather weak aquo acids. Typical are the trialkyltin(IV) species which behave as simple monoprotic acids, reaction [9]. These have

$$R_3Sn^+{}_{(aq)} + H_2O_{(l)} \rightleftharpoons R_3SnOH_{(aq)} + H_3O^+{}_{(aq)} \quad [9] (16)$$

pK_a values between 6 and 7, i.e. acid strengths comparable to the weak oxyacids, e.g. HClO. The pK_a values increase slightly with higher alkyl groups as also is observed with the RHg^+ ions and as would be expected from inductive effects. At neutral pH, the ratio $[R_3SnOH]:[R_3Sn^+]$ still favors the hydroxo complex as is illustrated in the distribution diagram for $(CH_3)_3Sn(IV)$ in Figure 1. Since the R_3SnOH is uncharged it should facilitate

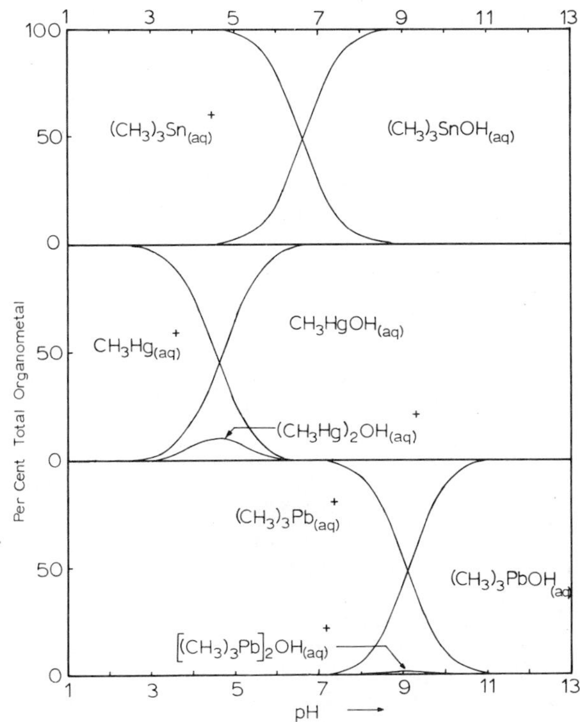

Figure 1. Species distributions as a function of pH for 10^{-3}M solutions of $(CH_3)_3Sn^+$, CH_3Hg^+, and $(CH_3)_3Pb^+$. The fraction of organometal in the binuclear complexes will increase with increasing organometal concentration. Distribution diagrams for more concentrated solutions of CH_3Hg^+ and $(CH_3)_3Pb^+$ can be found in Refs. 40 *and* 117, *respectively.*

the distribution of the organotin species into hydrophobic phases. This may be of importance for membrane transport of these species. There is little tendency to form polynuclear species. Data on the hydrolysis constants are collected in Table I.

As is to be expected, the trialkyllead(IV) cations are weaker acids than the tin analogs and will exist as the aquated cations at neutral pH. For reaction [10], the equilibrium

$$(CH_3)_3Pb^+_{(aq)} + H_2O_{(l)} \rightleftharpoons (CH_3)_3PbOH_{(aq)} + H_3O^+ \qquad [10]$$

constant log $*K_{11}$ is 9.1 (17). Some condensation has been observed with concentrated solutions, reaction [11].

$$(CH_3)_3PbOH_{(aq)} + (CH_3)_3Pb^+_{(aq)} \rightleftharpoons (CH_3)_3Pb(OH)Pb(CH_3)^+_{3(aq)} \qquad [11]$$

Figure 1 also illustrates the species distribution for a millimolar solution of $(CH_3)_3Pb^+$.

The dipositive dialkyltin(IV) ions are much stronger aquo acids than the trialkyls, and at concentrations above ca. 10^{-5} M they form significant fractions of polynuclear hydrolysis products (16,18,19). In general, their strength as acids is comparable to that of Sn^{2+}, and many solution properties of the R_2Sn^{2+} ions are similar to those of Sn^{2+}. The log $*K_{11}$ value of $(CH_3)_2Sn^{2+}$, $\leqslant$ 3.5, (16), is comparable to that for nitrous acid. This does not tell the whole story, however, because much of the hydrolyzed organotin is in the form of species such as $[((CH_3)_2Sn)_2(OH)_2]^{2+}$, $[((CH_3)_2Sn)_4(OH)_6]^{2+}$, etc. in more concentrated solutions. Figure 2 illustrates the effect of concentration on the species distribution, and Table II lists the hydrolysis constants. While these polynuclear complexes should be of little importance in most environmental or biological systems, they will make up a large fraction of the dialkyltin(IV) species in reactions designed to model these systems at $3 \leqslant pH \leqslant 8$ if the organotin concentration is as high as millimolar. It should also be noted that the sets of equilibrium constants such as those in Tables I and II normally do not include solubility product data. While a species distribution can be calculated with them for the entire pH range, it will be meaningless at higher concentrations if precipitation occurs. (See Figure 15.7 (a), ref. 13.) This effect is illustrated in Figure 2 for the 10 mM solution. This figure also clearly illustrates the amphoteric behavior of $(CH_3)_2Sn^{2+}$.

Again, the analogous dimethyllead(IV) ions are much weaker acids than the tin analogs (20), and the equilibrium constants are collected in Table II. The practical consequence of the magnitude of these constants is that very little $(CH_3)_2Pb^{2+}$ exists in solution at pH 7. Approximately one equivalent of acid must be titrated to reach pH 7.

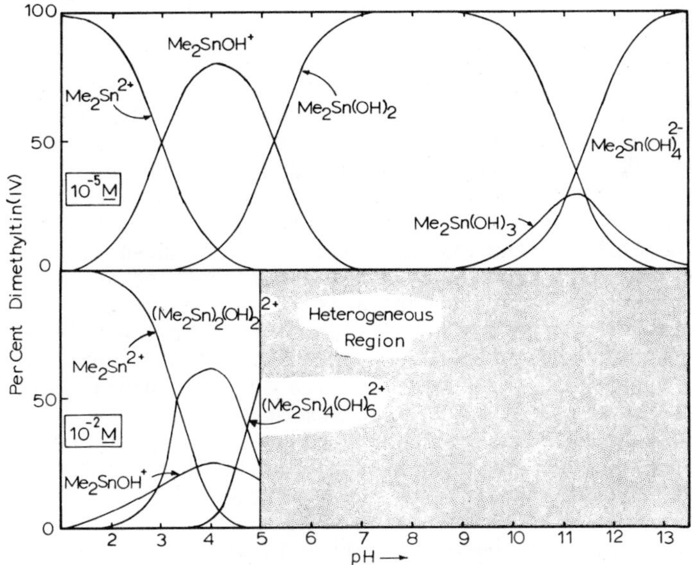

Figure 2. Species distribution as a function of pH for 10^{-5}M and 10^{-2}M solutions of $(CH_3)_2Sn^{2+}$

Table II. Summary of $(CH_3)_2Sn^{2+}$ and $(CH_3)_2Pb^{2+}$ Hydrolysis at 25°

Species	log $^*K_{pq}$,M=Sn(18)	log $^*K_{pq}$,M=Pb(20)
$(CH_3)_2M(OH)^+$	-3.00	$\leq$-7.4
$(CH_3)_2M(OH)_2$	-8.29	-15.54
$[(CH_3)_2M]_2(OH)_2^{2+}$	-4.83	-10.83
$[(CH_3)_2M]_3(OH)_4^{2+}$	-----	-24.31
$[(CH_3)_2M]_4(OH)_6^{2+}$	-16.17	------
$(CH_3)_2M(OH)_3^-$	-20.0	-28.52
$(CH_3)_2M(OH)_4^{2-}$	-32.2	------

B. Complexation of Organometallic Cations in Aqueous Solution. Information on stability constants of organometallic cations in aqueous solution makes it possible to predict a number of things. If particularly stable complexes are formed with a certain functional group that is present in environmental or biological systems, these groups are likely to be involved in the transport and physiological activity of the organometal. Often organometals are introduced, as in the case of tin(IV) compounds, as the halides; and a knowledge of the stability constants for complexes with halide ions will permit a prediction of whether the halide will persist or aquate in natural waters. With organometallic compounds, syntheses often are carried out with non-aqueous solvents, a practice that is less common with simple inorganic species. Stability data can allow predictions of whether model compounds have sufficient stability to be important species in dilute aqueous solution.

Trends in Stability: Hard and Soft Acid Character. The organometallic cation for which the most systematic studies on complex stability have been made is CH_3Hg^+. It is a very soft acid and tends to bind strongly to soft bases, i.e. to heavier donor atoms (21,22). This is demonstrated by the equilibrium constants for reaction [12] with X = halide ion. These increase markedly

$$CH_3Hg^+ + X^-_{(aq)} \rightleftharpoons CH_3HgX_{(aq)} \qquad [12]$$

with heavier halides: log K_1 = 1.50 (F^-), 5.25 (Cl^-), 6.62 (Br^-), 8.60 (I^-) (14). Because of the high stability of the chloride, bromide, and iodide complexes, these compounds have quite low aqueous solubility. The principal coordination number of mercury(II) is two, and the CH_3HgX molecules do not interact strongly with water. Addition of Cl^- or Br^- can be used to reverse the binding of CH_3Hg^+ to nucleic acids, e.g. in agarose electrophoresis with CH_3HgOH as a denaturing agent (23).

In contrast to the aqueous solubility, the CH_3HgX molecules should have reasonable lipid solubility, and chloro complexing may be important in facilitating transport of methylmercury(II) in biological systems. The distribution coefficient of CH_3HgCl between toluene and an aqueous phase was found to be 11.1, reaction [13] (24).

$$CH_3HgCl_{(aq)} \rightleftharpoons CH_3HgCl_{(toluene)} \qquad [13]$$

The alkyltin(IV) species behave as hard acids, and in this respect are very different from the alkylmercury(II) ions which are prototype soft acids (21). Experimentally this is demonstrated by the trend in the stability constants of the halide ion complexes, reactions [14]: log K_1 = 2.3 (F^-), -0.17 (Cl^-) (25). The constants with bromide and iodide are

$$[Me_3Sn]^+_{(aq)} + X^-_{(aq)} \rightleftharpoons \{[(CH_3)_3Sn(OH_2)_2]^+\cdot X\} \xrightleftharpoons{-H_2O} XSnMe_3\,(aq) \qquad [14]$$

so small that values have not been determined. It is likely that the complexation in solution is mainly of the outer sphere type, because the formation of inner sphere $(CH_3)_3SnX$ requires dehydration and a structural change from a planar to a pyramidal skeleton for $(CH_3)_3Sn^+$. These data together with the knowledge that substitution at the tin center occurs rapidly tell us that dissolution of $(CH_3)_3SnX$, X = Cl, Br, I, in water will yield the same aquated species within the time of mixing. Even in sea water, mean $[Cl^-] \simeq 0.5$ M or blood 0.103 M, $(CH_3)_3Sn^+$ will be present principally as the aquo cation at pH < 5 and $(CH_3)_3SnOH$ at pH 7. The low equilibrium concentration of $(CH_3)_3SnCl$ may be important in processes that involve extraction into lipid phases, e.g. in membrane transport, although the hydroxide will also be quite lipid soluble. Studies on the permeability of mitochondrial membranes show that R_3Sn^+ ions mediate the transport of Cl^- and OH^- across the membrane and provide a means for rapid equilibration of pH (26). The discussion of the chemistry in this reference is somewhat obscured by neglect of the hydrolysis of these ions which was discussed in the previous section. The R_3Pb^+ ions form only very unstable complexes with halide ions: K_1 (not log) $(CH_3)_3Pb^+ = 6.5$ (F^-), 2.1 (Cl^-); $(C_2H_5)_3Pb^+ = 3.5$ (F^-), 3.7 (Cl^-) (27,28). Also they are not so hard, since K_F/K_{Cl} is only 3.1 for $(CH_3)_3Pb^+$, while it is 3×10^2 with $(CH_3)_3Sn^+$.

The R_2Sn^{2+} ions form somewhat more stable complexes with anionic ligands because of their dipositive charge, but the halide complex stability constants indicate that they, too, are very hard acids. For reaction [15], the values log K_1 are 3.70 (F^-) (25), 0.38 (Cl^-) (29), < -0.5 (Br^-) (29). In the $(CH_3)_2Sn^{2+}$ - Cl^- system, Raman spectra demonstrate that the complexing is mainly of the outer sphere type, and inner sphere complexing only occurs with very concentrated solutions where the water activity is reduced to a low value.

$$[Me_2Sn]^{2+} + X^-_{(aq)} \rightleftharpoons \{[(CH_3)_2Sn(OH_2)_4]^{2+} \cdot X^-\} \underset{}{\overset{-H_2O}{\rightleftharpoons}} [(CH_3)_2SnX(OH_2)_n]^+ \qquad [15]$$

These data clearly show the much less favorable displacement of water by the heavier halides in the case of R_3Sn^+ and R_2Sn^{2+} compared to the RHg^+ ions. A practical consequence is that the organotin(IV) cations will tend to interact relatively more strongly with nitrogen and oxygen donors, particularly negatively charged ones, compared to sulfur donors, while the converse will be true for the alkylmercury(II) species. The R_2Pb^{2+} ions form much less stable complexes than the tin analogs: $(CH_3)_2Pb^{2+}$, 54.0 (F^-), 5.8 (Cl^-); $(C_2H_5)_2Pb^{2+}$ 35.0 (F^-), 9.2 (Cl^-) (27,28). They also are much less hard than the tin analogs. For $(CH_3)_2Sn^{2+}$ K_F/K_{Cl} is *ca.* 2.1×10^3, while the ratio for $(CH_3)_2Pb^{2+}$ is 9.3.

Stability with Biomolecules and Related Ligands. The hard or soft acid character is of use in predicting the types of donors that will give the most stable complexes. Of at least comparable importance is the intrinsic basicity of the donor; for example, a mercaptide will form much more stable complexes with CH_3Hg^+ than a thioether even though both ligands are sulfur donors. With a given type of donor, reasonably good linear free energy relations exist for protonation and methylmercuriation as observed originally by Simpson (30) for nitrogen donors and examined more generally by Erni and Geier (31,32). This is illustrated in Figure 3 for some oxygen and sulfur donors. As expected from the soft acid character of CH_3Hg^+, the intercept in the log K_{CH_3HgL} *vs.* log K_{HL} plot is much larger for the sulfur than the oxygen donors. Stability constants for a number of CH_3Hg^+ complexes are collected in Table III.

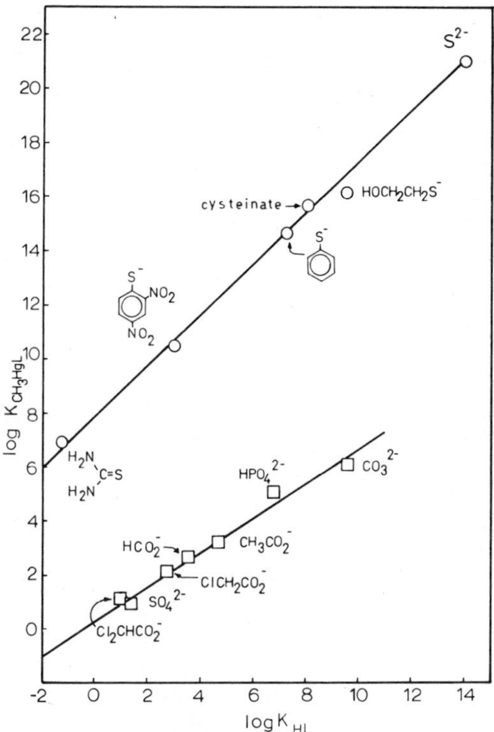

Figure 3. Stability constants for CH_3Hg^+ with sulfur and oxygen donors; log K_{CH_3HgL} vs. protonation constants log K_{HL} (= pKa)

Table III. Selected Stability Constants of Organometallic Cations with Organic Ligands.

Cation	Ligand	log K	Ref.
CH_3Hg^+	$HOCH_2CH_2S^-$	16.1	(14)
	$C_{10}H_{16}O_6N_3S^-$, glutathionate	15.9	(45)
	$C_3H_6O_2NS^-$, cysteinate	15.7	(45)
	$C_6H_5S^-$, thiophenolate	14.7	(46)
	$C_6H_3O_4N_2S^-$, 2,4-dinitrothiophenolate	10.5	(47)
	$(H_2N)_2C{=}S$	6.9	(32)
	$S(CH_2CH_2CO_2^-)2$	4.2	(47)
	$S(CH_2CH_2OH)_2$	3.1	(48)
	$C_7H_5O_2^-$ tropolonate	5.4	(49)
	$C_6H_5O^-$ phenolate	5.5	(32)
	$C_6H_4O_3N^-$ 4-nitrophenolate	3.8	(32)
	$CH_3CO_2^-$	3.2	(50)
	HCO_2^-	2.7	(50)
	$Cl_2HCO_2^-$	1.1	(50)
	$C_3H_3N_2^-$ imidazolate	11.8	(32)
	$C_3H_4N_2$ imidazole	7.1	(32)
	$H_2NCH_2CH_2NH_2$	8.2	(14)
	CH_3NH_2	7.6	(51)
	$(CH_3)_3N$	5.0	(51)
	$NH_2CH_2CO_2^-$ glycinate	7.9	(51)
	$C_{12}H_8N_2$ 1,10-phenanthroline	7.2	(52)
	C_5H_5N pyridine	4.7	(53)
$(CH_3)_3Pb^+$	$C_5H_9O_2^-$ pivalate	1.2	(17)
	$CH_3CO_2^-$	0.97,0.54	(17,54)
	HCO_2^-	0.86	(17)
	$ClCH_2CO_2^-$	0.52	(17)
$(CH_3)_2Sn^{2+}$	$C_{12}H_8N_2$ 1,10-phenanthroline	3.9	(35)
	$C_6H_4O_2N^-$ picolinate	5.3	(35)
	$C_5H_7O_2^-$ acetylacetonate	6.0	(35)
$(CH_3)_2Pb^{2+}$	$CH_3CO_2^-$	2.6	(54)
		(log K_2 1.0)	

Few data are available for stability constants of other organometallic ions with such ligands. Recently, some data for complexes of $(CH_3)_3Pb^+$ with carboxylates have been reported by Sayer <u>et</u> <u>al</u>. (<u>17</u>). A rough linear free energy relation is observed between complexation by $(CH_3)_3Pb^+$ and protonation, and this is illustrated in Figure 4. Similar behavior is to be expected for the R_3Sn^+ ions, although the stability constants with hard bases should be slightly larger. Consistent with this is the observation that phosphate buffers, pH 7.4, decrease the binding of $(C_2H_5)_3Sn^+$ to rat hemoglobin (<u>33</u>). Phosphate buffers also were observed to decrease the extraction of $(C_2H_5)_3{}^{113}Sn(IV)$ into a chloroform phase. In a qualitative study of complexing of R_3Sn^+ species with a number of biological molecules, no evidence could be found for complexes with ATP, glutathione, arginine, or lysine in a borate-EDTA buffer at pH 8.4 (<u>34</u>).

The R_2Sn^{2+} and R_2Pb^{2+} species probably form a much larger variety of complexes than the trialkyl species, but the extensive hydrolysis of the cations has made the accurate determination of stability constants by potentiometric titrations very difficult. While this is the classical technique for such determinations, it requires an accurate knowledge of all the hydrolysis constants as well as the ligand protonation constants. To complicate matters, mixed hydroxide-ligand complexes, e.g. $[R_2Sn(OH)_nL_m]$, often will be formed giving a large number of possible species and rendering the analysis of titration data even more difficult. While CH_3Hg^+ is extensively hydrolyzed in solutions with pH > 1, its essentially monofunctional character precludes the formation of mixed hydroxo-ligand complexes and permits the relatively straightforward determination of stability constants.

A few data have been obtained for complexes of $(CH_3)_2Sn^{2+}$ with bidentate ligands, using a computer to analyze the data (<u>35</u>,<u>36</u>). As is expected from the hard acid character of $(CH_3)_2Sn^{2+}$, the most stable complex is formed with the negative bidentate oxygen donor, acetylacetonate, log K_1 6.0; and the least stable complex was formed with the neutral bidentate nitrogen donor 1,10 phenanthroline, log K_1 3.9. The stability of the acetylacetonate complex is similar to that of the Ni^{2+}, complex, log $K_1 \sim 6.0$ (<u>37</u>), while the phenanthroline complex is much less stable than for Ni^{2+}, log $K_1 \sim 8.6$ (<u>37</u>).

<u>Future Prospects</u>

For the organometallic cations that are extensively hydrolyzed in solution, spectroscopic techniques can be used to determine both equilibrium constants and in many cases the binding sites. While neither the organometallic cations nor most of the ligands of interest have suitable electronic transitions, both Raman and nmr spectroscopy have been used successfully. Resonances of both the cation and

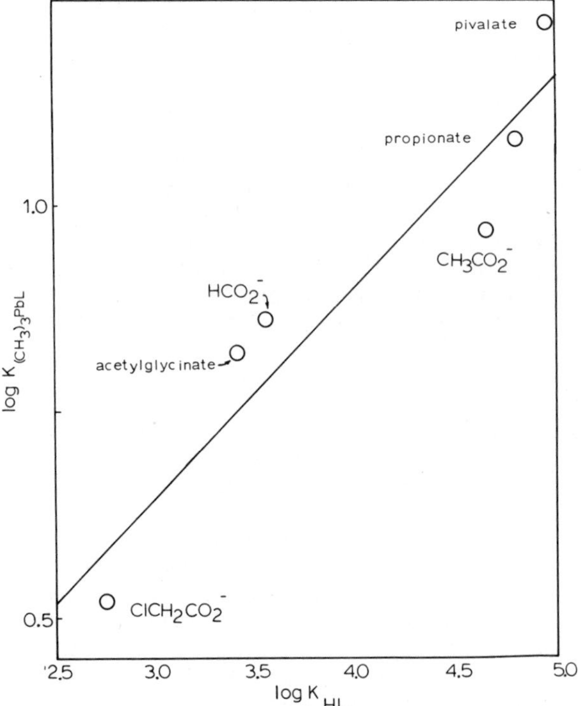

Figure 4. Stability constants for $(CH_3)_3Pb^+$ with carboxylic acids; $\log K_{(CH_3)_3PbL}$ vs. $\log K_{HL}$. Data from Ref. 17.

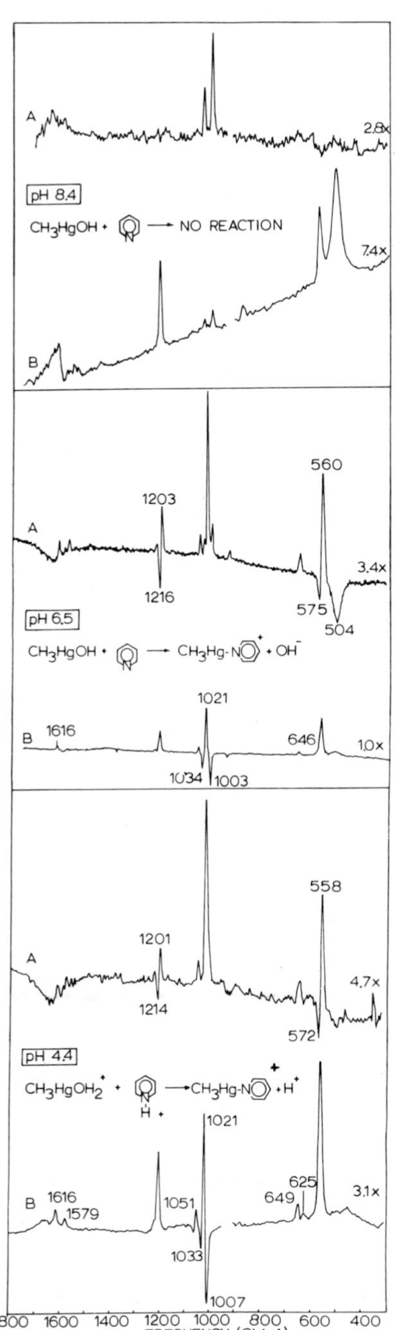

Figure 5. Raman perturbation difference spectra for the $CH_3Hg(II)$–pyridine system at pH 8.4, 6.5, and 4.4. In each set, A is the $CH_3Hg(II)$ + pyridine vs. $CH_3Hg(II)$ difference spectrum, while B is the $CH_3Hg(II)$ + pyridine vs. pyridine difference. Relative ordinate expansions are indicated at the right.

ligand protons have been used to study many reactions of CH_3Hg^+ with amino acids (38) and with the nucleoside inosine (39). Raman perturbation difference spectroscopy has been used to study the reactions of CH_3Hg^+ with pyrimidine (40,42,43) and purine (39,41,42) nucleotides as well as with calf thymus DNA (43).

Figure 5 illustrates the application of the Raman technique to the simple $CH_3Hg(II)$-pyridine system. At pH 8.4, the $CH_3Hg(II)$ + pyridine *vs.* $CH_3Hg(II)$ difference gives just the spectrum of pyridine showing that no reaction of the methylmercury cation has taken place. Similarly the $CH_3Hg(II)$+ pyridine *vs.* pyridine difference gives just the spectrum of CH_3HgOH. At pH 6.5 and 4.4 both spectra show extensive perturbations. The 1021 cm^{-1} band of the complex is sufficiently well resolved from the 1003 cm^{-1} band of pyridine or the 1008 cm^{-1} band of the pyridinium ion to permit measurements of the concentrations of unreacted ligand using the Raman intensities. While 100% is unreacted at pH 8.4, the values are 20% and *ca.* 0%, respectively at pH 6.5 and 4.4. This is in good agreement with values calculated from the known stability constant. The Raman difference technique is particularly sensitive in detecting small amounts of reaction, and both Raman and nmr can be used to determine the product distribution.

These spectroscopic techniques should be equally suitable for the study of R_3Sn^+ and R_2Sn^{2+} interactions, and a start already has been made with R_3Pb^+ chemistry (17).

Acknowledgements. This work has been supported, in part, by the National Science Foundation Grant CHE 76-18591 and by the Public Health Service, Grant AM-16101 from the National Institute for Arthritis, Metabolism, and Digestive Diseases. The author also would like to express his appreciation to the Office of Naval Research, as part of this material was presented at a workshop on organotin chemistry in February, 1978. Thanks are due Mary R. Moller for the $CH_3Hg(II)$-pyridine spectra.

Literature Cited

1. Tobias, R. S., *Organometal. Chem. Revs.* (1966), *1*, 93.
2. Coates, G. E. "Organometallic Compounds," p 72, Methuen, London, 1960.
3. Adedeji, F. A., Connor, J. A., Skinner, H. A., Galyer, L. and Wilkinson, G., *J. Chem. Soc. Chem. Commun.* (1976), 159.
4. Wood, J. M., Kennedy, F. S., and Rosén, C. G., *Nature* (London) (1968), *220*, 173.
5. Jensen, S. and Jernelöv, A., *Nature* (London) (1969), *223*, 753.

6. Schellenberg, M., Ph.D. Thesis, Eidgenössische Technische Hochschule Zürich, 1963.
7. Kochi, J. K., Factors Involved in the Stability of Alkyl-Metal Bonds, this volume.
8. Nugent, W. A. and Kochi, J. K., J. Amer. Chem. Soc. (1976), 98, 273.
9. Komiya S. and Kochi, J. K., J. Amer. Chem. Soc. (1976), 98, 7599.
10. Belluco, U., Giustiniani, M., and Graziani, M., J. Amer. Chem. Soc. (1967), 89, 6494.
11. Kuch, P. L. and Tobias, R. S., J. Organometal. Chem. (1976), 122, 429.
12. Tobias, R. S. and Freidline, C. E., Inorg. Chem. (1965), 4, 215.
13. Baes, C. F., Jr. and Mesmer, R. E. "The Hydrolysis of Cations," Wiley, New York, 1976.
14. Schwarzenbach, G. and Schellenberg, M., Helv. Chim. Acta. (1965), 48, 28.
15. Rabenstein, D. L., Evans, C. A., Tourangeau, M. C., and Fairhurst, M. T., Anal. Chem. (1975), 47, 338.
16. Tobias, R. S., Farrer, H. N., Hughes, M. B., and Nevett, B. A., Inorg. Chem. (1966), 5, 2052.
17. Sayer, T. L., Backs, S., Evans, C. A., Millar, E. K., and Rabenstein, D. L., Can. J. Chem. (1977), 55, 3255.
18. Tobias, R. S. and Yasuda, M., J. Phys. Chem. (1964), 68, 1820.
19. Tobias, R. S., Ogrins, I., and Nevett, B. A., Inorg. Chem. (1962), 1, 638.
20. Freidline, C. E. and Tobias, R. S., Inorg. Chem. (1966), 5, 354.
21. "Hard and Soft Acids and Bases," R. G. Pearson, Ed., Dowden, Hutchinson, and Ross, Strondsburg, Pa., 1973.
22. Pearson, R. G., J. Chem. Educ. (1968), 45, 581.
23. Bailey, J. M. and Davidson, N., Anal. Biochem. (1976), 70, 75.
24. Gruenwedel, D. W. and Davidson, N., J. Mol. Biol. (1966), 21, 129.
25. Cassol, A., Magon, L., and Barbieri, R., Inorg. Nuclear. Chem. Lett. (1967), 3, 25.
26. Selwyn, M. J., Dawson, A. P., Stockdale, M., and Gains, N., Eur. J. Biochem. (1970), 14, 120.
27. Pilloni, G. and Magno, F., Inorg. Chim. Acta. (1970), 4, 105.
28. Pilloni, G. and Magno, F., Inorg. Chim. Acta. (1971), 5, 30.
29. Farrer, H. N., McGrady, M. M., and Tobias, R. S., J. Amer. Chem. Soc. (1965), 87, 5019.
30. Simpson, R. B., J. Amer. Chem. Soc. (1964), 86, 2059.
31. Geier, G., Erni, J., and Steiner, R., Helv. Chim. Acta. (1977), 60, 9.

32. Erni, I. W., Ph.D. Thesis, Eidgenössische Technische Hochschule Zürich, 1977.
33. Rose, M. S., Biochem. J. (1969), 111, 129.
34. Aldridge, W. N. and Street, B. W., Biochem. J. (1964), 91, 287.
35. Tobias, R. S. and Yasuda, M., Inorg. Chem. (1963), 2, 1307.
36. Yasuda, M. and Tobias, R. S., Inorg. Chem. (1963), 2, 207.
37. "Stability Constants of Metal-Ion Complexes," L. G. Sillén and A. E. Martell, eds., Special Publication No. 17, 1964, and No. 25, 1971, The Chemical Society, London.
38. Rabenstein, D. L., Accts. Chem. Res. (1978), 11, 100.
39. Mansy, S. and Tobias, R. S., Biochemistry (1975), 14, 2952.
40. Mansy, S., Wood, T. E., Sprowles, J. C., and Tobias, R. S., J. Amer. Chem. Soc. (1974), 96, 1762.
41. Mansy, S. and Tobias, R. S., J. Amer. Chem. Soc. (1974), 96, 6874.
42. Mansy, S., Frick, J. P., and Tobias, R. S., Biochim. Biophys. Acta. (1975), 378, 319.
43. Chrisman, R. W., Mansy, S., Peresie, H. J., Ranade, A., Berg, T. A., and Tobias, R. S., Bioinorg. Chem. (1977) 7, 245.
44. Zanella, P., Plazzogna, G., and Tagliavini, G., Inorg. Chim. Acta. (1968), 2, 340.
45. Simpson, R. B., J. Amer. Chem. Soc. (1961), 83, 4711.
46. Schwarzenbach, G. and Karlen, U., quoted in ref. 32.
47. Gross, H., Diplomarbeit, ETH Zürich, quoted in ref. 32.
48. Fehr, R., Diplomarbeit, ETH Zürich, quoted in ref. 32.
49. Steiner, R., quoted in ref. 32.
50. Libich, S. and Rabenstein, D. L., Anal. Chem. (1973), 45, 118.
51. Rabenstein, D. L., Ozubko, R., Libich, S., Evans, C. A., Fairhurst, M. T., and Suvanprakorn, C., J. Coordn. Chem. (1974), 3, 263.
52. Anderegg, G., Helv. Chim. Acta. (1974), 57, 1340.
53. Hohl, H., Diplomarbeit, ETH Zürich, (1968), quoted in ref. 32.
54. Pilloni, G., Milani, F., Inorg. Chim. Acta. (1969), 3, 689.

RECEIVED August 28, 1978.

10

Organosilanes as Aquatic Alkylators of Metal Ions

RICHARD E. DESIMONE

Department of Chemistry, Wayne State University, Detroit, MI 48202

Much of the chemistry of concern to this symposium has in recent years received very broad recognition and consequently has been studied in depth from a variety of perspectives. Questions which are now being asked in these areas have become quite detailed and sophisticated, although admittedly much still remains to be learned. In contrast to this situation, that surrounding the element silicon, its environmental significance and relevant chemistry, is still in its infancy. Indeed the answer to the simple question "What, if any, is the environmental significance of organosilicon compounds?" is far from clear at this point in time. It is the purpose of what follows to point out some basic and potentially relevant chemistry of silicon and to try to focus to some small degree on its environmental import.

Silicon: Occurance and Distribution

It is commonly known that silicon is one of the most abundant of elements (TABLE I) and also one of the most widely distributed (1). The vast majority is tied up in silicate rocks

TABLE I. SILICON - HOW MUCH, AND WHERE?

Earth's surface - land, sea and air	
Oxygen -	53.3%
Silicon -	15.9%
Hydrogen -	15.1%
Aluminum -	4.8%
Average silicon content	
2.77×10^5	ppm in earth's crust
40	ppm in man
3	ppm in seawater

0-8412-0461-6/78/47-082-149$05.00/0

and minerals where it is relatively immobile and of little concern to us. In living organisms, silicon plays (and has played) an important role in nearly all stages of evolutionary development. From silicate bacteria to protozoa, algae, and the higher plant and animal organisms, nearly all contain and use silicon in one form or another. It is not the purpose of this article to survey the expanding area of bio-organosilicon chemistry (2) nor to consider the innumerable occurrences of silicon in living organisms (3). It is worth mentioning however, that among the myriad of known silicon compounds in a wide variety of biota, there is a conspicuous absence of true organosilicon molecules. In the higher animal organisms for example, silicon typically occurs as ortho- and oligosilicic acids and silicates, ortho and oligosilicic esters of carbohydrates, proteins, steroids, lipids and phospholipids, and as insoluble silicon polymers. While much has been learned in recent years about silicon in living organisms a great deal certainly remains to be discovered. We know very little about the occurrence of natural organosilicon compounds and we can by no means make the assumption that they do not exist. Much work is needed in this area.

At this point in time and for the purposes of this article it seems that what should be of concern is not the occurrence of toxic organosilicon compounds, but the possibility that reactions of silicon compounds with other,more generally troublesome metals or metalloids, will produce species which are in fact significantly damaging from an environmental viewpoint. In particular we will focus on organo group transfer reactions in aqueous media.

Silicon as an Organo Group Donor

Organo group transfer from silicon has been known for some time, the first reported instance being in 1896 with the formation of an organomercurial, p-dimethylaminophenylmercuric chloride (4). The reactivity of the chlorosilane is not surprising

$$\left((H_3C)_2N-C_6H_4 \right)_3 SiCl + HgCl_2 \longrightarrow (H_3C)_2N-C_6H_4-HgCl + \left((H_3C)_2N-C_6H_4 \right)_2 SiCl_2 \qquad (1)$$

since these are among the more reactive of silicon species. However, similar organo group transfer reactions occur in aqueous media with alkylarylsilanes, with transfer of the aryl group, and these proceed relatively quickly to substantial completion (5).

$$H_3C-C_6H_4-Si(CH_3)_3 + HgCl_2 \xrightarrow[H_2O]{EtOH}$$

$$H_3C-C_6H_4-HgCl + (CH_3)_3SiOH + HCl \quad (2)$$

$$(H_3C)C_6H_4-Si(CH_3)_3 + Hg(OAc)_2 \xrightarrow[H_2O]{HOAc}$$

$$(H_3C)C_6H_4-HgOAc + (CH_3)_3SiOH + HOAc \quad (3)$$

Such reactions are not limited to organo group transfer to mercury. Al, P, Ga, Fe and Sb are known organo group acceptors in both aqueous and non-aqueous media (6). A few representative examples are (7, 8)

$$(C_6H_5)_2SiCl_2 + SbCl_5 \longrightarrow C_6H_5SiCl_3 + C_6H_5SbCl_4$$

$$C_6H_5SbCl_4 \xrightarrow{H_2O} C_6H_5SbO(OH)_2 + HCl \quad (4)$$

$$(CH_3)_4Si + GaCl_3 \longrightarrow (CH_3)_3SiCl + CH_3GaCl_2 \quad (5)$$

$$(CH_3)_3Si-O-Si(CH_3)_3 + GaCl_3 \longrightarrow CH_3GaCl_2 + (CH_3)_3SiCl + ((CH_3)_2SiO)_n \quad (6)$$

(6) is noteworthy in that the silicon compound, hexamethyldisiloxane, is usually considered to be the simplest silicone. Thayer has reported that this molecule reacts with mercuric salts in a very complex reaction yielding ∿20 products (9).

The other major class of alkylating agents among silicon compounds is the organofluorosilicates. These have been extensively studied by Müller and co-workers (6, 10). Reaction with $HgCl_2$, in the presence of NH_4Cl to increase solubility, occurs readily to give monomethyl and dimethylmercuric species.

$$HgCl_2 + (NH_4)_2(CH_3SiF_5) \xrightarrow[H_2O/NH_4Cl]{20°C} CH_3HgCl + (NH_4)_2(SiF_5Cl)$$

$$CH_3HgCl + (NH_4)_2(CH_3SiF_5) \xrightarrow[H_2O]{100°C} (CH_3)_2Hg + (NH_4)_2(SiF_5Cl) \quad (7)$$

Similarly, elements such as Sb and Bi can be alkylated or arylated with substantial yields (11).

$$SbF_3 + 3(NH_4)_2(CH_3SiF_5) \xrightarrow[H_2O]{} Sb(CH_3)_3 + 3(NH_4)_2(SiF_6) \quad (8)$$

$$Bi(OH)_3 + 3\ C_6H_5{-}SiF_3 + 6NH_4F + 3HF \xrightarrow[H_2O]{}$$

$$(C_6H_5)_3Bi + 3H_2O + 3(NH_4)_2(SiF_6) \quad (9)$$

In view of the fact that Sb is one of the elements which man "moves" through the biosphere at a much greater rate than occurs naturally, reactions such as (8) are perhaps of more significance than they might outwardly appear to be.

The preceeding would appear to provide ample precedent for the ability of Si to function as an organo group donor. Unfortunately much of this work seems not to have received the recognition it deserves,especially among persons concerned with environmental problems.

In 1971, during a study of cobalmin-dependent methyl transfer to mercuric salts, it was observed that the common NMR reference compounds sodium 2,2-dimethyl-2-silapentane-5-sulfonate (DSS)(Eqn. 10) and sodium 3-trimethylsilylpropionate-d_4 (TSP) were capable of transfering CH_3^- to mercuric salts (12).

$$(CH_3)_3Si(CH_2)_3SO_3Na + HgX_2 \xrightarrow[H_2O]{}$$

$$(CH_3)_2\underset{\displaystyle X}{\underset{|}{Si}}(CH_2)_3SO_3Na + CH_3HgX \quad (10)$$

Qualitatively it was observed that CH_3Hg^+ formation occurred with several mercuric species, in the order $Hg(NO_3) > Hg(OAc)_2 \gg HgCl_2$. Complete demethylation of silicon was found to occur only with $Hg(NO_3)_2$. Thayer also finds that thallium (III) acetate and lead (IV) acetate react to give methylthallium and methyllead compounds (9). Recently, Bellama (13) and Thayer (9) have studied the kinetics of the reaction with various mercuric salts, finding second order rate constants of $9 \times 10^{-3}\ \underline{M}^{-1}\ sec^{-1}$ and $5 \times 10^{-6}\ \underline{M}^{-1}\ sec^{-1}$ for $HgCl_2$ with TSP and DSS respectively. The reaction of $Hg(CLO_4)_2$ proceeds with $k > 10^{-1}\underline{M}^{-1}sec^{-1}$for both DSS TSP, but more rapidly with TSP. For $Hg(OAc)_2$ and DSS $k=6.6\times10^{-3}\ \underline{M}^{-1}\ sec^{-1}$ (9).

Speculation on the mechanism has centered on two points; an intramolecular acid-base interaction between the silicon and a terminal oxygen of the anion, and the acidity of the attacking (and quite variable) mercuric species. The former (12) would be consistent with the fact that TSP always reacts faster than DSS, which has an extra carbon and would probably not adapt as well to the required conformation. The second point, the nature of the attacking mercuric species, is generally of great significance (for any metal). Anion dependent and pH dependent studies reveal dramatic effects on rate for these environmental parameters (14). The detailed understanding of any transmethylation reaction will require adequate speciation of all reactants and products under any given set of conditions. This has generally been one of the major weaknesses of most studies to date.

In some work currently in progress, Bellama has reported the facile transfer of organo groups to mercuric salts using a number of 1-organosilatranes in protic or aprotic media (15).

$$R\text{-}Si(OCH_2CH_2)_3\ddot{N} + HgX_2 \longrightarrow RHgX + X\text{-}Si(OCH_2CH_2)_3\ddot{N} \quad (11)$$

Interestingly, the parent organotrialkoxysilanes are inert under similar reaction conditions. Presumably the extra electron density at the silicon activates the Si-C bond to attack by Hg(II).

Environmental Considerations

We come now to the environmental significance of the chemistry just discussed. In the absence of significant knowledge of naturally occurring organosilicon compounds, it seems worthwhile to consider the impact of man-made organosilicon compounds. By far the largest group of such substances are the silicones, which have found literally hundreds of uses (Table II), almost all inspired because of the chemical inertness of the polymeric material (16, 17). The estimated silicone market in the United States in 1973 was 91 million pounds. Most applications involve complete release of the silicone into the environment. Most uses, except for the proposed use in electrical transformers as a replacement for PCB's, involve very small quantities (18).

A legitimate question at this point seems to be "What ultimately happens to all of this material?" There appears to be no published work on microbial demethylation of silicon(es) and it is not clear how various environmental factors affect the degradation of the polymeric material. Dow-Corning has reported that moist soil seems to be the most destructive environment to

TABLE II. APPLICATIONS OF SILICONES (16)

1) Waxes and polishes
2) Foaming and antifoaming agents
3) Release agents (in molding processes)
4) Protective coatings
5) Lubricants
6) Cosmetics
.
.
.
n) Cooling fluids in electrical transformers*

* a "bulk volume" application (pending)

the silicone molecule (18). It is postulated that the degradation process begins with a de-polymerization of the long siloxane chain to form volatile cyclic siloxanes of 4-5 Si-O units, a process with a half life of ∿10 days. Under the influence of moisture, oxygen and u.v. light, the volatile cyclic siloxanes are then presumed to degrade in the atmosphere to SiO_2, H_2O, and CO_2. Details on these studies are lacking. One is curious to know for example whether the soil in the experiments was sterile, whether the process could be interrupted or altered by the presence of other substances such as methyl acceptors, and whether ^{14}C labeling studies have been used to trace the methyl group carbon to product CO_2. Hopefully these details will be forthcoming.

In other experiments, Dow-Corning reports that a 15% emulsion of silicon fluid subjected to the action of activated sewage sludge for a period of 70 days showed no evidence of bio-degradation (18). However Bellama reports that relatively low viscosity silicone fluids will methylate Hg(II)(19).

In summary, we have barely scratched the surface of some very interesting and possibly significant chemistry. Much work remains to be done in areas mentioned and surely in others not yet discovered.

Literature Cited

1. Ochiai, E., "Bioinorganic Chemistry, An Introduction", pp. 5-12, Allyn and Bacon, Inc., Boston, 1977.
2. Voronkov, M.G., Chem. Brit.(1973) 9, 411.

3. Voronkov, M.G., Zelchan, G.I., Lukevitz, E.J., "Silicon and Life", Zinatne Publishing House, Riga, 1971.
4. Combes, C., Compt. Rend., (1896), 122, 622.
5. Eaborn, C., "Organosilicon Compounds", Butterworths, London, 1960.
6. Müller, R., Organometal. Chem. Revs., (1966), 1, 359.
7. Jakubowitch, A.J., and Mozarew, G.W., J. Gen. Chem. USSR, (1953), 23, 1414.
8. Schmidbaur, H., Angew. Chem., (1964), 76, 753.
9. Thayer, J.S., personal communication.
10. Müller, R. and Dathe, C., Chem. Ber. (1965), 98, 235.
11. Müller, R. and Dathe, C., Chem. Ber. (1966), 99, 1609.
12. DeSimone, R.E., J.C.S. Chem. Comm. (1972), 780.
13. Bellama, J.M. and Nies, J.D., personal communication.
14. Jewett, K.L., Brinckman, F.E. and Bellama, J.M., following paper.
15. Bellama, J.M. and Nies, J.D., J.C.S. Chem. Comm., submitted.
16. Howard, P.H., Durkin, P.R., and Hanchett, A., "Assessment of Liquid Siloxanes", Report # PB 247778, National Technical Information Service, U.S. Dept. of Commerce, 1974.
17. Calandra, J.C., Keplinger, M.L., Hobbs, E.J., and Tyler, L.J. Polymer Preprints, (1976), 17, 1.
18. Pollution Engineering, Aug., 1977, p. 41.
19. Bellama, J.M., personal communication.

Discussion

G. E. PARRIS (Food and Drug Administration): Have there been direct toxicity studies on silicones from the standpoint of environmental impact?

DeSIMONE: There have been many studies on direct toxicity, including intravenous studies. Except for minor eye irritation in certain cases, these appear harmless to just about everything.

PARRIS: The reactions you showed suggest that silicones may be new methylating agents in the environment.

DeSIMONE: It remains to be found out whether in fact enough organosilicon compounds do exist in the environment, where they exist, and whether what you just suggested is true.

PARRIS: One of the critical places will be if silicones find more use as dielectric fluids. I doubt that silicones will be the major replacement for PCB's. I think that most people who are dealing with the toxicity of silicones and the importance of environmental chemistry of silicones have not considered them as potential methylating agents.

C. FREY (Dow Corning Corporation): I would like to make a

couple of comments on the implications of the paper. The data there showing 91,000,000 lbs. of silicones gaining entry into the environment could very well be an item for some concern, but the great bulk of that material, as many people know, is polydimethylsiloxane. It's interesting to note that the subsequent work of Dr. Thayer [vide infra] seems to show that mercuric nitrate, which was just about the most potent mecurial cleaving agent, is without effect on cyclodimethylsiloxanes after several months of contact. So the notion that the organosilicon compounds, which are not naturally-occurring but find their way into the environment, will be cleaved by mercury is probably difficult to sustain.

DeSIMONE: The question is not whether mercury will cleave these compounds, but rather during the decomposition (in moist soil as you suggest is the most efficient way to do it), what happens if some methyl acceptor is present while you've got these cleavage products floating around.

FREY: I think it's important to differentiate between what might be called environmental chemistry and the chemistry of spills; that is, what happens if you have large concentrations of siloxane in contact with large concentrations of some metal. On an environmental scale, I think one has to be concerned with the reaction of methylsilicon compounds with naturally-occurring elements or compounds. As far as the soil is concerned, we'll present the details this summer at the 5th International Organosilicon Symposium in Karlsruhe [August, 1978]. There is no unusual chemistry there that involves the carbon-silicon bonds themselves.

J. J. ZUCKERMAN (University of Oklahoma): The half-life of 10 days, did that refer to the depolymerization-cyclization to the 5- or 4-membered rings?

DeSIMONE: That was the impression I got.

ZUCKERMAN: There was another reaction you mentioned which indicated the decomposition with respect to ultra-violet light giving CO_2.

DeSIMONE: I got this impression from an article in "Pollution Engineering" [Reference 18], an interview with John Ryan.

FREY: I think the question has something to do with the rates of soil-catalyzed reaction, whatever that happens to be. The details of that reaction, like all others, depend on the circumstances, and we'll be going into that detail in a paper that is in preparation.

J. M. BELLAMA (University of Maryland): About the nature of water-soluble organosilicon chemistry, the TSP and DSS to which you referred are extended chains; several people have proposed a head-to-tail interaction where the presence of an electron-rich species (an available Lewis base moiety somewhere remote in a molecule from the silicon) can bend around and join. We have looked at these kinds of interactions [Inorg. Chem. (1965), 14,

1618], and we have two feelings about them. One is that these interactions can be important when n (the number of intervening methylene groups) is 1 or 2. Models of these compounds show that the interacting atoms are essentially in contact. With longer chains (n = 3-5), there seem to be some interaction. When you get to n = 6, it looks like no interaction remains. Secondly, the nature of the group on the silicon seems to be very important. Craig's original paper [J. Chem. Soc. (1954), 332] postulates the necessity of inducing a positive charge on silicon. It would seem that organic groups such as methyls or phenyls are not going to be particularly good in this respect. If you have substituents on the silicon like a chlorine, or perhaps like a hydrogen, which is very electronegative with respect to the silicon, the chances for these kinds of interactions are going to be far greater. So, using water-soluble trimethylsilyl compounds suggests that these kinds of interactions are going to be less important in establishing a site for attack by an electrophilic mercury than would be the case if other substituents were present on the silicon.

DeSIMONE: The difference between the rates of DSS and TSP may be a reflection of that extra carbon; it is a factor of a couple orders of magnitude.

BELLAMA: Müller and Frey [Z. anorg. allg. Chem. (1969) 368, 113] postulated the necessity for ammonium fluoride addition to the methyltrichlorosilane plus mercuric chloride. They claim that it's necessary to have a methylpentafluorosilicate species as the intermediate and as the active methylating agent. We have done this reaction without adding the ammonium fluoride. We don't know what intermediate is present, or what is actually doing the methylating, but it's not necessary to add the ammonium fluoride in order to get this kind of reaction to occur.

J. S. THAYER (University of Cincinnati): We have found that the DeSimone reaction [equation 10] is not confined to mercury. One gets similar reactions with thallium triacetate, with lead tetraacetate and, we believe, with potassium hexachloroplatinate. We studied the kinetics of this reaction and our figures agree moderately well the ones that were quoted here. Secondly, the other reaction was the direct reaction between hexamethyldisiloxane and mecuric nitrate. The two dozen products alluded to are a series of linear and cyclic siloxanes, formed here by a disproportionation reaction, plus a variety of species we have not yet identified. Our belief is that this reaction proceeds by initial removal of a methyl group by mercury, from the siloxane moiety followed by rearrangement. We find that a similar reaction occurs with the germanium analog, hexamethyldigermoxane.

RECEIVED August 22, 1978.

11

Influence of Environmental Parameters on Transmethylation between Aquated Metal Ions

K. L. JEWETT and F. E. BRINCKMAN

Center for Materials Science, National Bureau of Standards, Washington, DC 20234

J. M. BELLAMA

Department of Chemistry, University of Maryland, College Park, MD 20742

While advances have occurred in detection of volatile biogenic methyl-metals and -metalloids, these procedures essentially have relied upon partition coefficients favoring degassing of permethylated species from aqueous media (1,2). Unfortunately, little work has been reported on the basic chemical features of relevant equilibria

$$Me_nM^{(z-n)+}{}_{(aq)} = Me_nM^{(z-n)+}{}_{(atm)},$$

or on determining partition coefficients.

Where neutral molecules are involved, *e.g.*, z = n, strong solvation by water is probably not important even though the central atom may be capable of greater than n-fold coordination. For example, in either fresh or salt water, Wasik *et al.* found that Me_2Hg partitions nearly equally between solution and atmosphere above (3). Preference for partition into fresh water was noted; presumably more favorable ligation by H_2O over Cl^- prevails, the so-called salting-out effect for hydrophobic molecules.

Though not yet resolved, there appears to be general agreement that biomethylation of metals and metalloids follows a stepwise process (4,5). A number of very important corollaries for assessing influences of environmental parameters hinge upon this basic viewpoint:

(a) polar, charged (n<z) intermediate methylelement metabolites are formed which probably display great affinity for aquation, hydrolysis, and ligation by naturally occurring solutes;

(b) transport of such solvated intermediate methylelements out of aquatic media by degassing is unfavorable;

(c) complicated transmethylation chemistry exists for such intermediate biogenic species which depends on comparative stabilities in aqueous solution for a variety of ligated methylelement ions; and

(d) elimination of toxic methylelement species, particularly by metathetical or reductive demethylation processes, likely de-

0-8412-0461-6/78/47-082-158$07.50/0

pends on highly specific coordination of $Me_nM^{(z-n)+}$ by certain charged or neutral ligands to form intermediates not sufficiently stable for further methylation.

Aquated Organoelement Ions. Recognition that *sigma*-bonded organometallic ions exhibit stable forms in aqueous solutions was shown many years ago. For example, one fundamental property of dissolved ions, that of electrolytic behavior, was demonstrated by conductivity measurements on $Me_2Tl^+{}_{aq}$ (6) and later polarographic studies with $MeHg^+{}_{aq}$ (7). Equally important considerations of structures of aquated organoelement species have received considerable attention by NMR (8,9) and laser Raman spectrometry (10, 11), particularly with the appearance of Tobias' classic review of the field over a decade ago (12).

Significant work has appeared providing methods and details of chemical processes which involve aquation numbers (12,13) and hydrolysis of water in coordination spheres of organometal ions (14). Interestingly enough, relatively scant work has been directed to kinetic studies of stable aquated organometal ions as these relate to transalkylation reactions (15,16), though analogous chemistry in aprotic solvents has preoccupied inorganic chemists for over a century. Nonetheless, a number of environmental events on a world-wide scale in recent times has focussed attention on the need to direct research to that aquatic chemistry which can elucidate formation and transport of organometal species of biological concern (4).

In this paper we deal with several organometal ions demonstrated as biogenic, but yet which also provide a chemical system amenable to treatment as classical aquated inorganic ions. Thereby, we can assess the influence of common environmental parameters, such as temperature, pH, salinity, and ionic strength, both in terms of the constitution of reacting organometallic species and their rates or modes of transmethylation.

Experimental Methods

Preparation of Solutions and Kinetic Runs. Solutions of commercial and synthesized (17) trimethyltin compounds and mercury(II) salts were prepared in distilled water, diluted, and mixed just prior to each kinetic run. Precautions for excluding air from solutions were unnecessary since reaction rates were not measurably different under anaerobic conditions. Sodium perchlorate provided gegenions of choice where high ionic strengths were required because of reduced likelihood of Hg^{2+} coordination by ClO_4^- (18).

Immediately prior to mixing of reactants, and thereafter, pH measurements were frequently taken for each kinetic run in NMR tubes or larger vessels with highly accurate microprobe equipment.

For this work, concentrations of reactants and ionic strengths were dictated both by solubility (hydrolysis) of reactants and rates of reactions amenable to the NMR measurement scheme employed (17). Generally, reactants were taken at about 0.025 M, permitting ionic strengths from 0.05 to 0.7 for observing second order rates between 1 x 10^{-4} to 1 x 10^{-2} $M^{-1}s^{-1}$.

Proton NMR spectra provided concentration data for all methylated metal species, following digital integration of peaks and suitable weighting of areas (17). These were taken for each kinetic run at intervals which insured uniform dispersion of input time and concentration data in regression analyses (19). The rate data were fitted by least squares to the simple second-order law

$$k_2(\text{obs})\ t = \frac{1}{(A-B)} \ln \left[\frac{B(A-x)}{A(B-x)}\right] ,$$

where A and B are initial reactant concentrations and x is extent of reaction at any given time t. Removal of significantly outlying points in fitted data (> 2σ) for N=13 to 20 points yielded smoother slopes [k_2(obs) values] within 95 percent confidence intervals or better.

Estimation of Reactant and Product Concentrations. A comprehensive computer program, CHEMSPECIES (20), was devised which permitted calculation of equilibrium concentrations of tin and mercury species. Input parameters were a set of formation constants (β) listed in Table I selected from literature. For specific kinetic runs, observed values for $[Me_3Sn^+]_{total}$ and $[Me_2Sn^{2+}]_{total}$ or $[MeHg^+]_{total}$ taken from NMR data, along with measured $[Cl^-]_{total}$ and pH, provided necessary boundary conditions for compositional analysis at any t. Output data computed included $[Hg^{2+}_{free}]$ and $[Cl^-_{free}]$, along with relative (mole fraction) or net concentrations of all methyltin, methylmercury, and mercury(II) species present exceeding a specified value (usually 0.01 mole percent).

Results and Discussion

Choice of Kinetic System. Transmethylation between Sn(IV) and Hg(II) in Water. The utility of neutral tetraalkyltins for forming corresponding alkylmercurials from HgX_2 salts in protic media was shown by Abraham *et al.* in a series of studies (31) conducted in methanol-water solutions. Similarly, later work of Dodd and Johnson examined cleavage of *sigma*-bonded RCH_2- groups (R = pyridino) on several transition metal centers by mercury or thallium salts in aqueous solutions (32). In both cases, the authors made reasonable assumptions that the methyl donors involved were relatively inert towards associative substitution, particularly towards pre-equilibrium distribution of ligands (anions) in competition with the electrophilic center. Thereby, especially in the latter studies, the relative reactivities of $HgCl_n^{2-n}$ (or $TlCl_n^{3-n}$) could be more simply evaluated as a function of changes

TABLE I

Selected Formation Constants for Tin(IV) and Mercury (II) Species in Water

Formation Reaction	log β	References
Reactants:		
$Hg^{2+} + Cl^- = HgCl^+$	6.72	21,22
$Hg^{2+} + 2Cl^- = HgCl_2^o$	13.23	21,22
$Hg^{2+} + 3Cl^- = HgCl_3^-$	14.23	21,22
$Hg^{2+} + 4Cl^- = HgCl_4^{2-}$	15.20	21,22
$Hg^{2+} + OH^- = HgOH^+$	10.03	22,23
$Hg^{2+} + 2OH^- = Hg(OH)_2^o$	21.16	22,23
$Hg^{2+} + 3OH^- = Hg(OH)_3^-$	20.71	22,23
$Hg^{2+} + 4OH^- = Hg(OH)_4^{2-}$	20.26	23
$Hg^{2+} + OH^- + Cl^- = Hg(OH)Cl^o$	17.43	22,24
$Hg^{2+} + 2OH^- + Cl^- = Hg(OH)_2Cl^-$	19.03	24
$Hg^{2+} + 3OH^- + Cl^- = Hg(OH)_3Cl^{2-}$	19.60	24
$Hg^{2+} + OH^- + 2Cl^- = Hg(OH)Cl_2^-$	16.87	24
$Hg^{2+} + 2OH^- + 2Cl^- = Hg(OH)_2Cl_2^{2-}$	18.51	24
$Hg^{2+} + OH^- + 3Cl^- = Hg(OH)Cl_3^{2-}$	17.07	24
$Me_3Sn^+ + Cl^- = Me_3SnCl^o$	- 0.17	25
$Me_3Sn^+ + 2Cl^- = Me_3SnCl_2^-$	- 1.74	25
$Me_3Sn^+ + OH^- = Me_3SnOH^o$	- 6.77	26
Products:		
$MeHg^+ + Cl^- = MeHgCl^o$	5.25	27
$MeHg^+ + OH^- = MeHgOH^o$	9.51	27
$2MeHg^+ + OH^- = (MeHg)_2OH^+$	11.74	28
$Me_2Sn^{2+} + Cl^- = Me_2SnCl^+$	0.37	29
$Me_2Sn^{2+} + 2Cl^- = Me_2SnCl_2^o$	0.14	29
$Me_2Sn^{2+} + 3Cl^- = Me_2SnCl_3^-$	- 1.31	29
$Me_2Sn^{2+} + OH^- = Me_2SnOH^+$	- 3.50	30
$Me_2Sn^{2+} + 2OH^- = Me_2Sn(OH)_2^o$	- 9.00	30

in the ionic properties, *e.g.*, [Cl^-], of the reaction medium.

As a model system for assessing effects of environmental parameters on aqueous transmethylation, we extended our earlier work (17) on the reaction,

$$Me_3Sn^+_{aq} + Hg^{2+}_{aq} = Me_2Sn^{2+}_{aq} + MeHg^+_{aq}, \qquad (1)$$

because of its relevance to the observed biomethylation of Sn(IV) (33) and involvement of a charged methyl donor offering greater similarity to polar intermediates currently viewed as important to biogenesis of organometals. As Huber (15) and we (34) have shown, an attractive alternative exists with the rapid transmethylation chemistry of aquated methyllead species, but kinetic limitations are imposed by the NMR measurement scheme employed. Moreover, though necessary equilibrium data for calculating kinetic contributions of methyltin species in chloride solutions have become adequate, sufficient formation constants for analogous lead ions have not.

In Tables II and III are collected, respectively, kinetic parameters obtained for reaction *1* under conditions which isolate effects of temperature or ionic strength (μ) and salinity ([Cl^-]).

Over the range of experimental conditions examined for reaction *1* in the present work, a simple biomolecular rate law was obeyed

$$\text{rate} = k_2(\text{obs})\ [Me_3Sn^+_{aq}]_{total}\ [Hg^{2+}_{aq}]_{total}. \qquad (2)$$

Excellent linear fits of kinetic data were obtained for all reactions studied, each typically exceeding 75 percent completion. Reaction *1* appears to be essentially irreversible: precipitation of Me_2Sn^{2+} or $MeHg^+$ does not alter the rate, nor do these products undergo further transmethylation reactions detectable by NMR.

Effects of Temperature. The long-familiar Arrhenius equation which relates the reaction rate constant with temperature defines a quantity, E*, regarded as the activation energy of a component or overall process:

$$\text{rate} = k = Ae^{-E^*/RT} \qquad (3a)$$

and,

$$\ln k = -E^*/RT + \ln A. \qquad (3b)$$

In aqueous solutions, especially, it is apparent (35,36) that the energetics (or activation parameters) of the reaction coordinate are profoundly affected by reorganization of solvent molecules surrounding the reactant molecules, particularly as this modifies their charge distribution. For an ionic biomolecular reaction

TABLE II

Effects of Temperature upon Transmethylation between $Me_3Sn^+_{aq}$ and Hg^{2+}_{aq} or Tl^{3+}_{aq}

RUN[a]	T, ^oC[b]	k_2(obs)±S.E.[c]	-ln k_2(obs)	-ln k_2(calc)	N	X[d]
(a) $Me_3Sn^+ + Hg^{2+}$						
1	10.5±1.1	1.72±0.06	6.37	6.32	20	57.4
2	20.1±0.6	4.33±0.10	5.44	5.50	14	71.6
3	29.0±0.3	9.33±0.21	4.67	4.78	13	79.5
4	40.0±0.2	15.56±0.32	4.16	3.95	14	88.9
5	49.8±0.4	42.44±1.15	3.16	3.26	12	83.5
	r = 0.944		$E^* = 14.2 \pm 0.9$	ln A = 18.9		
(b) $Me_3Sn^+ + Tl^{3+}$						
1	10.0±0.4	0.75±0.03	7.20	7.12	17	78.5
2	20.1±0.4	2.34±0.15	6.06	5.96	16	72.0
3	27.8±0.5	9.06±0.30	4.70	5.11	17	81.8
4	41.3±0.8	17.29±0.71	4.06	3.75	13	74.0
5	52.6±0.4	72.80±4.19	2.62	2.69	13	90.4
	r = 0.988		$E^* = 19.1 \pm 1.7$	ln A = 26.7		

[a]For series (a) $\mu = 0.051$; for (b) $\mu = 0.17$. [b]T ± standard deviation for N observations. [c]k_2(obs) x 10^3 M^{-1} s^{-1}. [d]Extent of reaction, %. E^* in kcal mol^{-1}.

TABLE III

Influence of Ionic Strength and Salinity on Reaction Rate[a]

RUN	Cl/Hg	k_2(obs) x 10^3	± S.E.[b]	log k_2(obs)	μ[c]	$\sqrt{\mu}$	$[Cl^-]$	$\sqrt{[Cl^-]}$
0	0.0[d]	<0.04[e]	--	<-4.4[e]	0.0836	0.288	0.0	0.0
1	1.25	3.56	0.35	-2.45	0.0828	0.288	0.0243	0.156
2a	2.00	7.77	0.38	-2.11	0.1038	0.322	0.0524	0.229
2b	2.00	7.45[f]	0.49	-2.13	0.1030	0.321	0.0504	0.224
3	2.93	9.04	0.36	-2.04	0.1028	0.321	0.0765	0.277
10	3.03	8.15	0.53	-2.09	0.4041	0.636	0.0767	0.277
11	3.06	9.23	0.27	-2.03	0.6186	0.786	0.0767	0.275
12	3.01	9.45	0.28	-2.02	0.1766	0.420	0.0759	0.275
13	3.02	9.16	0.24	-2.04	0.2764	0.526	0.0753	0.274
4	4.00	10.24	0.41	-1.99	0.2565	0.506	0.1023	0.320
5	5.00	9.56	0.67	-2.02	0.1499	0.387	0.1249	0.353
6	6.07	7.80	0.58	-2.11	0.1767	0.420	0.1517	0.389
7	10.03	7.77	0.38	-2.11	0.2761	0.525	0.251	0.501
8	22.13	4.40	0.25	-2.36	0.6211	0.788	0.594	0.768
9	23.03	4.61	0.19	-2.34	0.550	0.742	0.575	0.758

[a]Obtained at 26 ± 1°; [b]k_2 in $M^{-1}s^{-1}$; [c]Ionic strength = $\mu = 1/2\ \Sigma m_i z_i^2$, counterions from $NaClO_4$; [d]Reaction between $Me_3SnClO_4 + Hg(ClO_4)_2$, 1:1; [e]Estimated upper limit based on maximum NMR signal detected after 23 hr. run; [f]Reaction between $Me_3SnNO_3 + HgCl_2$, 1:1.

$$A^{z+} + B^{z-} \rightarrow [A\text{-}B]^{z\pm} \rightarrow \text{products} \qquad (4)$$

derivation of ΔF*, ΔH*, and ΔS* from experimental determinations of equation *3* can provide gross insights into mechanistic pathways accompanying solvent changes, or between related reactions systems.

TABLE IV

Comparison of Activation Parameters for Transmethylation between Metals in Protic Media

Parameter[a]	$Me_4Sn+HgCl_2$ 96% MeOH[b]	$Et_4Sn+HgCl_2$ 96% MeOH[b]	$Me_3Sn^++HgCl_2$ H_2O[c]	$Me_3Sn^++TlCl_3$ H_2O[d]
k_2(298) $M^{-1}s^{-1}$	2.59	0.00630	0.00602	0.00434
E_a kcal mol^{-1}	10.42	13.74	14.2	19.1
ΔF* kcal mol^{-1}	16.89	20.46	20.5	20.7
ΔH* kcal mol^{-1}	9.85	13.15	13.6	18.5
-TΔS* kcal mol^{-1}	6.97	7.18	6.86	2.20
ΔS* cal $deg^{-1}mol^{-1}$	-23.4	-24.1	-23.0	-7.38

[a]For comparison SI units are not used. [b]Reference 31. [c]This work.
[d]References 16 and 37.

In Table IV are compared E* values for demethylation of Me_3Sn^+ by Hg^{2+} and Tl^{3+} (37). Two important general observations can be made. A high degree of linearity prevails, indicating that neither does E* display temperature dependence nor do the expected compositional changes in reactant ions and $[A\text{-}B]^{z\pm}$ importantly alter activation energetics over the course of reaction *1*. Second, the E* values determined, particularly with the methyltin-mercury(II) reaction, are low, being comparable to transmethylation between neutral metal species in non-polar organic media (38) or tetraalkyltins and $HgCl_2$ in protic solutions (31). Table IV provides comparisons for these last reactions with activation parameters derived from the present work.

Abraham and Johnston (31) concluded that for the neutral R_4Sn reactions, a S_E2(open) transition state *5* was more likely than a four-center S_E2(cyclic) activated complex *6*,

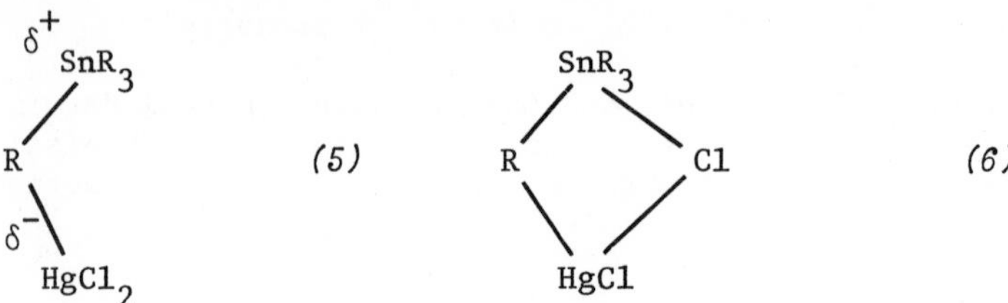

the former being more consistent with variations in ΔF^* and ΔH^*, or near constancy of ΔS^*, in solvent compositions from 100 percent methanol to 30 percent water in methanol. These authors suggested that the highly polar $[R_3Sn\text{--}R\text{--}HgCl_2]^{\pm}$ transition state *5* possesses salt-like or ion-pair-like properties. In this context it is interesting to note that reaction *1* did not proceed in methanol at a rate measurable by NMR, but with additions of only very small increments of water, measurable rates were observed (16,37). Presumably, recalling the activation parameters obtained for methylation of Hg^{2+} or Tl^{3+} by trimethyltin species, even limited availability of polar water molecules somehow serves to support necessary molecular rearrangements for formation of a charge-compensated activated complex which is energetically available to ground state solutes.

In sum, while we can presently presume that many associative or ion-pair-like interactions can occur over the course of reaction *1* (at all temperatures examined) which involve a number of polar transition states composed of $Me_3SnCl_n^{1-n}$ and $HgCl_m^{2-m}$ species, the low E^* values obtained require that solvent (water) effects mitigate any coulombically repulsive pathways. Water can provide such mitigation either by causing formation of new reactants with nearly neutral charge distributions, *e.g.*, $[HgCl_2^{o}] > [HgCl^+]$ or $[HgCl_3^-]$, or by providing appropriate assistance for forming low-energy ion-pair-like transition states (39). Fortunately, the matter can be put to several tests based upon classical theory of kinetic salt effects.

Effects of Ionic Strength and Salinity. The kinetic information for reaction *1* summarized in Table III can be broadly characterized by three separate relationships depicted graphically in Figure 1. For kinetic runs 1 through 3 (indicated by ⨁), wherein the ionic strength of the reaction medium was maintained at about $\sqrt{\mu} = 0.3$, almost three-fold increase in the bimolecular rate k_2(obs) resulted from increased relative concentration of chloride ion. Substitution of ClO_4^- (run 2a) by NO_3^- (run 2b) did not significantly alter the transmethylation rate, doubtless in consequence of very weak or nonexistent coordination between nitrate ion and Hg^{2+} (40) comparable to the situation for ClO_4^- (18). We conclude that available concentrations of reactive $HgCl_n$ species are not materially affected by such non-coordinating anion substitutions.

Exclusion of all chloride by perchlorate in run 0, however,

effectively reduced the observed reaction rate to a value below NMR detection. Consistent with this, a very good linear correlation between $\sqrt{\mu}$ or $\sqrt{[Cl^-]}$ and log k_2(obs) was obtained for runs 1-3 (r = 0.975 and 0.986, respectively), making it very clear that chloride ion exerts a specific ion effect (39) on reaction *1*.

Recognizing that we are dealing with metathesis of covalently bound ligands on aquated metal ions in a highly polar medium suggested the expedient of applying basic principles derived for reactions between simple inorganic electrolytes. According to the Brønsted and Bjerrum activity rate theory, ionic reactions proceed through formation of an activated complex. The rate of reaction is taken as proportional to the concentration of the complex which is in equilibrium with medium and reactant ions (36), *e.g.*, equation *4*. The Debye-Hückel limiting law for activity coefficients, γ, permits evaluation of charge, medium, and ionic size effects (39) associated with the activated complex by

$$\log_{10}\gamma_{\pm} = -0.509\ z_+z_-\sqrt{\mu} \qquad (7)$$

in water at 25° where the ionic strength, μ, is given

$$\mu = 1/2\sum_i c_i z_i^2 .$$

Thus, for the reaction between ion A of charge z_A at concentration c_A and ion B of charge z_B at concentration c_B, the effect of ionic strength upon the rate of reaction is approximately (α is a constant of theory dependent upon units)

$$\ln k/k_o = 2\ z_A z_B\ \alpha\sqrt{\mu}, \qquad (8a)$$

where k_o becomes the reaction rate at infinite dilution. This primary kinetic salt effect predicts generally that

if z_A and z_B have the same sign, k increases with μ
if z_A and z_B have opposite sign, k decreases with μ
if either z_A or z_B is zero, k is independent of μ.

Usually, it is assumed here that specific ion effects (*viz.*, Cl^- above) or large (highly delocalized) ions are not important in the reaction, and μ is small (< 0.01). In practice, a number of alternate relationships have been proposed (39), mainly employing empirical constants, to permit application to reactions treated at higher ionic strength (μ > 0.1), for example (36)

$$\ln k/k_o = 2\ z_A z_B\ \alpha\left[\frac{\sqrt{\mu}}{1+\sqrt{\mu}} - 0.2\ \mu\right] \qquad (8b)$$

Applications of either equation *8a* or *8b* to the two remaining sets of rates listed in Table III revealed several important points. For runs 3 and 10-13 where [Cl^-] was held constant with increasing μ (⊕ in Figure 1), reasonable correlations between log k_2(obs) and $\sqrt{\mu}$ are obtained with a small, slightly negative slope, μ = -0.0362 (equation *8a*). We regard this result as consistent with one major transmethylation process, or several minor parallel reactions, involved in the rate-determining step wherein reactant ions of opposite charge slightly contribute to the overall methyl exchange.

For the last case (◯) shown in Figure 1, where both [Cl^-] and μ were monotonically increased, a substantial reduction in k_2(obs) was measured. Again, correlation of runs 3 through 9 listed in Table III by either equation *8a* or *8b* indicated a significant linear relationship (r = 0.888), but the highly negative slope (m = 0.826) is now believed to result from two separate effects. The first, suggested by equation *8a*, is that a major pathway, or the sum of several important parallel routes, involves reaction of oppositely charged pairs of Me_3Sn- and Hg-substrates. This conclusion is consistent with our finding that a low energy of activation is characteristic for the overall process at similar conditions of [Cl^-] and μ, suggesting such species with minimal coulombic repulsions.

A predominant second effect operates, we believe, in this case as a consequence of extensive ligation of Hg^{2+}_{aq} by Cl^- (41). Kinetic studies already noted (32) show that increased concentrations of $HgCl_n^{2-n}$ species are responsible for reducing demethylation rates of a variety of methylmetal donors as much as 3 to 150 fold as n goes from 2 to 4 in the electrophile. At constant μ in the present cases for ClO_4^- and Cl^- dependence, we found over a 2-fold diminution in k_2(obs) which must be attributable to [Cl^-]. Further evidence that increasing [Cl^-] rather than increasing μ is the dominant factor in retarding transmethylation rate resulted from our correlation of log k_2(obs) with $\sqrt{[Cl^-]}$ in the form of equation *8a*. Here, we treated [Cl^-] as the dominant contributor to ionic strength, with the remaining ions behaving as they do in the [ClO_4^-] dependence case.

The problem confronting us at this point was to devise means by which the relative concentrations of Me_3Sn^+ and Hg^{2+} species could be determined at various [Cl^-] and pH conditions. Thereby, a kinetic model could be fashioned which featured the possible pair-wise interactions of all reactants.

Speciation of Reactants and Products in Aqueous Transmethylation between Tin(IV) and Mercury(II). Recent advances (21-30) in electrochemistry and spectrometry have spurred new and refined determinations of those equilibria (Table I) necessary to construct a reliable map of compositions for methylmetal and metal ions considered in the present work. We employed an extended version of programs discussed by Dryssen *et al.* (24) which simultaneously considers an array of formation constants ($\beta_{species}$) within experimentally observed boundary conditions of total chloride, pH, and total methylmetal ions. A representative plot of several CHEMSPECIES determinations of mercury and methyltin reactant species at t = 0 is depicted in Figure 2. It is obvious that substantial changes in the composition of $HgCl_n^{(2-n)+}$ occur over about a ten-fold increase in [Cl^-], while relatively smaller effects on Me_3Sn species occur. In similar calculations at lower ionic strengths and variations in pH (1-3 units), little overall change in the composition diagram was seen.

In this fashion, the CHEMSPECIES program was used to determine both the distribution of reactants for each kinetic run and to map the distribution of reactants plus products during the course of a reaction. Results for one such run at Cl/Hg = 3 are summarized in graphical form in Figure 3. Two trimethyltin and three chloromercury(II) species were found to account for greater than 99.9 percent of reactants (grouped as R in Figure 3) over the entire Cl/Hg range of 3 to 23 studied. Product (P) analyses by the CHEMSPECIES program indicated that Me_2SnCl_2 became significant only in late states of the reaction, especially at runs in higher [Cl^-]. None of the reaction conditions employed in the work generated significant concentrations (*e.g.*, < 0.1 mole percent) of hydroxy- or hydroxychloro-species.

The pH of many kinetic runs was monitored as a function of time, or the extent of reaction (x). Figure 4 illustrates a typical variation of [H^+] with extent of reaction, corresponding to the Cl/Hg = 3 run above. An acid-forming process was evident, just as expected from hydrolysis of the metal species involved (14). For example

$$HgCl_2 + H_2O = Hg(OH)Cl + Cl^- + H^+, \qquad (9)$$

yields an equilibrium constant, K', given by Table I as

$$K' = \frac{\beta_{Hg(OH)Cl} \cdot K_w}{\beta_{HgCl_2}} = 1.58 \times 10^{-10},$$

where $K_w = 10^{-14}$ is the water dissociation constant. Similarly, for product $MeHg^+$ species, hydrolysis is expected by

$$MeHgCl + H_2O = MeHg(OH) + Cl^- + H^+, \qquad (10)$$

for which Table I gives

$$K'' = \frac{\beta_{MeHg(OH)} \cdot K_w}{\beta_{MeHgCl}} = 2.09 \times 10^{-10}$$

Many such competitive hydrolysis reactions can occur, though individually less significant. Thus, similar treatment of the analogous Me_3SnCl and Me_2SnCl^+ hydrolysates indicated that these species contributed 10^{11} and 10^8 less to [H^+] than the respective Hg species considered in reactions *9* and *10*. All of these reactions, in sum, will contribute to observed pH changes during the

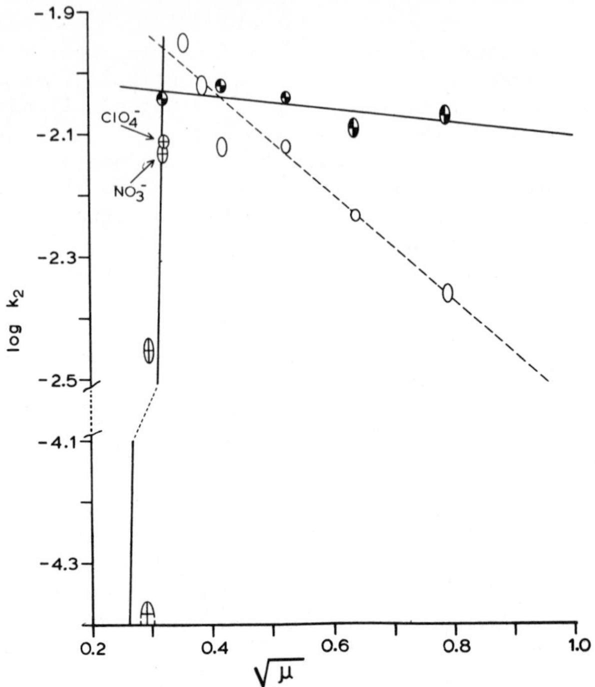

Figure 1. Plot of log k_2(obs) vs. $\sqrt{\mu}$ shows three classes of behavior for rates of transmethylation between Me_3Sn^+ and Hg^{2+}, depending upon specific availability of $[Cl^-]$ (⊕), high $[ClO_4^-]$ (◕), or high $[Cl^-]$ (○).

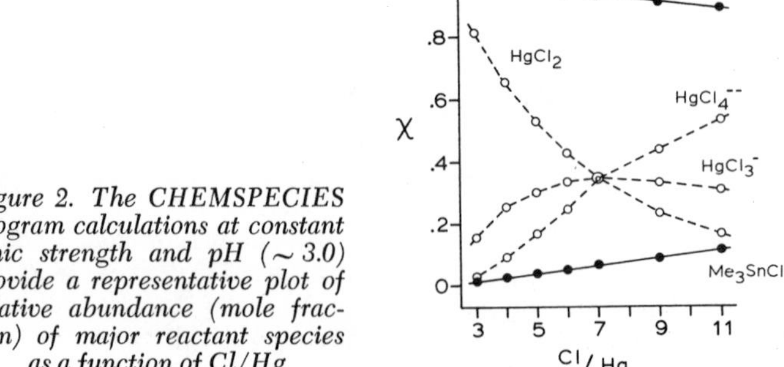

Figure 2. The CHEMSPECIES program calculations at constant ionic strength and pH (~ 3.0) provide a representative plot of relative abundance (mole fraction) of major reactant species as a function of Cl/Hg

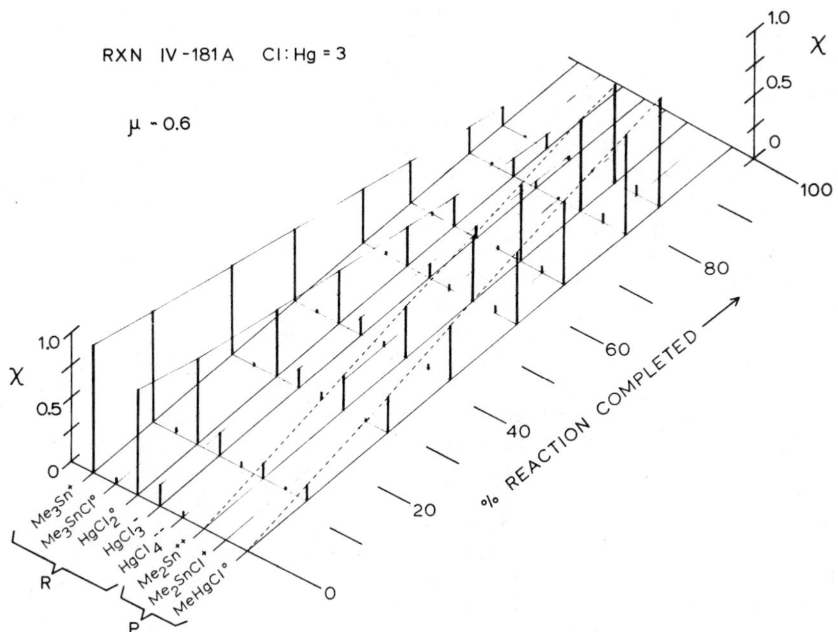

Figure 3. For a typical NMR kinetic run at Cl/Hg = 3, relative abundances of initial reactants, and subsequent abundances of both reactant and product species are compared as a function of extent of reaction

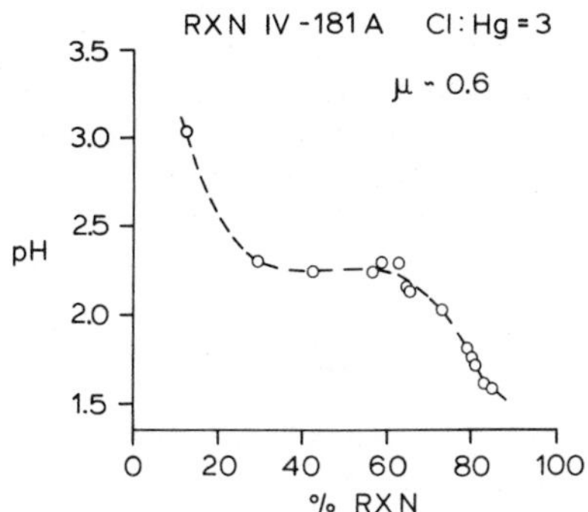

Figure 4. Changes in pH for a typical transmethylation reaction are followed as a function of extent of reaction. The [H^+] buffering is apparent for 30–70% of the reaction.

course of reaction *1*, but do not directly reflect the extent of transmethylation. CHEMSPECIES calculations with data from Table I can account for the large increase in H^+ during reaction in terms of < 500-fold changes in concentrations of Hg(OH)Cl and MeHgOH, all at < 10^{-6} M. Nonetheless, additional important (and unavailable) formation constants involving hydroxy- or oxo-species are needed for full interpretation of late reaction products. NMR spectra provided no evidence for new products, but such solutes, as with previous examples, could be at very low concentrations while yet manifesting large pH changes, *e.g.*, oligomerization of Me_2SnOH^+ (12,30). In any case, neither precipitation of such acid-forming products was observed, nor did significant deviation in the bimolecular rate (equation *2*) appear after substantial reaction which might implicate oxo-species in rate-determining steps.

Relative abundances of reactant and product mercurials may principally control the pH of the reaction. In Figure 4, where extent of reaction, that is $[MeHg^+]_{total}/[Hg^{2+}]_{total}$, versus pH is compared, this condition of buffering was apparent. In the region from about 30-70 percent reaction, we found that pH varied little more than ± 2.7 percent in either the low (Cl/Hg = 3) or high chloride (Cl/Hg = 23) situations. Over this range, mean pH increased from 2.26 to 2.38 with $[Cl^-]$, just as expected by shifting such controlling reactions as *9* and *10* to the left.

While complex, the general points raised here are typical for inorganic complexes in natural waters, and a number of numerical solutions have been described to treat such equilibria (24,42). In natural waters the types of transmethylation reactions of concern to us must occur at much slower rates, both on account of substantial reductions in $[Hg^{2+}]_{total}$ and $[Me_3Sn^+]_{total}$ and radical changes in $[H^+]$ or $[Cl^-]$.

Based on simulated pH, $[Cl^-]$, and metal concentrations encountered in natural waters, our preliminary CHEMSPECIES calculations indicate that involvement of hydroxy species in reaction *1* would be substantial. Nonetheless, in considering a transition from fresh water to sea water systems, the effects of chloride ion, of the type discussed here, become most important. The concept of metal ion buffering by addition of appropriate ligands to solutions has been amply discussed (42). In complete analogy to pH control, mixtures of acids and conjugate bases will resist change in $[\text{metal ion}]^{z+}$, as

$$[\text{metal ion}]^{z+} = \frac{K\ [ML]}{[L]} \quad . \qquad \begin{array}{l} ML = \text{metal ligand} \\ L = \text{ligand} \end{array}$$

In general, we must regard such chemistry as basic to moderating organometal ion concentrations in biota, where a substantial variety of "appropriate" ligands are available. Transmethylation reactions will presumably have their rates substantially dictated by such chloride "buffer systems".

Consideration of Reaction Pathways for Methylation of Hg^{2+} Species by $Me_3Sn^+_{aq}$. For the experimentally determined rate of reaction *1* we considered all possible "ion pair" associations implied by equation *4*. We are not concerned at this point whether these ionic associations possess substantial lifetimes in solution or are transitory complexes. In turn, the formation constants listed in Table I express concentrations of such possible ion-pair contributions

$$K_{association} = \frac{[A^{z+}][B^{z-}]}{[AB]^{\pm}} \frac{\gamma_A \gamma_B}{\gamma_{AB}} \qquad (11)$$

where γ_A, γ_B, and γ_{AB} express the usual activity coefficients for involved ions in terms of the Brønsted-Debye-Hückel equations (36). Consequently, for an overall bimolecular process, we rewrite equation *2* in terms of the principal six reacting species determined for all conditions of $[Cl^-]$ employed in this work:

$$\begin{aligned} rate = {} & k'_1[Me_3Sn^+][HgCl_2] + k'_2[Me_3Sn^+][HgCl_3^-]\gamma_1^2 \\ & + k'_3[Me_3Sn^+][HgCl_4^{2-}]\gamma_2 + k'_4[Me_3SnCl][HgCl_2] \\ & + k'_5[Me_3SnCl][HgCl_3^-] + k'_6[Me_3SnCl][HgCl_4^{2-}] \qquad (12) \end{aligned}$$

Here, we exclude reactants involving hydroxy-tin or -mercury species, also on the basis of our speciation results obtained with the CHEMSPECIES program discussed above. We therefore can reasonably presume a linear contribution to k_2(obs) by a number of pairwise (*i.e.*, bimolecular) reactions which proceed concurrently (32, 36).

In principle, other bimolecular (or n-order) events could contribute to demethylation of Me_3Sn^+ by Hg^{2+}. As described in the Experimental section, we have constrained pH, pCl, and concentrations of reactants to insure sufficient reaction rates for NMR measurements, while avoiding solubility difficulties. By reducing excessive hydrolysis to insoluble hydroxides, involvement of non-bimolecular reactions, particularly with polymeric species, was probably eliminated (30,41).

It is possible to experimentally determine the activity coefficients γ_1 and γ_2 by use of the Debye-Hückel extensions of the Brønsted equation cited before. Normally this is performed at ionic strengths considerably less than those employed in the present study. In his elegant early work, Davis showed procedures by which such ion-pair reactions contributing to an overall bimolecular process could be linearly evaluated in terms of the k_o and z values (43). An application to the complex tin-mercury system also seems possible, but is beyond the scope of the present paper.

From equation *12* it is seen that quantitative evaluation of specific rate constants for the six pair-wise reaction pathways noted required a multivariate analysis involving treatment of k_2(obs) as a variable dependent upon six k' values and $[Cl^-]$.

Statistical Analysis of the Dependence of k_2(obs) on $[Cl^-]$. In their treatment of cleavage rates of *sigma*-bonded pyridinomethyl derivatives of transition metals by $HgCl_n$ or $TlCl_n$, Dodd and Johnston successfully employed (32) an extension of Davis' linear sum method for concurrent reaction pairs. Basically, this approach assumes kinetically independent existence and reactivity for each discrete electrophilic metal species, whether charged or neutral. Thus, relative concentrations of $HgCl_n^{2-n}$ reactants (taken as mole fractions, χ) were used to weight linear combinations of associated specific rate constants.

It seemed reasonable to adopt this procedure, inasmuch as available data indicates that likely pre-equilibrium associative processes involving chloride exchange (44) and aquation (8) probably all occur at rates greater than 10^6 times the transmethylation rates under study.

Consequently, expressing reaction *1* in more general terms by means of equations *2* and *12*, we obtained

$$k_2(\text{obs}) = \sum_{\substack{n=0\\m=2}}^{\substack{n=1\\m=4}} k_{nm} \cdot \chi_n \chi_m \qquad (13)$$

in which

$$k_{02} \cdot \chi_{[Me_3Sn^+][HgCl_2^o]}$$

$$k_{03} \cdot \chi_{[Me_3Sn^+][HgCl_3^-]}$$

$$k_{04} \cdot \chi_{[Me_3Sn^+][HgCl_4^{2-}]}$$

$$k_{12} \cdot \chi_{[Me_3SnCl^o][HgCl_2^o]}$$

$$k_{13} \cdot \chi_{[Me_3SnCl^o][HgCl_3^-]}$$

$$k_{14} \cdot \chi_{[Me_3SnCl^o][HgCl_4^{2-}]}$$

A series of CHEMSPECIES calculations for initial reaction conditions (t = 0 seconds) was performed using starting total reactant concentrations, pH, and total [Cl^-]. Ionic strengths (μ = 0.556) were maintained constant for all Cl/Hg ratios examined. A table of individual concentrations for each of the reacting species was constructed and the cross products of these values, expressed as mole fractions, were generated in accord with individual rate coefficients k_{01} through k_{14} above. A necessary and sufficient condition for equation *13* was that the sum of crossproduct mole fractions for each set of k_{nm}'s taken for each Cl/Hg taken equalled unity. For each Cl/Hg case k_2(obs) was determined by methods given in the Experimental Section.

A multi-regression analysis program was employed which provided complete analyses of variance, from the standpoint of tests for F-ratio, multiple correlation coefficient, t-test for regression coefficients, and scatter of residuals (19). For completeness, a step-wise regression was performed for all 2^6-1 = 63 combinations of the six independent variables, although several statistical tests quickly revealed that two significant limiting conditions were characteristic to the data (45).

The evolution of the analytical results are summarized in Table V, where it is seen that very satisfactory fits of over 99.5 percent of the kinetic information could be obtained with uniform dispersion of residuals for ten k_2(obs) at various Cl/Hg from 3 to 23. Basically, we find that the specific rate coefficient k_{14} for the reaction pair [Me_3Sn-Cl^o] and [$HgCl_4^{2-}$] approaches zero ($10^{-6}M^{-1}s^{-1}$). In addition, severe non-orthogonality (conversely, high degree of correlation) occurred between two pairs of reactants k_{03}-k_{12} and k_{04}-k_{13}. This condition was predicted by the correlation matrix for regression (19)

	k_{02}	k_{03}	k_{04}	k_{12}	k_{13}	k_{14}
k_{02}	---	0.013	-0.987	0.010	-0.988	-0.824
k_{03}		---	-0.160	1.000	-0.159	-0.546
k_{04}			---	-0.157	1.000	0.881
k_{12}				---	-0.156	-0.543
k_{13}					---	0.881
k_{14}						---

and the special equilibrium relationships inherent with formation constants listed in Table I. One of these equilibria can be evaluated for the $k_{03}k_{12}$ case as

TABLE V

Summary of Best Linear Fit for k_2(obs) *versus* [Cl^-] by Six Independent Pair-wise Rate Constants

Variable nm	Estimated Rate[a] k_{nm}	± S.E.	Significance of t-test[b]	Relative Reaction Rate
02	17.37	1.87	< 0.001	1.00
03	7.83	0.63	< 0.001 (03 vs. 12)	0.45
04	9.41	1.25	< 0.001 (04 vs. 13)	0.54
12	115.8	9.3		6.7
13	129.8	17.3		7.5
14	<0.01[c]	<0.01[c]	n.s.	∿ 0.0

[a]k_{nm} 10^{-3} M^{-1} s^{-1}; [b]Reference 19; [c]Estimated upper value derived from mean sum of squares deviant from regression.

STATISTICAL ANALYSIS[b]:

Intercept = $-9.25 \pm 0.10 \times 10^{-3}$ M^{-1} s^{-1}

Variance $R = 0.998$ where $R_{0.01} = 0.911$

$R^2 = 0.995$ ($100\ R^2$ = "percent fit")

$F = 419.0$ (df = 6) where $F_{0.005} = 12.95$

$$\frac{[Me_3Sn^+][HgCl_3^-]}{[Me_3SnCl][HgCl_2]} = \frac{\beta_{HgCl_3}}{\beta_{Me_3SnCl}} = 10^{1.17} = 14.79 \qquad (14)$$

For the $k_{04}k_{13}$ case, the following numerical relationship can be derived from

$$\frac{[Me_3Sn^+[[HgCl_4^{2-}]}{[Me_3SnCl][HgCl_3^-]} = \frac{\beta_{HgCl_4}}{\beta_{Me_3SnCl}} = 10^{1.14} = 13.80 \qquad (15)$$

By substitution of $k_{04} = 13.80\ k_{13}$ and $k_{03} = 14.79\ k_{12}$, two new independent variables were generated, *e.g.*, k_{41} and k_{31}, respectively, which yielded a correlation matrix showing satisfactory orthogonality (19). Multiple regression of this new set of three variables (again, it was shown that k_{14} was not significant) provided stable coefficients assignable as the individual or partial second-order rate coefficients signifying the relative contribution of each pair-wise reaction path considered.

Mechanistic Considerations of Several Pathways for Methylation of Hg^{2+} by Me_3Sn^+ in Natural Waters. In chloride solutions of constant ionic strength approximating those found in marine waters (42), where $\mu \simeq 0.6$ and $[Cl^-] \simeq 0.5M$, observed bimolecular transmethylation rates were accurately accounted for by a linear combination of certain tin and mercury reaction pairs. Our observed and estimated rates are summarized in Table VI and compared in the error plot depicted in Figure 5.

Of considerable interest is the possibility of applying these kinetic results to estimating rates for reaction *1* in natural aquatic situations where large $[Cl^-]$ and ionic strength gradients occur. Such commonplace environmental settings are found in estuarine interfaces with river fresh waters, where it should also be noted that likelihood of higher concentrations of bioactive metal ions is greater from disposal of industrial effluents (46). For our discussion, we can expect that reactant metal species in reaction *1* would be available only at greatly reduced concentrations *i.e.*, $< 10^{-8}M$. On the other hand, we can envision inter-tidal water conditions favoring distributions of methyltin and mercury chloro-species over hydroxy-species, as implied by CHEMSPECIES calculations at pH $\simeq$ 7-8 and pCl $\simeq$ 2. We are thus confronted with a main question of whether reduced, but buffered, $[H^+]$ and $[Cl^-]$, coupled with lowered and variable ionic strengths materially change the kinetic contribution of reaction pairs discerned in the laboratory.

We have shown that for those reaction pairs involving a neutral partner in the activated complex or rate-determining step, the Brønsted-Debye-Hückel relationship (equations *8a*, *8b*) predicts that $\ln k = \ln k_o$. That is, k_2(obs) may become independent of

TABLE VI

Estimation of k_2 Based on the Reaction Pair Model

Run	k_2(obs)	± S.E.[a]	k_2(calc)[a,b]	μ	Cl/Hg
16	4.40	0.25	4.49	0.562	22.16
9	4.61	0.18	4.53	0.550	23.03
17	5.85	0.27	5.79	0.382	14.77
7	6.25	0.48	6.39	0.550	11.00
6	6.75	0.23	6.66	0.550	8.97
4	7.11	0.32	7.04	0.550	6.00
3	7.11	0.36	7.19	0.550	4.99
2	7.40	0.42	7.38	0.550	4.01
1	7.63	0.19	7.66	0.550	3.00
8	7.70	0.39	7.67	0.550	3.00
10	8.06	0.29	8.11[c] 8.10[d]	0.051	2.00
11	9.12	0.20	7.69[c] 9.06[d]	0.051	3.01
12	10.24	0.41	7.38[c] 10.40[d]	0.077	4.00
13	10.58	0.41	7.20[c] 11.50[d]	0.100	5.00

[a]k_2 in 10^{-3} $M^{-1}s^{-1}$; [b]Given by k_{pm} values in Table V substituted into equation *13*. [c]Calculated from regression equation given in Table V; [d]See text and equations *13, 16*.

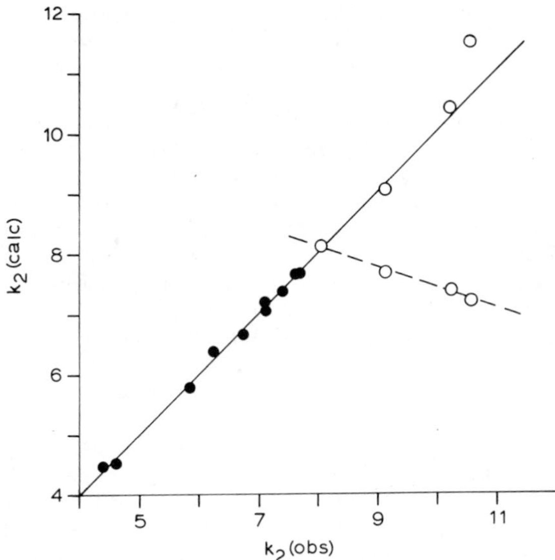

Figure 5. Observed and calculated rate constants ($x\ 10^3 M^{-1} s^{-1}$) are compared in an error (residual) diagram. Kinetics parameters estimated for various Cl/Hg values in solutions at high ionic strengths accurately predict k_2(calc) in such conditions (—●—), but fail for rates observed in solutions of low ionic strength (- -○- -). Introduction of a correction factor, based upon changes in ionic strength, into appropriate kinetic parameters permits fairly accurate extension of original parameters to faster rates in solutions of low ion concentrations (—○—).

ionic strength if solution conditions favor only formation of neutral tin and/or mercury reactants. We have tested this idea in two ways.

First, by comparing the relative rate contribution for reaction pairs in equation *1* listed in Table V, we see the most important are $Me_3Sn^+ + HgCl_2^o$; $Me_3SnCl^o + HgCl_2^o$; and $Me_3SnCl^o + HgCl_3^-$. All of these pathways are expected to be independent of μ. We now reexamine the effects of increasing μ discussed with Table III and Figure 1 employing a mechanistic picture: (a) with increased $[ClO_4^-]$, the observed near-zero slope results because virtually no compositional change occurred in the three principal reaction pairs over the range studied, hence $\ln k/k_o$ = constant $\simeq 0$; (b) where μ was increased with $[Cl^-]$, substantial compositional changes were effected in favor of $HgCl_4^{2-}$ concentrations. Thereby, reactions occurring by $Me_3Sn^+ + HgCl_4^{2-}$ and $Me_3SnCl + HgCl_4^{2-}$ either yielded a negative slope for equation *8* or materially retarded k_2(obs) as kinetically minor pathways.

A second important test involved a fortuitous opportunity to experimentally "isolate" a reaction pathway. We determined k_2(obs) for reaction *1* where Cl/Hg = 2 and μ = 0.051. Here, at reactant concentrations amenable to NMR measurements, CHEMSPECIES calculations showed that Me_3Sn^+ and $HgCl_2^o$ comprised 99.9 percent of total reactants. Therefore, equation *13* reduced to

$$k_2(\text{obs}) = \chi_{[Me_3Sn^+][HgCl_2^o]}\, k_{02} = 8.06 \times 10^{-3} M^{-1} s^{-1}$$

This result agrees well with the predicted value of 8.11×10^{-3} $M^{-1}s^{-1}$ which was based on kinetic parameters obtained for reactions conducted at the significantly higher ionic strength of 0.55. Were oppositely charged reaction pairs to become compositionally important at the lower ionic strength, we expect that these could contribute increased relative rates to the overall k_2(obs). Under ideal circumstances, where ion-pair associations and even mechanistic paths were independent of μ, individual linear contributions from such z_+z_- pairs should alter by some factor proportional to the $\Delta\mu$ studied here.

We examined this possibility for reaction *1* at the low ionic strength noted. A striking reversal in the relationship of k_2(obs) with Cl/Hg or $[Cl^-]$ was observed, that of mutual increase, as summarized in Table VI. Plotting these k_2(obs) obtained at low ionic strength against k_2(calc) obtained with kinetic parameters derived at high ionic strength, again as an error graph in Figure 5, clearly demonstrated inapplicability of certain terms in equation *13* without a $\Delta\mu$ correction factor for charged pathways (39).

To account for acceleration of k_2(obs) by Cl^- at low ionic strength, we can derive a simple correction based upon several important assumptions. For equation *8a* or *8b*, we take $\ln k = \ln k_o$ where $z = 0$ and/or $\mu = 0$; in practice our multiple regression solution to equation *13* also produces a concentration-independent term (intercept) which probably bears a physical relationship to k_o. Further, we presume that the concurrent k_{nm} values estimated in the high ionic strength solutions are valid at low μ. This is saying that no important mechanistic changes occur in the ten-fold change of ionic strength examined. With these points in mind, we rewrite equation *8a* solving for k_i corresponding to μ_i with elimination of k_o and Brønsted's proportionality constant α

$$\ln k_1 - 2\, z_+ z_- \sqrt{\mu_1} = \ln k_2 - 2\, z_+ z_- \sqrt{\mu_2}$$

More conveniently,

$$\ln k_1/k_2 = 2\, z_+ z_-\, [\sqrt{\mu_1} - \sqrt{\mu_2}]. \qquad (16)$$

We applied such a $\Delta\sqrt{\mu}$ correction for each of the four runs cited in Table VI. For run 10, all contributors to k_2(obs) except k_{02} fall out as previously described; for runs 11-13, reaction pair k_{14} was eliminated because of its small contribution, hence only reaction pairs k_{03} and k_{04} required correction with equation *16* using the two ionic strengths employed at μ_1 and μ_2, *i.e.*, 0.550 and 0.051, 0.077, or 0.100, respectively. Following this, the corrected terms were summed with appropriate weighting by the cross-product mole fractions as specified in equation *13*.

The new k_2(calc) are compared in Table VI, where it is seen that remarkable fits (<1-9%) are obtained with our elementary assumptions. It should be noted that even though a straight-line relationship is suggested from Figure 5 and the fits tabulated in Table VI, some divergence occurs. This must be expected, inasmuch as apparent ion strengths used in this work are only approximate, having been calculated from stoichiometries of reactants and gegenions (moles L^{-1}) rather than in molarity for calculated distributions of ligated tin and mercury ions. In the worst case of runs 9 or 16, apparent μ exceeds true ionic strength by about 3 percent, according to CHEMSPECIES calculations. Other sources of error include standard deviations in k_2(obs) of about 4 or 5 percent.

It is tempting to extend these simple ideas developed from Brønsted-Debye-Hückel concepts for transmethylation between tin and mercury ions to natural aquatic solutions. We have already noted the metal ion buffering and specific ligand effects which we expect to loom as important in environmental situations. Seemingly, if we can assess the significance of those factors, it would appear that major goals for the organometallic chemist will now be to establish comparative reaction rates for those concurrent ionic processes which, in sum, lead to appearance of toxic methylmetal species.

Plainly, for the system defined by reaction *1*, chloride ion and ionic strength dependence are now well established. Over gradients of 20- and 10-fold, respectively, there appears to be no discernable change in the major reaction pairs, either in their constitution or relative contributions to k_2(obs). In one sense this is surprising, particularly for the ion-molecule or ion-ion partners, since it has been shown in related reactions that a 10-fold change in [Cl^-] can alter k_{nm} to a much greater extent than that expected from equation *16* where Cl/Hg is large (32). At much reduced Cl/Hg values, formation constants for chloroelectrophiles far exceed those for ion-pairing between Cl^- and methyl donors (8,32), so reducing possibilities for altered reactivities in the latter.

With either $Me_3Sn^+_{aq}$ or Hg^{2+}_{aq}, substantial evidence is available indicating that either species forms well-defined structures in water (8,10,11,12,41,47), for which a distinctive stereochemical pathway to transmethylation should be reckoned. The three fastest reaction pairs measured in this work all involve tin or mercury species which are regarded on the basis of spectroscopic evidence as coordinately "unsaturated". That is, methyl or chloro-ligands do not occupy all available sites: 6 for tin and four or six for mercury (11,12,41,47). Therefore, labile water molecules are available for associative bridging complexes of the type shown with *5* and *6*.

Finally, a statement is in order, concerning the "true" chemical nature of the equilibrium pairs defined by equations *14* and *15*. These should not be regarded as mathematical devices to serve requirements of orthogonality in our regression analysis. Their existence is as real as any of the formation constants in Table I, and the subtle structural or electronic differences which energetically separate the k_{03} pair from the k_{12} pair, or k_{04} from k_{13}, should represent bases by which other kinetic properties can be distinguished in a manner similar to the present work. Their respective equilibrium constants suggest that we treat them as ion-pairs (46,48).

Acknowledgements

The authors thank Mr. R. A. Thompson and Ms. M. Crocker for assistance with NMR spectra and kinetic runs. Discussions with Dr. T. D. Coyle concerning kinetic and mechanistic interpretations of data were very helpful. We are especially grateful to Mr. M. R. Cordes of the NBS Applied Mathematics Division for his talented assembly of critical subroutines required for the CHEMSPECIES program. Continuing encouragement and partial financial support from the NBS Office of Environmental Measurements and the U.S. Naval Ship Research and Development Center (Annapolis) are gratefully acknowledged by K. L. Jewett and F. E. Brinckman. J. M. Bellama wishes to express appreciation for financial support by the National Science Foundation Grant OCE-76-23595 and by the University of Maryland Sea Grant Program.

Literature Cited

1. Brinckman, F.E., Iverson, W.P., and Blair, W., In: "Proceedings of the Third International Biodegradation Symposium," J.M. Sharpley and A.M. Kaplan, Eds., 919, Applied Science Publ., London, 1976.
2. Chau, Y.K. and Wong, P.T.S., In: "Occurrence and Fate of Organometals and Organometalloids in the Environment", American Chemical Society Symposium Series, Washington, D.C., F.E. Brinckman and J.M. Bellama, Eds., 1978.
3. Wasik, S.P., Brown, R.L., and Minor, J.I., J. Environ. Sci. Health, (1976) A 11, 99.
4. Ridley, W.P., Dizikes, L.J., and Wood, J.M., Science, (1977) 197, 329.
5. Carty, A.J., In: "Occurrence and Fate of Organometals and Organometalloids in the Environment", American Chemical Society Symposium Series, Washington, D.C., F.E. Brinckman and J.M. Bellama, Eds., 1978.
6. Hein, F. and Meininger, H., Z. anorg. allgem. Chem., (1925) 145, 95.
7. Benesch, R. and Benesch, R.E., J. Amer. Chem. Soc., (1951) 73, 3391.
8. Fratiello, A., Prog. Inorg. Chem., (1972) 17 Part II, 57.
9. Including estimation of sp-hybridization of metal-carbon bonds; cf. Holmes, J.R. and Kaesz, H.D., J. Amer. Chem. Soc. (1961) 83, 3903.
10. Hall, J.R., In: "Essays in Structural Chemistry", A.J. Downs, D.A. Long, and L.A.K. Staveley, Eds., 443, Plenum Press, New York, 1972.
11. Brinckman, F.E., Parris, G.E., Blair, W.R., Jewett, K.L., Iverson, W.P., and Bellama, J.M., Environ. Health Perspectives, (1977) 19, 11.
12. Tobias, R.S., Organometal. Chem. Rev., (1966) 1, 93.
13. Glass, G.E., Schwabacher, W.B., and Tobias, R.S., Inorg. Chem., (1968) 7, 2471.
14. Baes, C.F., Jr. and Mesmer, R.E., "The Hydrolysis of Cations", John Wiley and Sons, New York, 1976.
15. Haupt, H.-J., Huber, F., and Gmehling, J., Z. anorg. allgem. Chem., (1972) 390, 31.
16. Jewett, K.L., Dissertation, University of Maryland, 1978.
17. Jewett, K.L., Brinckman, F.E., and Bellama, J.M., In: "Marine Chemistry in the Coastal Environment", 304, American Chemical Society Symposium Series No. 18, Washington, D.C., T. Church, Ed., 1975.
18. Johansson, L., Coord. Chem. Rev., (1974) 12, 241 and references cited therein.
19. Chatterjee, S. and Price, B., "Regression Analysis by Example", John Wiley and Sons, New York, 1977.
20. A Fortran V listing of CHEMSPECIES program is available from the authors on a limited basis.

21. Ciavatta, I.L. and Grimaldi, M., J. Inorg. Nucl. Chem., (1968) 30, 197.
22. Helper, L.G. and Olofsson, G., Chem. Rev., (1975) 75, 585.
23. Sillen, L.G. and Martell, A.E., "Stability Constants of Metal Ion Complexes", The Chemical Society, London, Special Publication No. 17, 1964.
24. Dryssen, D., Jagner, D., and Wengelin, F., "Computer Calculation of Ionic Equilibria and Titration Procedures", Almqvist and Wiksell, Stockholm, 1968.
25. Cassol, A., Portanova, R., and Magon, L., La Ricerca Scientifica, (1966) 36, 1180.
26. Janssen, M.J. and Luijten, J.G.A., Rec. Trav. Chim., (1963) 82, 1008.
27. Geier, G. and Erni, I.W., Chimia, (1973) 27, 635.
28. Schwarzenbach, G. and Schellenberg, M., Helv. Chim. Act., (1965) 48, 28.
29. Cassol, A., Magon, L., and Barbieri, R., Inorg. Nucl. Letters, (1967) 3, 25.
30. Tobias, R.S. and Yasuda, M., J. Phys. Chem., (1964) 68, 1820.
31. Abraham, M.H. and Johnston, G.F., J. Chem. Soc. (A), (1970), 193.
32. Dodd, D., Johnson, M.D., and Vamplew, D., J. Chem. Soc. (B), (1971), 1841.
33. Huey, C.W., Brinckman, F.E., Grim, S.O., and Iverson, W.P., In: "Proceedings of the International Conference on Transport of Persistent Chemicals in Aquatic Ecosystems", II-73, National Research Council, Ottawa, 1974.
34. Jewett, K.L. and Brinckman, F.E., Preprints of Papers (ACS Div. Environ. Chem.), (1974) 14, 218.
35. Blandamer, M.J. and Burgess, J., Chem. Soc. Rev., (1975) 4, 55.
36. Davis, C.W., Progr. Reaction Kinetics, (1961) 1, 161.
37. Jewett, K.L., Brinckman, F.E., and Bellama, J.M., Abstr. Papers, 172nd National Mtg. ACS, San Francisco, CA, Sept., 1976, INOR 137.
38. Lockhart, J.E., "Redistribution Reactions", Academic Press, New York, 1970.
39. Pethybridge, A.D. and Prue, J.E., Progr. Inorg. Chem., (1972) 17 Part II, 327.
40. Hester, R.E. and Plane, R.A., Inorg. Chem., (1964) 3, 769.
41. Sandström, M., Acta Chem. Scand., (1977) A 31, 141.
42. Stumm, W. and Morgan, J.J., "Aquatic Chemistry", Wiley-Interscience, New York, 1970.
43. Davies, C.W. and Wyatt, P.A.H., Trans. Faraday Soc., (1938) 45, 770; 774.
44. For example with $HgCl_n$ see: Eigen, M. and Eyring, E.M., Inorg. Chem., (1963) 2, 636.
45. Davies, O.L. and Goldsmith, P.L., Eds., "Statistical Methods in Research and Production", Hafner Publ. Co., New York, 1972

46. Purves, D., "Trace-Element Contamination of the Environment", Elsevier Scientific Publ. Co., Amsterdam 1977. Especially see Chapter 7.
47. Deacon, G.B., Rev. Pure Appl. Chem., (1963) 13, 189.
48. An elegant spectroscopic definition is given by Waters, D.N., in reference 10.

Discussion

J. J. ZUCKERMAN (University of Oklahoma): I assume that these reactions occur in homogeneous solution?

BELLAMA: Yes.

ZUCKERMAN: In Chesapeake Bay water, might there be micelles where, in that organic medium, if the ionic strength of the aqueous portion rose, one would get a salting out into the micelle? Micelles perhaps containing or composed of chlorinated hydrocarbons or chlorination agents? Could you comment on it?

BELLAMA: These are very complex systems; we and others are just beginning to decipher them. The laboratory situations are complex enough; when you get out into the real world the situation becomes fantastically complex, not only in terms of the chemistry question that you have addressed, but also in terms of the ubiquitous microorganisms. So, yes, that effect is very possible; such effects very likely do occur. This is the work of the future: to try to discover what the individual pairwise interactions can tell us, if we really can restrict the systems to two separate interacting moieties.

F. E. BRINCKMAN (National Bureau of Standards): The real world confronts us and makes what is already a complicated solution to these multicomponent kinetic analyses even more complicated because of phase considerations. In Chesapeake water, the situation is far worse than even in micelles because of abundant microglobules which are of larger size, and in terms of surface activity are very great. These are fecal products, lipids, etc., that involve reacting species with partitioning. Unfortunately, we'll not get into it in this symposium with the exception of papers by Drs. Wasik and Good. We did some work based on the idea of the cybotactic volume [E.M. Kosower, J. Amer. Chem. Soc., (1958) 80, 3261]. We found that a great number of common bacterial products, organic materials, which will form homogeneous solutions can be compared as a solvent media for these transmethylation rate studies we showed here. It is the amount of water which is determining the rate. This leads to the idea of Kosower's concept of the cybotactic volume, which is that in the water or in a strong electrolyte medium, there are long-range structures called micelles if you like. In a bacterial product medium such as an acetone, *i.e.*, acetone-water solution, the acetone will order itself into a cybotactic volume which represents, if you like a "micro-micro micelle". It is in those volumes that these reactions do not happen. That's about all we know. We reported on that in San Francisco [*Abstr. Papers*, 172nd Nat'l ACS Mtg., Sept. 1976, INOR-136].

BELLAMA: One might add here that in the Baltimore Harbor, which must be one of the least-clean areas that we could examine, there is sewage outflow, the concentration of metals is about enough that you can walk on water wtihout external assistance, and you have organics as well; so the situation becomes very complex. On the other hand, there are portions of the Bay (some of the shellfishing areas) that are still very clean areas, and so one has a control.

F. HUBER (University of Dortmund): In homogeneous solution we measured the reaction of dimethylthallium acetate and mercury acetate, and we obtained the same order of magnitude for the k_2 values. We found an activation energy of about 18 kilocalories per mole, and I think it's about the same as you found with the tin and thallium compounds. In our paper [this volume], we postulate a mechanism for these transalkylation reactions between dimethylthallium compounds and mercury compounds. It's about the same as the mechanism for the transalkylation reaction we observed between dimethylthallium and thallium(III) acetate.

C. J. SODERQUIST (University of California, Davis): Have you tried an experiment where you took Chesapeake Bay water and some sediment and put in trimethyltin and mercury at levels you would encounter there (0.1 ppm)? Does the reaction take place? If it does, does it match what you would predict from your data?

BELLAMA: We have taken sediment or taken the organisms isolated from the Bay by Professor R.R. Colwell's group (University of Maryland) or by Dr. W.P. Iverson (National Bureau of Standards) and spiked with one metal or several metals to see whether or not the systems would promote methylation of a particular metal [C.W. Huey, *Ph.D. Dissertation*, University of Maryland, 1976]. The results are favorable.

BRINCKMAN: Yes, we have done that experiment and transmethylation occurs, but there is a very substantial uptake of the organotin by substrate. This is true in the growth medium for *in vitro* experiments dependent upon container effect. Either a polycarbonate Petri dish effect, or a sediment effect occurs, and the uptake of tin is very substantial. The methylation of mercury occurs, but we think it is slower.

M. O. ANDREAE (Scripps Institute of Oceanography): You

showed the production of trimethyltin in sediments. Is that a model experiment, or is that actually trimethyltin that you found in the real sediment?

BELLAMA: No, we haven't found it in sediment, although some organotins apparently have been found. [R.S. Braman *et al*. report trace methyltin species in fresh and marine waters [S.E. Regional ACS Meeting, Tampa 1977; submitted, *Anal. Chem*.]. Their precise identity is not really quite clear to us. It's the question of speciation, which is obviously an extremely important one, and one that at environmental concentrations become very difficult to do by some of our current analytical techniques. Methylation of metals seems to occur with virtually every element in the periodic table. We found that the transmethylation reactions seem to occur much more readily than transalkylations where you have a higher group than a methyl, with perhaps one exception that Dr. DeSimone touched on in his paper [equation 11]. That is, in our work on silatranes, we find that transalkylation reactions all seem to proceed very readily; indeed, many other R groups can be transferred at essentially the same rate as methyl. With regard to your question, a good starting place for transmethylation reaction is the trimethyltin cation, which is the classical water-soluble, water-stable, methyl-donating species that would be the reaction system of choice to most organometal chemists.

J. S. THAYER (University of Cincinnati): Most of the reactions that you cite had methyl transferred to mercury from some other organometal. Have you observed any reactions where a methyl group is transferred from methylmercury to some other metal?

BELLAMA: In general, mercury seems to be near the bottom of the activity series; you can take many methyl donors and transfer the methyl group to mercury, but there are exceptions.

BRINCKMAN: Jewett showed [K.L. Jewett, *Ph.D. Dissertation*, University of Maryland (1978)], and it has been independently shown [N.F. Gol'dshleger *et al*., *Dokl. Acad. Nauk SSSR* (1972) *206*, 106] that the methyl group can be transferred from methylmercury to Au(III), Pd(II), and Pt(II) with, for example, formation of intermediate transitory methylpalladium species. We found the second order rate constant to be quite slow, about 10^{-4} $M^{-1}s^{-1}$, or less.

RECEIVED August 22, 1978.

12

Demethylation of Methylcobalamin: Some Comparative Rate Studies

JOHN S. THAYER

Department of Chemistry, University of Cincinnati, Cincinnati, OH 45221

Introduction

Various studies have been reported on the kinetics and mechanism of the reaction between inorganic salts and methylcobalamin(1-9). Rate constants have been reported for CH_3HgX(4,5) and $PdCl_4{}^{2-}$(6). Yet there has been no systematic wide-ranging study into the various factors affecting the rate of this reaction. This paper gives a first report on such a study.

Experimental

All reactions were run in 0.10M acetic acid-0.10M sodium acetate solution (pH=4.55) at 25°C. The rate of disappearance of methylcobalamin was followed by standard spectrophotometric techniques(1,3,10) usually until 90% or more conversion occurred. Initial reactions followed second-order kinetics (first-order in each reactant); occasionally the metal substrate was in great excess, in which case pseudo-first-order kinetics were observed.

Results

A. Metathetical Exchange

Tables I and II list the reaction rate constants for various substrates. Occasionally a metal substrate was only partially soluble in the reaction mixture; the listed value is then the minimum value for the rate constant. For those systems where the reaction had to be studied for many days, controls (methylcobalamin solutions without metal compound) were run

0-8412-0461-6/78/47-082-188$05.00/0

Table I. Rate Constants for Metathetical Reaction of Methylcobalamin with Organometals

Compound	# Runs	$K(s^{-1}M^{-1})$
Mercury		
CH_3HgOAc	6	6.68×10^{-2}
C_6H_5HgOAc	3	$*3.5 \times 10^{-3}$
Thallium		
$(CH_3)_2TlOAc$	3	2.42×10^{-4}
$CH_3Tl(OAc)_2$	3	1.40×10^{-2}
Lead		
$(CH_3)_3PbOAc$	4	3.60×10^{-4}
$(C_2H_5)_3PbOAc$	3	2.00×10^{-3}
$(C_6H_5)_3PbOAc$	3	$*3.6 \times 10^{-3}$
$(C_2H_5)_2PbCl_2$	1	2.32×10^{-4}
$(C_6H_5)_2Pb(OAc)_2$	2	$*1.0 \times 10^{-4}$
$C_6H_5Pb(OAc)_3$	4	1.08×10^{-2}
Tin		
$(CH_3)_3SnOAc$	3	1.50×10^{-5}
$(C_2H_5)_3SnOAc$	4	2.42×10^{-5}
$(C_6H_5)_3SnOAc$	6	$*5.0 \times 10^{-4}$
$(CH_3)_2SnCl_2$	3	2.50×10^{-4}
Metalloids		
$(CH_3)_4As^+I^-$	3	1.95×10^{-3}
$(CH_3)_4Sb^+I^-$	3	1.35×10^{-4}
$(CH_3)_3Te^+I^-$	3	1.73×10^{-4}

Table II. Rate Constants for Reaction of Methylcobalamin with Inorganic Oxy Compounds

Compound	# Runs	$K(s^{-1}M^{-1})$
Au_2O_3	2	$*2.4 \times 10^{-2}$
In_2O_3	1	$*7.0 \times 10^{-6}$
SnO_2	3	$*1.3 \times 10^{-6}$
@$Pb(OAc)_4$	2	*0.40
Na_2HAsO_4	1	2.30×10^{-5}
Sb_2O_5	1	$*4.2 \times 10^{-5}$
$NaBiO_3$	6	0.44

*Minimum values for partially soluble compounds. OAc= acetate. @Compound introduced as lead tetraacetate, but hydrolysis occurs upon contact with solvent.

simultaneously. These controls showed no reaction during this time period.

Comparison of reaction rates indicate that the heaviest metals are the most reactive. Compounds of the Sixth Row metals from Pt through Bi react readily with methylcobalamin, though in many cases the initially formed monomethylmetal compound decomposes. Lighter metals are noticeably less reactive, except for Pd(6). Substitution of organic groups onto the metal reduces reactivity very sharply, with the extent of reduction depending on the number of organic groups. Thus, in the Pb series, reactivity towards methylcobalamin will vary in the order

$$RPbX_3 \; > \; R_2PbX_2 \; > \; R_3PbX$$

The nature of the inorganic groups present on the metal will also change the rate of reaction. Table III shows the effect of the anion upon the reactivity of methylmercuric ion.

Table III. Effect of Anion on Reaction Rate Constant for Methylmercuric Salt-Methylcobalamin Reaction*

Anion	Anion Conc.[a]	$k(s^{-1})$	Ratio
OAc^-	0	3.60×10^{-4}	1.000
Cl^-	1.36×10^{-2}	3.01×10^{-4}	0.856
Br^-	7.55×10^{-3}	1.37×10^{-4}	0.381
I^-	9.54×10^{-3}	Salt Precipitates	
SCN^-	6.57×10^{-3}	7.15×10^{-5}	0.199
CN^-	7.17×10^{-3}	3.44×10^{-5}	0.096

*In all cases, the concentrations of CH_3HgOAc and methylcobalamin were 5.75×10^{-3} and 1.87×10^{-4} molar respectively.

[a]Concentration of extra anion added. The calculated K for this system is $6.99 \times 10^{-2}\ s^{-1}M^{-1}$.

A similar repressant effect is noted for inorganic mercury salts (1,5) and also occurs for organothallium compounds, where the halides are less reactive than the acetates (10). By contrast, organotin chlorides seem to be somewhat more reactive than the corresponding acetates.

B. Successive Reactions

The great difference in the reaction rate constants for $Hg(OAc)_2$ and CH_3HgOAc enabled the two methylation reactions to be studied separately (1,3,4). This could be extended to other systems. As Figure 1 shows, K_2PtCl_6 clearly undergoes successive methylations. The report of Taylor and Hanna, that this compound only accepts a single methyl group (9), is incorrect simply because they did not study the system for a sufficient length of time. Our observations do support their claim, however, that there is a time lag between the mixing of the reagents and the start of the reaction(9). Table IV lists the rate constants for a number of compounds that undergo polymethylation.

Table IV. Rate Constants for Successive Reactions with Methylcobalamin

Compound	$K_1 (s^{-1}M^{-1})$	$K_2 (s^{-1}M^{-1})$
$Tl(OAc)_3$	1.60	$^*1.3 \times 10^{-2}$
$CH_3Tl(OAc)_2$	$^*2.0 \times 10^{-2}$	$^*1.3 \times 10^{-4}$
$C_6H_5Pb(OAc)_3$	9.5×10^{-2}	1.1×10^{-3}
K_2PtCl_6	0.965	8.4×10^{-3}

*Calculated on the assumption that methylthallium diacetate does not decompose in solution.

The first rate constant for K_2PtCl_6 agrees well with estimated values obtained from the data of Taylor and Hanna (9). Methylthallium diacetate has been reported to decompose in polar solvents (11):

$$CH_3Tl(OAc)_2 \rightarrow CH_3OAc + TlOAc \quad (1)$$

This decomposition has to be taken into account when calculating the rate constant for methylation. By an elaborate calculation (10), we found that the rate

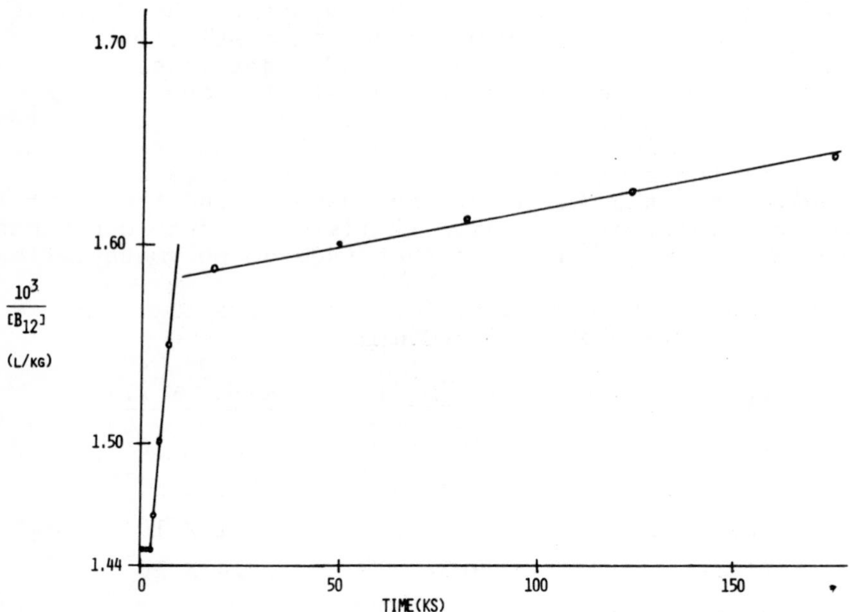

Figure 1. Kinetics plot of K_2PtCl_6-methylcobalamin reaction. Initial concentrations were 5.07 × 10^{-5} and 6.91 × 10^{-4}M, respectively.

constant for the reaction of methylthallium diacetate with methylcobalamin to be 5.65 X 10^{-2} $s^{-1}M^{-1}$. Under these conditions, methylation and reductive elimination are competitive reactions, and only a fraction of methylthallium diacetate is converted to the dimethyl compound.

Three compounds that might have been expected to undergo successive methylation did not, in fact, do so. These are shown in Table V.

Table V

Compound	*K*($s^{-1}M^{-1}$)	% Reaction
$KAuCl_4$	2.08	98.1
$Pb(OAc)_4$	*0.40	25.0
$NaBiO_3$	0.44	25.0

*Minimum value for slightly soluble compound

In excess methylcobalamin over an extended time period , $KAuCl_4$ only accepted a single methyl group. Since dimethylgold(III) compounds are known and stable, the inability may be due to the facile decomposition

$$CH_3AuCl_3^- \rightarrow CH_3Cl + AuCl_2^- \qquad (2)$$

This reaction has been reported for the Pd analog (6). No metallic Au mirror was formed during the reaction. Unlike K_2PtCl_6, $KAuCl_4$ showed no initial time lag; reaction began upon contact.

Neither $Pb(OAc)_4$ nor $NaBiO_3$ showed signs of successive methylation in the presence of excess methylcobalamin. In fact, only a portion of the metal compound actually reacted. Taylor and Hanna had a similar observation for $Pb(OAc)_4$ (8). Monomethyllead (IV) and monomethylbismuth (V) compounds are totally unknown. If they form at all, they must decompose so fast that methylation cannot proceed to any extent. Since the rate constant for $CH_3Pb(OAc)_3$ would be expected to be about 0.10 $s^{-1}M^{-1}$, the constant for the reductive elimination reaction must be at least comparable in magnitude. Since both of these compounds are strong oxidizing agents, it is possible that the reaction does not involve methyl transfer but rather direct oxidation of the methyl group.

C. Reaction of K_2PtCl_4 with Methylcobalamin

Taylor and Hanna reported essentially no reaction between potassium tetrachloroplatinate and methylcobalamin over a short period of time (9). Examining this system over a longer time period, we found that a reaction does occur. In fact, the reaction is autocatalytic, as shown in Table VI.

Table VI. Data for K_2PtCl_4 - Methylcobalamin Reaction

Time(ks)	$[CH_3B_{12}]$ Reacted	10^3 K*	$[CH_3B_{12}]$ Cal.@ Reacted
2.22	0.01×10^{-4}	3.22	0.01×10^{-4}
3.30	0.02	4.60	0.02
12.84	0.08	4.94	0.08
77.64	0.49	5.78	0.49
175.44	1.32	10.76	1.29
287.41	1.99+	38.06	2.27

*Rate constant ($s^{-1}M^{-1}$) calculated from second-order rate law, with initial concentrations K_2PtCl_4 = 2.00×10^{-4}M and CH_3B_{12} = 6.51×10^{-4}M.

@Concentration of methylcobalamin calculated using a rate constant 4.80×10^{-3} $s^{-1}M^{-1}$ in the autocatalytic equation (12):

$$x + A_o = \frac{A_o + B_o}{1 + \frac{B_o}{A_o} e^{-(A_o + B_o)kt}}$$

where x is the amount of methylcobalamin reacted after time t, and A_o and B_o are the initial concentrations of K_2PtCl_4 and methylcobalamin respectively.

The autocatalysis might have come from the presence of metallic platinum, formed by the reaction

$$PtCl_4^{2-} + CH_3B_{12} \rightarrow CH_3PtCl_3^{2-} + Cl^- \quad (3a)$$

$$CH_3PtCl_3^{2-} \rightarrow CH_3Cl + Pt + 2Cl^- \quad (3b)$$

We tested this by running the reaction in the presence of finely powdered platinum metal. The reaction shows a marked increase in rate, and followed standard second order kinetics(10).

D. Oxidative Reaction with Methylcobalamin

The report that Sn(II) species can be converted to methyltin(IV) compounds (13) prompted us to look at a variety of substrates that might undergo oxidative methylation. Table VII lists some of these, with their reaction rate constants.

Table VII. Rate Constants for Oxidative Reaction of Methylcobalamin with Metal Compounds

Compound	# Runs	*K($s^{-1}M^{-1}$)
Hg	3	1.0×10^{-5}
C_5H_5Tl	2	3.3×10^{-3}
$CsGeCl_3$	2	5.6×10^{-4}
$(CH_3)_6Sn_2$	2	1.2×10^{-4}
$(C_6H_5)_3P$	2	4.0×10^{-4}

*All compounds only partly soluble in reaction medium; hence, numbers listed are minimum values.

In the case of $CsGeCl_3$, one possibility is chlorine-methyl exchange, followed by disproportionation (Eqn. 4). The other compounds

$$2CH_3GeCl_2^- \rightarrow (CH_3)_2GeCl_2 + Ge + 2Cl^- \quad (4)$$

cannot proceed by this pathway. Investigations are now under way to determine the mechanism(s) by which these compounds react.

Discussion

Early workers quickly recognized that at least two reaction pathways existed for methylcobalamin reactions (7). The methyl-cobalt bond may cleave in any of three ways:

$$CH_3 - Co^{III} \rightarrow :CH_3^{\ominus} + {}^{\oplus}Co^{III} \qquad (5a)$$

$$CH_3 - Co^{III} \rightarrow \cdot CH_3 + \cdot Co^{II} \qquad (5b)$$

$$CH_3 - Co^{III} \rightarrow CH_3^{\oplus} + {}^{\ominus}:Co^{I} \qquad (5c)$$

In the latter two cases, the cobalt is being reduced, and the products (B_{12_r} and B_{12_s}) have quite different absorption spectra, making them easy to distinguish, at least in principle. The first cleavage occurs without oxidation, and the product is aquocobalamin, also with a unique absorption spectrum. Similar mechanisms could be proposed for other species, e.g. $R_2CH_3S^+$, that have been proposed as methylating agents.

A. Metathetical Demethylation

This is the most common mechanism found, and is essentially a transfer of a methyl carbanion to a metal substrate. The substrate is almost always cationic in nature; hence, the mechanism of the reaction is apparently S_E2 (14), proceeding through an intermediate of the type

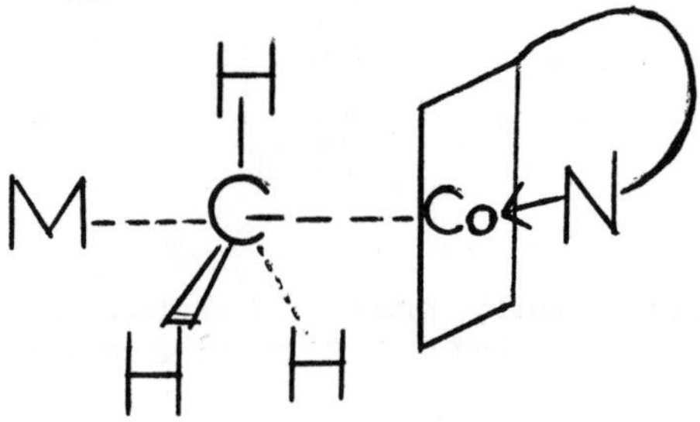

By corollary, anything that affects the electrophilicity of the metal will affect the rate of reaction. Organic groups, being good electron donors, should lower the electrophilicity of a metal; this is consistent with the observed decrease in reaction rates as the number of organic groups increases. The rate suppression by anions (Table III) results from electron-rich species competing for the same metal cation. Hence, any species that is electrophilic has the potential of reacting with methylcobalamin by this pathway, even if the initially formed product subsequently decomposes.

Workers studying the reaction of mercuric acetate with methylcobalamin found that the mercury interacts with the benzimidazole nitrogen of methylcobalamin to form the "base-off" complex (1,3,5) leading to more complex kinetics. This competing reaction, which can be eliminated by addition of halide ions (5), was also observed for $PdCl_4^{2-}$ (6). We did not observe it in the reactions of organometals with methylcobalamin, perhaps reflecting their reduced electrophilic character relative to $Hg(OAc)_2$.

B. Redox Demethylation

The most thoroughly studied system of this type is the $Sn(II)-CH_3B_{12}$ reaction (13). The Sn requires an oxidizing agent, such as aquocobalamin or Fe(III), for reaction to occur, and the proposed mechanism has homolytic cleavage of the methyl-cobalt bond as its crucial step (13,15).

Arsenic, selenium, and tellurium are methylated by a different pathway, which involves reduction of the substrate by removal of oxygen. The mechanism proposed by Challenger (16) and elaborated on by Cullen *et al.* (17) involves transfer of a methyl carbonium ion:

$$(HO)_3As: \quad + \quad CH_3^+ \rightarrow [CH_3As(OH)_3]^+ \qquad (6a)$$

$$[CH_3As(OH)_3]^+ \rightarrow CH_3AsO(OH)_2 + H^+ \qquad (6b)$$

$$CH_3AsO(OH)_2 + 2\,e^- \rightarrow CH_3(HO)_2As: + O^{2-} \qquad (6c)$$

$$CH_3(HO)_2As: + CH_3^+ \rightarrow [(CH_3)_2As(OH)_2]^+ \text{etc.} \qquad (6d)$$

In these systems methylcobalamin does not seem to be the methylating agent, at least not directly; instead, the methyl group comes from a sulfonium salt (16,17). Nevertheless, a similar mechanism could be written for methylcobalamin reactions.

The reaction between Pt(IV) or Au(III) and methylcobalamin has been alleged to require the presence of the lower oxidation state of the metal (7). This has been termed a "redox switch" mechanism, and has the following form (9):

$$PtCl_4^{2-} + CH_3B_{12} \rightarrow \text{complex} \qquad (7a)$$

$$\text{complex} + PtCl_6^{2-} \rightarrow CH_3PtCl_5^{2-} + Cl^- + B_{12} \qquad (7b)$$

The complex, reported by Wood and coworkers (15) to involve interaction of $PtCl_4^{2-}$ with a side chain of the corrin ring system, is characterized by having a more labile methyl-carbon bond than methylcobalamin itself. Consequently, it reacts more readily. Side chain complexes between metals and B_{12} derivatives have been proposed (18). The metals following the "redox switch" mechanism fall near the electrode potential for molecular oxygen (E^o = +0.68 v) (15). It should be noted that the $PtCl_4^{2-}$/Pt potential is +0.75 v (19). Our observation that Pt catalyses that reaction of $PtCl_4^{2-}$ with methylcobalamin suggests that this couple also falls into the "redox switch" category. Whether the "redox switch" mechanism is limited solely to platinum and gold or applies to a wider range of metals remains to be determined.

C. *In vivo* kinetics

The validity with which most reported kinetic and mechanistic studies can be extended to biological processes is quite uncertain. Very few rate studies have been reported for methylation by natural organisms. Trimethylarsine formed at the rate of 3.9×10^{-12} Ms^{-1} when *Candida humicola* methylated 5×10^{-3} M As_2O_3 solution (17). When one gram of rat cecal contents were treated with 2.0 ml 7.4×10^{-6}M $HgCl_2$ solution, 17.6 ng CH_3HgCl were formed over a period of twenty hours (20). Similar slow reactions were found in the biological methylation of Pb(II) and Tl (I) (21). Obviously, rates of methylation under biological conditions are going to be much slower than under test tube conditions. One reason is the much lower concentrations involved. Another reason, proposed by Robinson et al. (5), is that methylcobalamin in biological systems exists in hydrophobic environments, and that surfactants will greatly inhibit rates of reactions. Much remains to be done in this area.

Acknowledgments

We wish to thank Dr. John Wood and the Freshwater Biological Laboratory (Navarre, Minnesota) for providing the means to inaugurate this project. Drs. M. Orchin, J. B. Smart, and R. S. Tobias kindly provided some of the more exotic compounds used in this study. Dr. Estel Sprague provided some very helpful discussion on the kinetics involved in these reactions.

Literature Cited

1. DeSimone, R. E., Penley, M. W. Charbonneau, L., Smith, S. G., Wood, J. M., Hill, H.A.O., Pratt, J. M., Ridsdale, S., Williams, R. J. P., Biochim. Biophys. Acta (1973), 304, 851.

2. Bertilsson, L., Neujahr, H. Y., Biochem. (1971), 10, 2805.

3. Chu, V. C. W., Gruenwedel, D. W., Bioinorg. Chem. (1977), 7, 169.

4. Chu, V. C. W., Gruenwedel, D. W., Z. Naturforsch. (1976), 31C, 753.

5. Robinson, G. C., Nome, F., Fendler, J. H., J. Amer. Chem. Soc. (1977), 99, 4969.

6. Scovell, W. M., J. Amer. Chem. Soc. (1974), 96, 3451.

7. Agnes, G., Bendle, S. Hill, H. A. O., Williams, F. R., Williams, R. J. P., Chem. Comm. (1971), 850.

8. Taylor, R. T., Hanna, M. L., Environ. Sci. Eng. (1976), All, 201.

9. Taylor, R. T., Hanna, M. L., Bioinorg. Chem. (1976), 6, 281.

10. Thayer, J., unpublished results.

11. Pohl, U., Huber, F., J. Organometal. Chem. (1976), 116, 141.

12. Capellos, C., Bielski, B. H. J., "Kinetic Systems," 59-62, Wiley-Interscience, New York, 1972.

13. Dizikes, L. J., Ridley, W. P., Wood, J. M., J. Amer. Chem. Soc. (1978), 100, 1010.

14. Matteson, D. S., "Organometallic Reaction Mechanisms," Academic Press, New York, 1974.

15. Fanchiang, Y. T., Ridley, W. P., Wood, J. M., Chapter , this volume.

16. Challenger, F., Chapter 1, this volume.

17. Cullen, W. T., Froese, C. L., Lui, A., McBride, B. C. Patmore, D. J., and Reimer, M., J. Organometal. Chem. (1977), 139, 61.

18. Pratt, J. M., Inorganic Chemistry of Vitamin B_{12}," 189-190, Academic Press, New York, 1972.

19. Hartley, F. R., "The Chemistry of Platinum and Palladium," 13, Applied Sciences, London, 1973.

20. Rowland, I., Davies, M., Grasso, P., Arch. Env. Health (1977), 24.

21. Huber, F., Schmidt, U., Kirchmann, H., Chapter this volume.

Discussion

J. J. ZUCKERMAN (University of Oklahoma): Concerning the list of insoluble species reacting with the methylating agent: Is reaction occurring with soluble species (although the Ksp value for SnO_2 is a very small number), or is it in fact occurring on the surface? If so, are all these species in a slurry of similar paricle size?

THAYER: In some cases there is a partial solubility, and we expect that many of these are reacting with the species actually in solution, even though we could not get complete solubility. In some other cases, e.g., lead tetraacetate, which is hydrolyzed to lead dioxide or some sort of oxy-acetato-lead species, there is not appreciable solubility. There you are probably getting a surface reaction. This may be true for some other species as well.

ZUCKERMAN: Was there some standardization in particle size in the list that you gave?

THAYER: I tried to make these solids as finely divided and as homogeneous as I could, but other than that I'm afraid there wasn't any.

J. M. WOOD (University of Minnesota): When you are studying these reactions you also must look at problems related to competing reactions for the displacement of benzimidazole. You can have equilibria with the base-on, with base-off or protonated base-off. Just recently, we have been able to show that by making a number of corrin-ring derivatives and comparing complexation with a corrin macro-cycle, one can now form outer-sphere complexes with propionamide side chains. Consequently, the kinetic interpretation of your reactions can be extremely difficult because you can be looking at four or five alternate equilibria in these systems.

THAYER: I agree. Not much can be said yet about mechanisms for these reactions. But for most organometals I don't think complexation either with the benzimidazole or the ring is important. With many inorganic species, it is.

ZUCKERMAN: If the metal species finds itself in a complex with the benzimidazole, then it is also on the wrong side of the ring to effect methylation. Is there some easy way to get around that?

WOOD: It has a profound effect on the kinetics for the reaction. If you are looking at electrophilic attack on the cobalt-carbon bond and you displace the benzimidazole, taking mercury as a simple example, you get about two orders of magnitude difference in the reaction rate because it's difficult to displace the methyl group as a carbanion. But if you consider the reaction with a single-electron oxidant in tin(II), you look at a free radical displacement of the methyl group. This seems to be insensitive to coordination by benzimidazole and is very insensitive to pH in these systems. Depending on whether the mechanism is electrophilic attack or single-electron oxidative addition, an enormous change in the reaction rates is seen for this reaction. However, the chemistry is not trivial because the benzimidazole can come off, it can be protonated, or it can react with a metal ion. The formation of a complex with a corrin macrocycle which changes the electron density on the cobalt-carbon bond can greatly change the reactivity of the methyl group to either electrophilic attack or possibly free radical attack.

ZUCKERMAN: That would change the molecularity of the rate-determining step, if one metal were to complex with the benzimidazole and a second metal species were to attack an apical site. So that would be revealed in the data, provided one could interpret those data in terms of molecularity.

W. R. CULLEN (University of British Columbia): We are brainwashed by the idea that methyl B_{12} is involved in all of these reactions. We firmly believe that there is no B_{12} involved in arsenic methylation. Professor Wood pointed out that the nature of the methylating species had to be established in nature. Much of our discussion now involves base-on, base-off details, yet the compound may not in fact be the methylating agent. I suggest that the most important thing we can do is to investigate these reactions in biological systems, to find out what is going on and what is doing the methylating. There is one easy way of getting oxidation: that is to use CH_3^+ as an attacking agent. Use of CH_3^- with oxidation requires additional manipulation of electrons. CH_3^+ is a good methylating agent from nature and is readily available. I suggest we should think more about CH_3^+; it can come from methyl B_{12} as well.

WOOD: I agree. The value of this model chemistry is to point up possible mechanisms for transmethylation reactions in biological systems. But the important thing is to establish precisely which coenzyme intermediate is involved in the transmethylation reaction. That's why experiments with isotopes are necessary in complicated systems like lake sediments and mixed microbial communities. It's important to determine what the methyl donor is. Then you can start thinking in terms of mechanism. For the metalloids, you are primarily looking at nucleophilic attack on the carbon-sulfur bond of something like S-adenosylmethionine, or else something like methylated co-enzyme M in anaerobic systems. You can get methyl transfer to arsenic in a high oxidation state if you activate the corrin ring by a reaction with a platinum complex. That may not have very much significance to biological systems, but from nmr data you see that platinum complex activates the system. It is the same sort of chemical shift that you see for B_{12} when it binds to the enzymes, where you have to supply 70 kilocalories to break that bond for substrate rearrangement reactions in the enzymes. So, we don't really know what is happening in the B_{12} proteins yet.

R. H. FISH (University of California, Berkeley): I don't think you ever identified any compounds in these rate studies. Is that correct?

THAYER: We have not identified any products.

FISH: The results seems to be tenuous unless you identify the compounds. How trimethyltin forms in the environment is still controversial. Professor Wood shows that tin(II) goes to methyltin(IV), but what happens after that? We have looked at methyl- and dimethyltin species reacting with methyl B_{12} and couldn't find any reaction occurring. Again, you must identify the products before you can talk about rates of reaction pertaining to methylation.

THAYER: Not necessarily. In the case of the tin compounds, it seems there is a substantial difference in rates of reaction depending on what inorganic groups are present. With methyltin trichloride, I couldn't see any reaction. With methyltin triacetate virtually no reaction occurred. However, dimethyltin dichloride reacts considerably faster than dimethyltin diacetate. It is well known that tin-oxy species will form intermolecular bridging bonds resulting in polymeric species particularly where you have 2 oxy groups per tin, or even more where you have three. This may influence the reaction of organotin compounds with whatever methylation agent there may happen to be. There was one case where we could identify the product, the trimethyltelluronium ion. I found tellurium metal and dimethyltelluride. For tetramethylarsonium ion and methylcobalamim a very unpleasant odor which quite probably was trimethylarsine was given off.

F. HUBER (University of Dortmund): We have preliminary results on the reaction of tin(IV), not with methylcobalamin, but with methylcobaloxime. From nmr measurements we found that we probably get the methyltin compounds when we react tin(IV) with methylcobaloxime. We also get signals for trimethyltin and monomethyltin. We have a problem because the positions of the nmr signals are not very constant. We have to look more at the coordination situation in these solutions. We have been looking at oxygen systems with these reactions, but we have to consider the natural systems which certainly involve sulfur ligands. When we make compounds of organolead and organotin compounds with the carboxylic acids containing SH groups, we find a preference for lead bonding to sulfur, and not to oxygen. We can make carboxylates which are bonded through sulfur to lead, but it is different with tin compounds. This is a very interesting field and we should try to investigate this field of sulfur coordination to organometal compounds.

F. E. BRINCKMAN (National Bureau of Standards): This concerns the pressing question of trace metal flux; I'm struck that there are so many corrinoids available in environmental detritus, either of anthropogenic or natural biogenic origin, which could act as solubilizing agents. With Ksp values for many of the oxides that you indicated, such solubilization might be important to metal fluxes. I'm particularly concerned with the tin case; I'm concerned about its bioavailability. Admittedly, methylcobalamin might be the wrong model, but it is an attempt to look at it as a solubilizing agent, not necessarily as a methylating agent, for uptake into food webs. Have you looked at some other tin mineral solubilities in your solutions of methylcobalamin?

THAYER: No, we haven't. I've confined my attention to tin(IV) since oxy-metal species seem to undergo this reaction more readily than species with other groups present. There is a report of tin tetraacetate, but under these conditions I'm sure it's immediately hydrolyzed.

BRINCKMAN: Have you looked at any tin sulfides or any other metal sulfides?

THAYER: There is only species with a metal-sulfur bond that I have looked at. This was chloromercury-thiomethyl and -thio-t-butyl. We got a very rapid reaction with methylcobalamin, but it was the other group, the chloride, that was being removed. In acidic media, with something like bis(methylthio)mercury in the presence of methylcobalamin, a reaction probably would be rather slow, but I think you could observe it. A little bit of bis-thiol derivative of mercury dissolves in dilute acid (~ pH=4) and will react.

CULLEN: I want to emphasize that it's very important that we look at sulfur involvement in these compounds, especially in the case of the metalloids, and indeed, in all metals with a high sulfur affinity. Also, nature actually doesn't do all these rather complicated reactions in one step. I think we ought to look for mechanisms that go slower, using a large number of steps to do so. In other words, we don't require oxidative reductions. In case of arsenic there is an oxidation and then of course a reduction. These are two different things.

WOOD: In 1953, after the Minimata (Japan) disaster, when people started looking for the cause, the molecule that was isolated from shellfish in Minimata Bay was methylmercurythiomethyl. I think it's interesting that government agencies are still not analyzing for it in shellfish, as far as I'm aware. It took almost four years to persuade the EPA to stop analyzing for total mercury and start analyzing for methylmercury.

RECEIVED August 22, 1978.

13

Mechanisms for Alkyl Transfers in Organometals

JAY K. KOCHI

Department of Chemistry, Indiana University, Bloomington, IN 47401

The class of vitamin B_{12} (methylcorrinoid) derivatives is one of three major coenzymes which is available for methyl transfer in biological systems, being particularly effective with inorganic substrates. Thus methylcobalamin has been implicated in methyl transfers to a variety of metal ions including mercury, lead, tin, and thallium, as well as platinum, palladium, and gold (1).

The mechanism of methyl transfer from cobalt to another metal center, that is, transmetallation, is the subject of extensive study. Two general mechanisms have been proposed for the methylation of metal ions by methylcobalamin (2). In type **I** reactions, the metal ion acts as an electrophile during the transfer of a methyl anion equivalent. In type **II** reactions, methyl radicals are transferred between metal centers. Thus type **I** reactions are considered to involve heterolytic cleavage of the cobalt-carbon bond of methylcobalamin, whereas the homolytic cleavage of the same bond in type **II** reactions is induced by a reduced member of a redox couple. In a more general sense, type **I** and type **II** reactions can be considered as two-equivalent and one-equivalent processes, respectively.

The concept of electron transfer relates two-equivalent processes with their one-equivalent counterparts. For example, consider the carbonium ion as the key reactive intermediate in solvolysis reactions, which historically have served as prototypes for numerous ionic processes. Electron transfer by one-equivalent reduction produces alkyl radicals,

$$CH_3^+ \underset{}{\overset{+\epsilon}{\rightleftharpoons}} CH_3\cdot$$

which are crucial to homolytic processes. The same interchange between ionic and radical species applies to electron transfer processes between carbanions and alkyl radicals.

$$CH_3^- \underset{}{\overset{-\epsilon}{\rightleftharpoons}} CH_3\cdot$$

0-8412-0461-6/78/47-082-205$07.50/0

Finally, an overall two-equivalent change interrelates carbonium ions and carbanions.

$$CH_3^+ \underset{}{\overset{2\varepsilon}{\rightleftharpoons}} CH_3^-$$

Viewed in this way, this hypothetical transformation is better considered as a two-step process involving successive electron transfers. Thus, in electrochemical processes, only one electron is transferred in a single act since the simultaneous transfer of two electrons, like a biphotonic process, is a much less probable event.

Although the observation of electron transfer processes between alkyl radicals and carbon-centered ions are as yet relatively rare, those between metal complexes of course represent a well-established part of inorganic chemistry. Outer-sphere and inner-sphere mechanisms developed by Taube serve as an excellent basis for electron transfer processes between two metal complexes (3). The presence of a bimetallic intermediate either as a precursor or successor (postcursor) complex can play an important role in inner-sphere electron transfer processes. Outer- and inner-sphere mechanisms have also been applied to the oxidation-reduction reactions of alkyl radicals with metal complexes (4). However, the detailed examination of the mechanism of oxidation-reduction processes with alkyl radicals is made difficult by their transient nature. As a result, the mechanisms have been derived heretofore from deductions based on kinetic observations and product analyses, and a few intermediates have only recently been detected. For example, the inner-sphere complex between methyl radicals and copper(II),

$$CH_3\cdot + Cu^{II} \longrightarrow CH_3\text{-}Cu^{III}$$

has been observed by flash photolytic techniques and found to decay with first-order kinetics ($k = 7 \times 10^2\ sec^{-1}$) in aqueous solutions at 25° (5). Indeed, the association of alkyl radicals with metal complexes through inner-sphere complexes may be the route by which most, if not all, oxidation-reduction reactions of alkyl radicals with metal complexes occur, irrespective of whether they have been previously classified as inner- or outer-sphere processes. In this regard, the root of the difference between wholly inorganic systems and the hybrid alkylmetal systems probably lies in the great propensity of the carbon-centered ions, both carbonium ions and carbanions, to be highly solvated, either as ion-pairs or inner-sphere complexes.

An inner-sphere alkylmetal intermediate such as that in the above equation may be derived by an alternative route involving either oxidation or reduction of a stable alkylmetal complex, e.g.,

$$CH_3\text{-}Cu^{II} \xrightarrow{-\varepsilon} CH_3\text{-}Cu^{III}$$

In this instance, the precursor itself is unstable. However, there are a variety of other stable alkylmetal complexes extant from which electron transfer is possible. Reversible dissociation of such intermediates then relates alkyl radicals to organometals derived by conventional two-equivalent processes, i.e.,

$$R^- + M^+ \rightleftharpoons R\text{-}M \underset{}{\overset{-\epsilon}{\rightleftharpoons}} R\text{-}M^{+\cdot} \rightleftharpoons R\cdot + M^+$$

This dichotomy is inherent in all of these processes, and there is a severe problem of rigorously demonstrating how each may participate in a particular organometal reaction.

Several major questions arise in the treatment of alkylmetals as intermediates in transmetallations: (a) the mechanistic distinction between electrophilic and electron transfer mechanisms in the cleavage of an alkyl-metal bond, (b) the separation of concerted, two-equivalent processes from successive, one-equivalent processes, and (c) the labilization of alkylmetals especially with regard to electron transfer processes. Binary alkylmetals of mercury, lead and tin are useful models for the study of these questions since they are substitution-stable compounds and generally well behaved in solution for kinetic studies. In this report, we shall describe how various alkyl derivatives can be examined systematically to allow direct photoelectron spectroscopic study of the bonding orbitals in these organometals. Two series of well delineated cleavages of alkylmetals will then be discussed. These include:

(a) the electrophilic protonolysis of dialkylmercury compounds,

$$R'HgR + HOAc \xrightarrow{k_E} RH + R'HgOAc$$

and (b) the electron transfer oxidation of the same organomercurials with hexachloroiridate(IV).

$$R'HgR + IrCl_6^{2-} \xrightarrow{k_{ET}} R'HgR^{+\cdot} + IrCl_6^{3-}$$

$$R'HgR^{+\cdot} \xrightarrow{fast} R\cdot + R'Hg^+ \text{ , etc.}$$

Using these systems, we will compare and contrast electrophilic and electron transfer mechanisms in the cleavages of alkyl-metal bonds in both main group and transition metal complexes.

I. Organometals as Electron Donors—Ionization Potentials

A. Dialkylmercury Compounds. The He(I) photoelectron spectra of dialkylmercury compounds show two principal bands of interest (6). The first ionization potential, I_D, occurring in a range between 7.57 eV (di-t-butylmercury) and 9.33 eV (dimethylmercury) is included in a fairly broad, unsymmetrical band. A second, weaker band occurring between 14.4 and 15.0 eV is due to ionization from the mercury $5d^{10}$ shell. The ionization energies for these two bands are tabulated in Table I. Repre-

Table I. The First and $5d^{10}$ Vertical Ionization Potentials (eV) of Dialkylmercury Compounds.

R—Hg—R'		First I_D	$5d^{10}$ I_D
CH_3	CH_3	9.33	14.93
CH_3	C_2H_5	8.84	14.85
C_2H_5	C_2H_5	8.45	14.71
CH_3	i-C_3H_7	8.47	14.86
CH_3	i-C_4H_9	8.75	14.74
CH_3	t-C_4H_9	8.32	
C_2H_5	i-C_3H_7	8.18	14.61
n-C_3H_7	n-C_3H_7	8.29	14.63
i-C_3H_7	i-C_3H_7	8.03	14.46
C_2H_5	t-C_4H_9	8.06	
i-C_3H_7	t-C_4H_9	7.73	
n-C_4H_9	n-C_4H_9	8.35	
i-C_4H_9	i-C_4H_9	8.30	14.47
t-C_4H_9	t-C_4H_9	7.57	
i-C_4H_9	neo-C_5H_{11}	8.33	14.49
neo-C_5H_{11}	neo-C_5H_{11}	8.30	14.41

sentative spectra of only the first band are reproduced in Figure 1 for one series of alkylmethylmercury compounds, i.e., R–$HgCH_3$.

The effect of alkyl substitution on the first ionization potential of a series of alkyl derivatives is attributed primarily to polarization effects in the molecular ion final state. It has been recognized that such electronic effects are additive along the series: Me, Et, i-Pr, and t-Bu. Thus, the energy effect of replacing Me by Et is expected to equal that of replacing Et by i-Pr, or of replacing i-Pr by t-Bu. In each case, α-hydrogens in CH_3 are being sequentially replaced by methyl groups. Additive energy effects have been used by Taft (7) as a criterion for identifying polar effects as denoted by the empirical substituent constant σ^*:

$$(\sigma^*) \qquad \text{Me : Et : i-Pr : t-Bu} \;=\; 0 : 0.10 : 0.19 : 0.30$$

It has been shown that the Taft relationship holds for the ionization potentials of alcohols and other alkyl compounds. Figure 2 illustrates the linear relationship between σ^* values and the ionization potentials of a series of alcohols, alkyl bromides, alkylhydrazines, and aldehydes.

The correlation of the ionization potentials of a series of alkylmercurials R-$HgCH_3$ also plotted in Figure 2 is distinctly nonlinear. The incremental changes in energies become progressively smaller or "saturated" as one proceeds from methyl to *tert*-butyl. Moreover, the same pattern of saturation obtains for the analogous series of Grignard reagents R-MgX and tri-

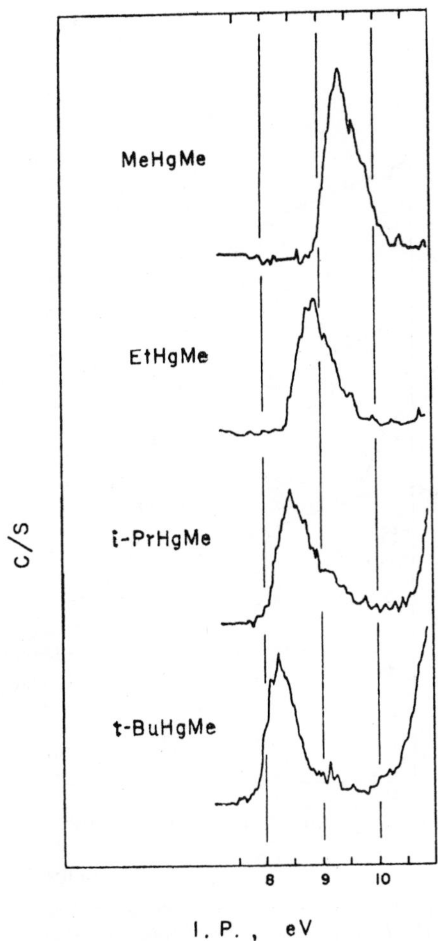

Figure 1. He(I) photoelectron spectra of the first bands in a series of organomercurials, MeHgR, where R = Me, Et, 1-Pr, and tert-*Bu. Units in the ordinate are arbitrary* (6).

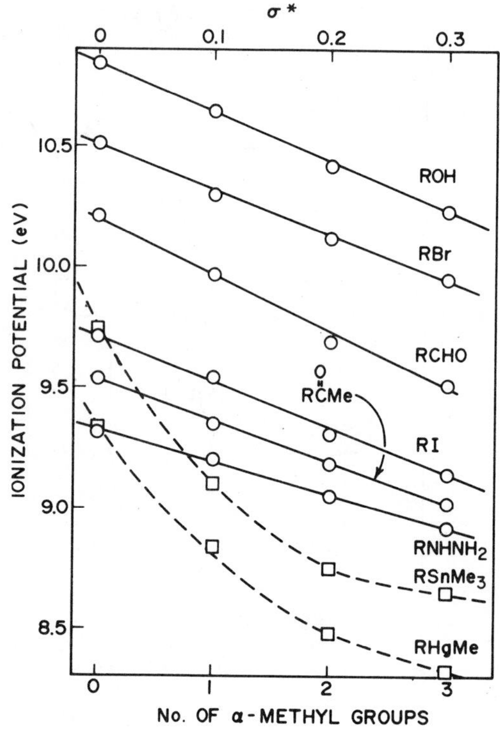

Inorganic Chemistry

Figure 2. Correlation of the first ionization potential of various alkyl-substituted compounds vs. Taft's σ^* constant (top scale) and number of α-methyl groups (bottom scale) (6).

methyltin compounds $RSnMe_3$ measured independently (8,9).

The difference between energy effects which are saturated and those that are additive may be related to the nature of the highest occupied molecular orbital (HOMO). For those systems containing nonbonding electrons, the ionization proceeds from a HOMO which is largely orthogonal to the σ bonding orbitals, particularly those associated with the bonding of the heteroatom to carbon. Thus, although interactions between the alkyl group and the heteroatom can be observed, ionization from the HOMO is only weakly coupled to the bonding system and additivity is observed. In contrast, the ionization process in organometals such as Me_2Hg proceeds from a molecular orbital that has substantial metal-carbon bonding character. This conclusion is portrayed with a simple LCBO diagram as:

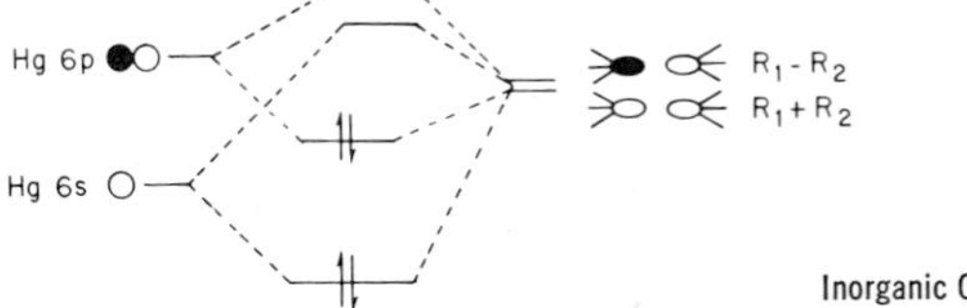

Inorganic Chemistry

where the HOMO is formed from the bonding combination of the antisymmetric combination of the σ type group orbitals of the alkyl fragments with the mercury 6p atomic orbital. In fact, ionization of the HOMO of alkyl radicals (i.e., $R\cdot \rightarrow R^+ + e$) serves as a reasonable model for the ionization of the HOMO of organomercurials. Significantly, both the measured ionization potentials of alkyl radicals (10) and those obtained from SCF-MO calculations listed in Table II (11) show the characteristic saturation effect described above, and they correlate well with the first vertical ionization potentials of the organomercurials listed in Table I. Thus, in this case, ionization from the HOMO is strongly coupled to the bonding system. In such a situation, changes in electron repulsion may well be large and not effectively constant as the alkyl group is systematically varied from Me to t-Bu.

Table II. Experimental and Calculated Ionization Potentials of Alkyl Radicals.

Alkyl Radical	Experimental (eV)	Calculated (eV)
Methyl	9.84	9.95
Ethyl	8.38	8.56
iso-Propyl	7.55	7.60
tert-Butyl	6.93	6.83

The variation in the 5d ionization potential through the series of dialkylmercury compounds arises primarily from changes in the ground state charge on the mercury atom. The

$I_D(5d)$ data can be used to calculate the charge on the mercury atom as presented in Table III, together with the calculated

Table III. Calculated Charges (q) on the Mercury Atom and Approximate Electronegativities of the Alkyl Groups.

Dialkylmercury	q	Electronegativity
$(CH_3)_2Hg$	+0.0234	1.98 (CH_3)
$(C_2H_5)_2Hg$	-0.0338	1.81 (C_2H_5)
$(n\text{-}C_3H_7)_2Hg$	-0.0547	1.75 ($n\text{-}C_3H_7$)
$(i\text{-}C_3H_7)_2Hg$	-0.0990	1.62 ($i\text{-}C_3H_7$)
$(i\text{-}C_4H_9)_2Hg$	-0.0703	1.70 ($i\text{-}C_4H_9$)
$(neo\text{-}C_5H_{11})_2Hg$	-0.112	1.58 ($neo\text{-}C_5H_{11}$)

values of the electronegativities of the alkyl groups, since the $5d^{10}$ bands could not be measured with a high degree of accuracy. Therefore, calculated charges and electronegativities of the alkyl groups must be regarded as having more qualitative rather than quantitative significance at this point. Nevertheless, the approximately equal changes in electronegativity on proceeding from methyl to ethyl (1.98 to 1.81) and from ethyl to isopropyl (1.81 to 1.62) are at least consistent with the additivity cited above.

Among binary mercury(II) derivatives, dialkylmercury compounds possess the lowest ionization potentials as shown in Table IV.

Table IV. First Vertical Ionization Potentials (eV) of Binary Mercury(II) Derivatives.

ClHgCl (11.37)	MeHgCl (10.88)	MeHgMe (9.33)
BrHgBr (10.62)	MeHgBr (10.16)	
IHgI (9.50)	MeHgI (9.25)	

B. Tetraalkyllead Compounds. The vapor phase photoelectron spectra of the Group IVb tetramethyls, i.e., neopentane, tetramethylsilane, tetramethylgermane, tetramethylstannane and tetramethylplumbane, have been scrutinized. With the exception of neopentane, the spectra are all similar, each showing two broad bands. The higher energy band occurs at about 14 eV, and comparison with simple compounds suggests that this band is associated with ionization from the σ(C-H) bonding molecular orbital mainly localized on the methyl fragment. The threshold or adiabatic ionization energy of the lower energy band decreases in the order: $(CH_3)_4C > (CH_3)_4Si > (CH_3)_4Ge > (CH_3)_4Sn > (CH_3)_4Pb$ from 10.25, 9.42, 9.38, 8.85, to 8.38 eV, respectively, indicating that ionization is associated with electrons localized relatively close to the metal atom. The lower band has been assigned to ionization from the $3t_2$ orbital derived principally from the σ(M-C) bonding orbitals. In tetramethylplumbane it is split into two well-resolved bands. Examination of the pes for the series of methyl/ethyllead compounds in Table V reveals that

Table V. Ionization and Oxidation Potentials of Organolead Compounds.

$Et_{4-n}PbMe_n$	Oxidation Potential (V)	Ionization Potential (eV)
Et_4Pb	1.67	8.13
Et_3PbMe	1.75	8.26
Et_2PbMe_2	1.84	8.45
$EtPbMe_3$	2.01	8.65
Me_4Pb	2.13	8.90

the ionization potentials decrease monotonically with increasing substitution of ethyl for methyl groups around the lead nucleus. The regular trend noted over the entire series of methyl/ethyllead compounds suggests that substitution of an ethyl group for a methyl group is largely an electronic effect, and that steric interactions between alkyl groups around the lead atom are not large. Moreover, it is interesting to note that the cumulative effects of α-methyl groups are similar in a comparison of the series of dialkylmercury compounds listed in Table I with those tetraalkyllead compounds listed in Table V.

The anodic oxidation of the same series of methyl/ethyllead compounds has also been examined in acetonitrile solutions with lithium fluoroborate as a supporting electrolyte (12). The number of electrons involved in the anodic process was determined to be 1.0 by thin layer chronopotentiometry using a platinum electrode for all the tetraalkyllead compounds examined. The anodic oxidation of each of the tetraalkyllead compounds was found to be irreversible by current-reversal chronopotentiometry, suggesting that the alkyllead cation-radical is unstable. Eq 1 represents its mode of decomposition.

$$R_4Pb^{+\cdot} \longrightarrow R_3Pb^{+} + R\cdot \qquad [1]$$

The potential in a magnetically-stirred solution of tetraalkyllead depended on the identity of the lead compound. Since each of these oxidative processes is irreversible, the observed potentials shifted to more positive values as the current density increased, and no theoretical significance can be placed on the absolute values of the measured potentials. At a given current density, however, the potentials reflect the relative ease of removal of a single electron from the series of methyl/ethyllead compounds examined in this study. The latter rests on the presumption that anodic oxidation of these closely related compounds proceeds via a common mechanism. Indeed, electron detachment from tetraalkyllead measured in the gas phase by photoelectron spectroscopy also shows a striking relationship with the electrochemical oxidation potentials in acetonitrile solution.

II. Electrophilic Cleavage of Organometals—Quantitative Effects of Alkyl Groups

Acetolysis of dialkylmercury compounds liberates one equivalent of alkane and of alkylmercury acetate according to eqs 2 and 3 (13).

$$R\text{-}Hg\text{-}R' + HOAc \xrightarrow{k} RH + R'HgOAc \qquad [2]$$

$$R\text{-}Hg\text{-}R' + HOAc \xrightarrow{k'} R'H + RHgOAc \qquad [3]$$

The further cleavage of alkylmercury acetate is too slow to interfere with the acetolysis study of dialkylmercury. The pseudo first-order rate constants for acetolysis, k and k', in eqs 2 and 3 were determined from the rates of liberation of alkanes RH and R'H, respectively.

The protonolysis of dialkylmercury in acetic acid solutions proceeds by a rate-limiting proton transfer. The experimental values of the kinetic isotope effect of 9-11 are close to the theoretical maximum expected for the transfer of deuterium relative to proton in this system. The large values of k_H/k_D also suggest a rather linear transition state for proton transfer in which the contribution from the symmetric stretching mode is small. Such a transfer of a proton halfway in the transition state places a considerable positive charge on mercury. These results together with the retention of configuration during proto-demercuration are consistent with a 3-center transition state of the type depicted below.

$$\left[\begin{array}{l} R \\ \vdots \quad H\text{---}OAc \\ Hg \\ | \\ R' \end{array}\right]^{\ddagger}$$

The more or less triangular array of carbon, mercury and the proton in the transition state for protonolysis was originally proposed by Kreevoy and Hansen (14). The extent to which there is nucleophilic assistance during acetolysis of dialkylmercury is not treated explicitly. Instead, for the acetolysis in eq 2, the effects of alkyl groups on the cleavage reaction can be classified into two categories—namely, leaving group (HgR') effects and cleaved group (R) effects. Steric effects due to the leaving groups are unimportant in acetolysis, and it helps to limit the mechanistic considerations to the immediate locus of the reaction site.

Leaving Group Effects (HgR'). The effect of leaving groups HgR' on the cleavage of a particular alkyl-mercury bond accelerates in the order: R' = Me < Et < i-Pr < t-Bu. This reactivity sequence represents the increasing ability of these alkyl groups to accommodate a positive charge when they are attached to the departing cationic mercury (HgR'). Electron release by various alkyl groups in response to a fixed electron demand in the ground state of CH_3HgR' is also reflected in the magnitudes of the methyl proton coupling constants, $J(^{199}Hg\text{-}H)$. More appropriately, electron release by alkyl groups in the transition state for acetolysis may be modeled by the cation-radical of dialkylmercury. The latter is probed independently by measuring the energetics of electron detachment from a homologous series of RHgR', e.g.,

$$CH_3Hg\text{-}R' \longrightarrow [CH_3Hg\text{-}R']^{+} + \epsilon$$

as described in the foregoing section. Indeed, there is a linear correlation of log k for acetolysis and the vertical ionization potentials of a series of $CH_3Hg\text{-}R'$. Since photoelectron ionization is a vertical process, it is electronic in origin and must be largely free of steric factors. The correlation thus supports the conclusion that steric effects of leaving groups are unimportant in acetolysis of this series of organomercurials. The correlation between rates of cleavage and ionization potential is not restricted to organomercurials. The same relationship is also obtained in 4-coordinate organolead compounds (15,16).

$$Me_nEt_{4-n}Pb + HOAc \xrightarrow{k(Me)} CH_4 + Me_{n-1}Et_{4-n}PbOAc \quad [4]$$

$$Me_nEt_{4-n}Pb + HOAc \xrightarrow{k(Et)} CH_3CH_3 + Me_nEt_{3-n}PbOAc \quad [5]$$

Thus, the rate of acetolysis of the Me-Pb bond [i.e., log k(Me)] decreases linearly with the ionization potential of $Me_nEt_{4-n}Pb$, where n = 0, 1, 2, 3, 4, and a parallel relationship is obtained during the concomitant cleavage of the Et-Pb bond [log k(Et)]. Increasing steric factors in the leaving group (i.e., trialkyllead) prevent extension to higher homologs.

In the acetolysis of dialkylmercury, the leaving group (HgR') effects [under conditions of a constant cleaved (R) group] can be expressed quantitatively by the linear free energy relationship in eq 6,

$$\log \kappa/\kappa_0 = \mathbf{L} \quad [6]$$

where κ_0 is the rate constant for acetolysis of R-HgR' in eq 2 when R' = Me, and κ is that for R' = Et, i-Pr or t-Bu. **L** is a leaving group constant which has a characteristic value for each HgR', and it does not depend on the nature of the cleaved group (R). Normalizations of **L** to **L**(Et) = 0.10 are listed in Table VI as **L'** to allow direct comparison with the Taft σ^* constants.

There is a striking difference between the values of **L'** and σ^*, although both are due to electronic or polar effects. Thus, there is a "saturation" in incremental changes in energy for **L** as each hydrogen in $RHgCH_3$ is sequentially replaced by methyl groups in the series: $RHgCH_3$, $RHgCH_2CH_3$, $RHgCH(CH_3)_2$, and $RHgC(CH_3)_3$. On the other hand, the corresponding changes in σ^*

Table VI. Leaving Group Parameters in Acetolysis of Dialkylmercury. Comparison with Taft σ^*.

Leaving Group (HgR')	L	L'	σ^*
$HgCH_3$	0	0	0
$HgCH_2CH_3$	0.76	0.10	0.10
$HgCH(CH_3)_2$	1.28	0.17	0.19
$HgC(CH_3)_3$	1.44	0.19	0.30

are "additive," increasing linearly from CH_3, CH_2CH_3, $(CH_3)_2CH$ to $(CH_3)_3C$. Indeed, Taft has employed the additivity requirement for identifying polar effects.

The difference between energy effects which are saturated and those that are additive provides the key to the understanding of substituent effects in electrophilic cleavages. There exists a strong linear correlation between σ^* values and the ionization potentials of a series of alcohols, alkylhydrazines, aldehydes and alkyl halides represented by the process: $RX \rightarrow RX^{\cdot+} + \epsilon$. On the other hand, the ionization potentials of a series of organomercurials CH_3HgR' also plotted against σ^* show a saturation effect equivalent to that in acetolysis. The saturation also obtains for ionization from the same series of Grignard reagents and alkyltrimethyltin compounds, $(CH_3)_3SnR'$. The difference between saturation and additivity effects can be explained by considering the highest occupied molecular orbital (HOMO) in each series. For those compounds containing nonbonding electrons, the ionization proceeds from a HOMO which is largely orthogonal to the orbital involved in the bonding of X to carbon in R-X, and its effect on the electron density in the bond is minimal. In contrast, the ionization process in organometals such as Me_2Hg proceeds from a bonding molecular orbital with a node at mercury. Consequently, the electron density in the bond to carbon is diminished substantially, and the cationic character of the α-carbon is accompanied by a decrease in electron repulsion, which is not effectively constant as the alkyl group is systematically varied from Me to t-Bu. Indeed, the validity of this description is shown by values of the ionization potentials of alkyl radicals [i.e., $R\cdot \rightarrow R^+ + \epsilon$], which agree remarkably well with those obtained from SCF-MO calculations (17,18). Significantly, the ionization potentials of alkyl radicals show the characteristic saturation effect described above, and they also correlate well with log k for acetolysis and the ionization potentials of the organomercurials.

It is noteworthy that leaving group (HgR') effects due to substitution of methyl groups in the β-position of the alkyl chain R' are highly attenuated relative to that accompanying α-substitution. For instance, the pseudo first-order rate constant for methane evolution from MeHgEt is 2.35×10^{-6} sec^{-1} and that for MeHg–i-Bu is 2.39×10^{-6} sec^{-1}. Thus, leaving group effects are not simply related to the size of the alkyl group (R').

Cleaved Group Effects (R). The rates of acetolysis of alkyl

groups from dialkylmercury decrease in the order: R = Et > i-Pr > Me > t-Bu. The cleaved alkyl group effects [under conditions in which the leaving group (HgR') is constant] can be expressed quantitatively by eq 7,

$$\log \kappa'/\kappa'_0 = \mathbf{C} \qquad [7]$$

where κ'_0 is the rate constant for acetolysis of R-HgR' in eq 2 when R = Me, and κ' is that for R = Et or i-Pr. **C** is a cleaved alkyl group constant which has a characteristic value for each R, and it does not depend on the leaving group HgR'.

The non-systematic trend in the values of **C** in Table VII

Table VII. Cleaved Group Parameters in Acetolysis of Dialkylmercury.

Cleaved Alkyl Group (R)	**C**
CH_3	0
CH_3CH_2	0.55
$(CH_3)_2CH$	0.29
$(CH_3)_3C$	~-0.9

suggests that there are at least two opposing effects present in the acetolysis of an alkyl-mercury bond. The decrease observed in proceeding from Et, i-Pr to t-Bu follows from the increase in steric bulk at the site of protonation. On the other hand, the increase from Me to Et (and to i-Pr) is in accord with electron release from these R groups accompanying protonolysis, as described earlier. (The value of **C** for the t-butyl group is approximate.)

The rates of protonolysis of alkylmercury iodides (14) in aqueous perchloric and sulfuric acid follow the order expected from a dominance of steric factors, viz., Me : Et : i-Pr : t-Bu in the relative order: 123 : 49 : 16 : 1.0. The small kinetic isotope effect (k_H/k_D) measured from the protonolysis of methylmercury iodide may reflect either a transition state in which the bond to carbon is poorly formed or one in which it is almost complete. The latter could account for the reactivity pattern, but other uncertainties in this system discourage further discussion.

A similar effect, however, can be observed during the acetolysis of a series of well-behaved methyl-ethyllead compounds [$Me_nEt_{4-n}Pb$, when n = 0, 1, 2, 3] (15,16). If leaving group effects are taken into account, the cleavage of Me-Pb is consistently 8.6 times more facile than Et-Pb cleavage in all three intramolecular competitions as well as in the intermolecular competition using $Me_nEt_{4-n}Pb$ (n = 1, 2, 3) and Me_4Pb/Et_4Pb, respectively. It is noteworthy that the Me/Et reactivity in tetraalkyllead is reversed from that in dialkylmercury [k(Me)/k(Et) = 0.30], although the kinetic isotope effect of 9 in the acetolysis of tetraethyllead (19) is comparable to that observed with diethylmercury. The difference is due to increased steric hindrance in the 4-coordinate organolead compounds compared to the more accessible 2-coordinate organomercury analogs. The severe steric restrictions imposed on tetraalkyllead compounds is also borne out by the failure to extend the linear free energy relationships to higher alkyl homologs in protonolysis studies.

Generalized Equation for Protonolysis of Dialkylmercury. The linear free energy relationships in eqs 6 and 7 for leaving group effects and cleaved group effects, respectively, during acetolysis of dialkylmercury suggest that a generalized relationship is possible which correlates all the rates using the empirical parameters in Tables VI and VII, i.e.,

$$\log k/k_0 = \mathbf{L} + \mathbf{C} \qquad [8]$$

where k_0 represents the rate constant for acetolysis of Me_2Hg and k is that for any other RHgR'. The validity of eq 8 is shown by comparing the experimental rate constants with those calculated from the equation.

The empirical constant **C** in the generalized eq 8 for acetolysis of dialkylmercury takes into account any steric interactions due to the cleaved group (R) at the reaction site. In the absence of such steric effects, the cleaved group effect in electrophilic substitution should be influenced primarily by electron release and thus parallel the leaving group (HgR') effects as described in eq 6 and Table VI.

Significantly, the relative reactivities of the cleaved alkyl groups in electron transfer cleavages follow a pattern in which the incremental changes in energy for R = Me, Et, i-Pr and t-Bu show a "saturation" effect, which is the same as that observed in the leaving group effect (HgR') denoted by **L** in Table VI. Indeed, there is a linear correlation between the oxidation or ionization potentials and **L**. Thus, the saturation pattern for alkyl groups is independent of whether they are involved as cleaved (R) groups or as leaving (HgR') groups. It clearly relates to an intrinsic property of the alkyl-mercury bonds and reflects the manner in which an alkyl group responds to the presence of a positive charge on mercury. The saturation pattern for alkyl groups can be used as a diagnostic probe for the mechanism of cleavage in organometals.

Acetolysis studies on dialkylmercury have demonstrated not only the importance of the cleaved group (R) but also the leaving group (HgR') in electrophilic substitution. Alkyl groups are excellent probes for measuring these electronic effects quantitatively, and the correlations of the rates with the ionization potentials show that a positive charge is developed on both the leaving group (R) and the cleaved group (HgR'). These facets of the reactivity of dialkylmercurials toward protonic electrophiles may be extended more generally, since it has long been recognized that nucleophilic reactivity is influenced by the "polarizability" of the nucleophile. Thus, the Edwards oxybase equation contains both a term related to the oxidation potential of the nucleophile as well as a term related to its basicity (20,21). The molecular orbital analog of the Edwards equation has been developed by Klopman, in which electrostatic and covalent terms are the counterparts to basicity and polarizability, respectively (22, 23). Organometallic nucleophiles are σ-donors and have negligible basicity in the Edwards sense. Thus, the nucleophilic reactivity of organometals using either the Edwards or Klopman model should reduce to an equation such as eq 8, in which electron release by alkyl groups is the important consideration. The latter, in essence, represents a "virtual" ionization of the carbon-metal bond by the electrophile since it can be directly related to the energetics of electron detachment. The 3-center transition state represented earlier is an adequate model at this juncture.

III. Electron Transfer Cleavage of Organometals with Hexachloroiridate(IV)

A. Tetraalkyllead Compounds. Tetraalkyllead compounds react rapidly with hexachloroiridate(IV) at 25°C in acetonitrile or acetic acid solution (12). For example, the addition of tetramethyllead to a solution of $IrCl_6^{2-}$ in acetic acid results in the immediate discharge of the dark red-brown color. Methyl chloride, trimethyllead acetate and the two reduced iridium(III) products are formed from tetramethyllead and hexachloroiridate(IV) in acetic acid solutions according to the stoichiometry:

$$Me_4Pb + 2\,IrCl_6^{2-} \xrightarrow{HOAc} Me_3PbOAc + CH_3Cl + IrCl_6^{3-} + IrCl_5(S)^{2-} \quad [9]$$

where S = solvent

Only one alkyl group is readily cleaved from each tetraalkyllead compound. The mixed methyl/ethyllead derivatives afford mixtures of methyl and ethyl chlorides, the yields of which depend on the organolead compound.

$$\begin{matrix}Me \\ \quad\searrow \\ \quad\quad PbR_2 + 2\,IrCl_6^{2-} \\ \quad\nearrow \\ Et\end{matrix} \longrightarrow \begin{cases} Me\text{-}Cl + EtPbR_2^{+}, \text{ etc.} \\ Et\text{-}Cl + MePbR_2^{+}, \text{ etc.}\end{cases} \quad [10]$$

After normalization for each type of alkyl group in the reactant, the relative yields of ethyl chloride and methyl chloride are rather constant at about 25 in acetonitrile, but vary somewhat in acetic acid.

Tetraalkyllead compounds react with hexachloroiridate(IV) at differing rates, which were followed spectrophotometrically by the disappearance of the absorption bands at 490 and 585 nm. The kinetics showed a first-order dependence on tetraalkyllead and hexachloroiridate(IV) in both acetonitrile and acetic acid solution. The second-order rate constants determined in aceto-

$$-d[IrCl_6^{2-}]/dt = 2k[R_4Pb][IrCl_6^{2-}] \quad [11]$$

nitrile solutions increase progressively from Me_4Pb, Me_3PbEt, Me_2PbEt_2, $MePbEt_3$, to Et_4Pb as listed in Table VIII.

Table VIII. The Correlation of Selectivities and Rates of Oxidative Cleavage of Tetraalkylleads by Hexachloroiridate(IV) with the Energetics of Electron Detachment Processes.

$PbMe_nEt_{4-n}$	k (M^{-1} sec^{-1})	EtCl/MeCl	I_D (eV)	E (V)	ν_{CT} (cm^{-1})
$PbEt_4$	26	–	8.13	1.67	–
$PbEt_3Me$	11	24	8.26	1.75	20,400
$PbEt_2Me_2$	3.3	25	8.45	1.80	22,600
$PbEtMe_3$	0.57	24	8.65	2.01	23,200
$PbMe_4$	0.02	–	8.90	2.13	24,300

Two important criteria can be used to distinguish the reaction of tetraalkyllead with hexachloroiridate(IV) from the more conventional electrophilic processes, e.g., those involving Brönsted acids, silver(I), copper(I) or copper(II) complexes, etc. (15,19,24,25). First, the rate of reaction of $Et_{4-n}PbMe_n$ with hexachloroiridate(IV) increases successively as methyl is replaced by ethyl groups [see n = 4 to 0 in Table VIII, column 2]. Second, a given ethyl group is cleaved approximately 25 times faster than a methyl group [column 3]. Both of these reactivity trends are diametrically opposed to an electrophilic cleavage which occurs directly at the less hindered methyl site faster than at an ethyl site under equivalent conditions.

These results suggest that the rate-limiting step with hexachloroiridate(IV) occurs prior to alkyl transfer. The mechanism given in Scheme I involves electron transfer in eq 12 as the rate-limiting process.

Scheme I:

$$R_4Pb + Ir^{IV}Cl_6^{2-} \xrightarrow{k} R_4Pb^{+\cdot} + Ir^{III}Cl_6^{3-} \quad [12]$$

$$R_4Pb^{+\cdot} \xrightarrow{fast} R\cdot + R_3Pb^+ \quad [13]$$

$$R\cdot + Ir^{IV}Cl_6^{2-} \xrightarrow{fast} RCl + Ir^{III}Cl_5^{2-}, \text{ etc.} \quad [14]$$

Indeed, there is a good linear correlation of the rates (log k) with the one-electron oxidation or ionization potentials of tetraalkyllead compounds presented in Table V. Selectivity in the cleavage of alkyl groups from organolead according to Scheme I occurs during fragmentation of the cation-radical in a fast subsequent step 13, which is consistent with the mass spectral study. Thus, a quantitative determination of the cracking patterns of the series of $PbMe_nEt_{4-n}$ showed that scission of the Et-Pb bond is favored over the Me-Pb bond in the parent molecular ions, largely due to bond energy differences.

$$\begin{matrix} Me \\ Et \end{matrix} > PbR_2^{+\cdot} \longrightarrow \begin{cases} Me\cdot + EtPbR_2^{+} & [15] \\ Et\cdot + MePbR_2^{+} & [16] \end{cases}$$

Examination of the electron spin resonance spectrum during the reaction with hexachloroiridate(IV) did not reveal the presence of the cation-radical $PbEt_4^{+\cdot}$, which must be highly unstable even at temperatures as low as -20°C. Nonetheless, the formation of ethyl radicals in high yields was evident from spin-trapping experiments with nitrosoisobutane and phenyl-t-butylnitrone, in which the well-resolved spectrum of the ethyl adduct could be obtained.

The use of hexachloroiridate(IV) as an efficient scavenger for alkyl radicals is implied in Scheme I by the isolation of alkyl chlorides in high yields. In support, separate experiments do indeed show that ethyl radicals generated unambiguously from the thermolysis of propionyl peroxide are quantitatively converted by hexachloroiridate(IV) to ethyl chloride in eq 14. There is an alternative possibility that alkyl halide is formed directly from the cation-radical without the intermediacy of an alkyl radical.

$$R_4Pb^{+\cdot} + Ir^{IV}Cl_6^{2-} \longrightarrow R_3Pb^{+} + R\text{-}Cl + Ir^{III}Cl_5^{2-}\text{, etc.}$$

The difference between this formulation and that presented in eq 13 of Scheme I rests on the degree of metastability of the cation-radical toward fragmentation. The failure to observe the esr spectra of $R_4Pb^{+\cdot}$ and the irreversibility of the oxidation wave in chronopotentiometry suggests that its lifetime is short.

B. Dialkylmercury Compounds. Hexachloroiridate(IV) also readily cleaves dialkylmercury compounds by second-order kinetics similar to eq 15 for tetraalkyllead (26). Moreover, the products, both organic and iridium(III), as well as the stoichiometry of the reaction are also equivalent to that given in eq 9, viz.,

$$Me_2Hg + 2\,IrCl_6^{2-} \longrightarrow MeCl + MeHg^{+} + IrCl_6^{3-} + IrCl_5(S)^{2-}$$

Nmr studies indicate that $MeHg^{+}$ is bound to $IrCl_5(S)^{2-}$ in solution. The same stoichiometry applies to the higher homologs; the only difference lies in the complexion of the products, increasing amounts of alkenes and alkyl acetates being formed at the expense of alkyl chlorides on going from ethyl, isopropyl to t-butyl.

The cleavage of dialkylmercury by hexachloroiridate(IV) is highly dependent on the structure of the alkyl groups. Thus, in the homologous series of $RHgCH_3$, the relative rates of cleavage increase from R = methyl : ethyl : isopropyl : tert-butyl, roughly in the order of $10^0 : 10^3 : 10^5 : 10^6$. These results run counter to the pattern observed in the electrophilic cleavage of the same mercurials described in the foregoing section, or to expectations based on increasing steric hindrance. Instead, it suggests that the rate-limiting step occurs prior to alkyl transfer.

Scheme II:

$$R_2Hg + Ir^{IV}Cl_6^{2-} \xrightarrow{k} R_2Hg^{+\cdot} + Ir^{III}Cl_6^{3-} \quad [17]$$

$$R_2Hg^{+\cdot} \xrightarrow{fast} RHg^+ + R\cdot \quad [18]$$

$$R\cdot + Ir^{IV}Cl_6^{2-} \xrightarrow{fast} R_{ox} + Ir^{III}Cl_5X^{n-} \quad [19]$$

where X = Cl or solvent

The kinetics, products and selectivity as well as spin trapping with t-BuNO and O_2 all accord with the mechanism in Scheme II. The observation of paramagnetic intermediates by spin trapping indicates that alkyl radicals are formed during the cleavage of R_2Hg by $IrCl_6^{2-}$. In fact, the quantitative accounting of the alkyl fragments as alkylperoxy products, when the reaction is carried out in the presence of oxygen, shows that all of the alkyl groups must depart from mercury as free radicals according to eq 18. The latter is strongly supported by the observation that $IrCl_6^{2-}$ disappears under these conditions at just one-half the rate observed in an inert atmosphere, as predicted by Scheme II.

The failure to observe directly the electron spin resonance spectrum of $R_2Hg^{+\cdot}$ suggests that its lifetime is very short. It is present as one of the principal species during electron impact of R_2Hg in the gas phase, and a mercury(III) species has been observed as transient in the electrochemical oxidation of Hg-(cyclam)$^{2+}$. Selectivity in the cleavage of alkyl groups from unsymmetrical dialkylmercury by $IrCl_6^{2-}$ according to Scheme II occurs during fragmentation of $R_2Hg^{+\cdot}$ radical-cation subsequent to the rate-limiting step. The unimolecular decomposition of $(CH_3)_2Hg^+$ in the gas phase has been examined by photoelectron-photoion coincidence spectroscopy (27).

$$CH_3HgCH_3^{+\cdot} \longrightarrow CH_3Hg^+ + CH_3\cdot \quad [20]$$

$$CH_3HgCH_3^{+\cdot} \longrightarrow CH_3Hg\cdot + CH_3^+ \quad [21]$$

The threshold energy for fragmentation in eq 20 is found to be nearly 2.5 volts lower than that for eq 21. The exclusive cleavage of R = t-Bu and i-Pr and preferential cleavage of R = Et in the homologous series of $RHgCH_3$ is in accord with a weaker alkyl-mercury compared to a methyl-mercury bond. The predominant factor which determines alkyl vs. methyl cleavage are the strengths of the relevant C-Hg bonds. These values can be evaluated from the average bond energies for Me_2Hg, Et_2Hg, and i-Pr_2Hg which are 58, 48, and 42 kcal mol^{-1}, respectively (28,29).

According to Scheme II, the isolation of alkyl chlorides in high yields implies that hexachloroiridate(IV) is an efficient scavenger of alkyl radicals in eq 19 (R_{OX} = RCl, X = S). However, in addition to the redox transfer of chlorine from hexachloroiridate-(IV) in eq 14, an additional redox step is required, especially for R = isopropyl and t-butyl. The observation of isobutylene and tert-butyl acetate from tert-butyl radicals and hexachloroiridate-(IV) is analogous to electron transfer oxidation of alkyl radicals (30).

$$(CH_3)_3C\cdot + Ir^{IV}Cl_6^{2-} \longrightarrow Ir^{III}Cl_6^{3-} + (CH_3)_3C^+ \text{ , etc.}$$

The tert-butyl cation formed under such circumstances will undergo solvation, for example to tert-butyl acetate, or loss of a β-proton to isobutylene. The resultant iridium(III) product must then maintain its coordination sphere intact as $IrCl_6^{3-}$. Indeed, the distribution of $IrCl_6^{3-}$ and $IrCl_5(CH_3CN)^{2-}$ among reduced iridium(III) products formed from various alkylmercurials is precisely in accord with this formulation. Thus, the results clearly indicate that methyl and ethyl radicals react with $IrCl_6^{2-}$ in acetonitrile, exclusively by chlorine transfer. For isopropyl and tert-butyl radicals, approximately 85 and 50%, respectively, of the reaction proceeds by chlorine transfer and the remainder by electron transfer. The latter becomes more important in acetic acid solutions. The decreasing trend of alkyl radicals to react with $IrCl_6^{2-}$ by electron transfer in the order: t-Bu > i-Pr ≫ Et > Me follows the ease of ionization of the radical in the order: t-Bu < i-Pr < Et < Me as listed in Table II. Furthermore, the opposed trend in the yields of alkyl chlorides is consistent with the generally decreasing alkyl-chlorine bond energies from MeCl through t-BuCl. Whether chlorine transfer and carbonium ion formation represent inner- and outer-sphere redox processes, respectively, forms an interesting speculation.

Inner- and outer-sphere mechanisms merit consideration for the process by which electron transfer occurs from R_2Hg to $IrCl_6^{2-}$ in the rate-limiting step in eq 17. A linear free energy relationship between log k of reaction and I_D of R_2Hg is expected for this system if electron transfer occurs by an outer-sphere process. However, the negative deviation of di-tert-butyl-, diisopropyl-, and diethylmercury from the linear plot suggests that steric factors can be important in the electron transfer to $IrCl_6^{2-}$.

C. Dialkyl(bis-phosphine)platinum(II) Complexes. The cleavage of organoplatinum(II) complexes with outer-sphere oxidants was carried out as a comparison for the alkyls of the main group elements, lead and mercury, described above. Indeed, cis-dialkyl(bis-phosphine)platinum(II) complexes are readily oxidized by hexachloroiridate(IV) to afford two principal types of products depending on the structure of the alkyl group and the coordinated phosphine (31). Thus, the diethyl analog, cis-Et_2-$Pt(II)(PMe_2Ph)_2$, affords EtCl and ethylplatinum(II) species by oxidative cleavage of the Et-Pt bond,

$$Et_2Pt^{II}L_2 + 2\,IrCl_6^{2-} \longrightarrow EtPt^{II}L_2S^+ + EtCl + IrCl_6^{3-} + IrCl_5S^{2-}$$

where L = PMe_2Ph, PPh_3; S = solvent

whereas $Me_2Pt(PMe_2Ph)_2$ undergoes oxidation to dimethylplatinum(IV) species.

$$Me_2Pt^{II}(PMe_2Ph)_2 + 2IrCl_6^{2-} \xrightarrow{CH_3CN} Me_2Pt^{IV}(PMe_2Ph)_2X_2 + IrCl_6^{3-} + IrCl_5^{2-}$$

where X = Cl, $Ir^{III}Cl_5(CH_3CN)$ or CH_3CN

In the latter example, the cleavage of the Me-Pt bond can be induced by replacement of the phosphine to $Me_2Pt(PPh_3)_2$, which competitively undergoes oxidation to dimethylplatinum(IV) species as well as oxidative cleavage to MeCl and methylplatinum(II) species.

$$Me_2Pt^{II}(PPh_3)_2 \xrightarrow{2\,IrCl_6^{2-}} \xrightarrow{(54\%)} Me_2Pt^{IV}(PPh_3)_2X_2 \quad [22]$$

$$\xrightarrow{(46\%)} MePt^{II}(PPh_3)_2X + MeCl \quad [23]$$

The stoichiometric requirement of 2 equivalents of hexachloroiridate(IV) remains invariant for each dialkylplatinum(II), independent of the products of oxidation.

The rates of reactions of dialkylplatinum(II) complexes with hexachloroiridate(IV) in acetonitrile obeyed second-order kinetics, being first-order in each reactant according to eq 24.

$$-d[IrCl_6^{2-}]/dt = 2k[IrCl_6^{2-}][R_2PtL_2] \quad [24]$$

The second-order rate constants for the oxidation of R_2PtL_2 depend in an interesting manner on the nature of the alkyl group as well as the phosphine ligand as shown in Table IX.

Table IX. Rate Constants for the Oxidation of R_2PtL_2 by Hexachloroiridate(IV).

R_2PtL_2	k (l mol^{-1} sec^{-1})	k_{rel}
$Me_2Pt(PMe_2Ph)_2$	4.4×10^2	2.2×10^4
$Et_2Pt(PMe_2Ph)_2$	16×10^2	8×10^4
$Me_2Pt(PPh_3)_2$	2×10^{-2}	1.00
$Et_2Pt(PPh_3)_2$	7.6×10^{-1}	38

The energetics and kinetics as well as the observation of alkyl radicals by spin trapping and oxygen scavenging support a mechanism involving the rate-limiting electron transfer from dialkylplatinum(II) to hexachloroiridate(IV) similar to Schemes I and

II for tetraalkyllead and dialkylmercury, respectively.

Scheme III:

$$R_2PtL_2 + IrCl_6^{2-} \xrightarrow{k} R_2PtL_2^{+} + IrCl_6^{3-} \quad [25]$$

$$R_2PtL_2^{+} \longrightarrow RPtL_2^{+} + R\cdot \quad [26]$$

$$R\cdot + IrCl_6^{2-} \longrightarrow RCl + IrCl_5^{2-} \quad [27]$$

$$R_2PtL_2^{+} + IrCl_6^{2-} \longrightarrow R_2PtL_2Cl^{+} + IrCl_5^{2-}$$

According to Scheme III, the activation process is represented by the electron transfer step in eq 25. The greater reactivity of ethyl derivatives compared to the methyl analogs in the two series of R_2PtL_2 is in accord with their ability to act as donor ligands in electron transfer reactions. Similarly, the difference between the phosphine ligands, PMe_2Ph and PPh_3, can be attributed to their varying donor properties. It is unlikely that steric factors are dominant (or even important), since the difference in methyl/ethyl reactivity increases ten-fold from 3.6 to 38 when the ligand is changed from PMe_2Ph to the more bulky PPh_3. The latter reflects a saturation of electronic effects which is also seen in the change in the reactivity of the methyl derivative by a factor of 22,000 when PMe_2Ph is replaced by PPh_3, compared to a corresponding change of only 2100 observed with the ethyl derivatives.

The failure to observe the esr spectrum of the paramagnetic intermediate $R_2PtL_2^{+}$ indicates that its lifetime is short, consistent with the irreversibility observed in the cyclic voltammetry of R_2PtL_2. Other evidence for the formation of transient Pt(III) species have been advanced in the photolysis of $PtCl_4^{2-}$ and the pulse radiolysis of $PtCl_4^{2-}$ and $PtCl_6^{2-}$ (32,33). There is also kinetic evidence for the existence of metastable Pt(III) species in the oxidation of $PtCl_4^{2-}$ by hexachloroiridate(IV) (34).

According to Scheme III, the paramagnetic $R_2PtL_2^{+}$ suffers at least two principal fates: cleavage of an alkyl radical or further oxidation to dialkylplatinum(IV) by unimolecular and bimolecular routes, respectively. The competition between these pathways depends not only on the concentration of $IrCl_6^{2-}$, but more importantly on the stability of $R_2PtL_2^{+}$ as reflected in the strength of the alkyl-platinum bond. All else being the same, ethyl cleavage is more facile than methyl cleavage, in general accord with the trend in bond strengths. The tendency for methyl cleavage to occur more readily in the presence of coordinated dimethylphenylphosphine compared to triphenylphosphine may be attributed to greater electron release by the former. Difference in steric bulk (e.g., cone angles) of the phosphines may also be a factor.

The fragmentation of the alkyl group from $R_2PtL_2^{+}$ as a free radical in eq 26 during oxidative cleavage of Et_2PtL_2 and $Me_2Pt(PPh_3)_2$ is supported by spin trapping and oxygen scavenging, as well as the ability of $IrCl_6^{2-}$ to convert alkyl radicals to alkyl chlorides efficiently. The latter leads concomitantly to the reduction of hexachloroiridate(IV) to pentachloroiridate(III) as $IrCl_5(CH_3CN)^{2-}$ according to eq 27. Indeed, analysis of the reduced iridium products shows the presence of equimolar amounts of $IrCl_5(CH_3CN)^{2-}$ and $IrCl_6^{3-}$, the latter arising from the rate-limiting electron transfer step according to Scheme III.

IV. Delineation of Electrophilic and Electron Transfer Mechanisms

The examples cited show that both electrophilic and electron transfer processes can readily participate in the cleavage of organometals. Often the distinction is not cleanly delineated. Basically, an electrophilic cleavage of a carbon-metal bond is mechanistically distinguished from one proceeding via electron transfer as outlined in eqs 28 and 29, respectively.

$$R\text{-}M + E^{+} \xrightarrow{k_E} \left[R \overset{E^{+}}{\cdots} M \right]^{\ddagger} \longrightarrow R\text{-}E + M^{+}, \text{ etc.} \quad [28]$$

$$R\text{-}M + E^{+} \xrightarrow{k_{ET}} \left[R\text{-}M^{+\bullet}\ E^{\bullet} \right] \longrightarrow R\text{-}E + M^{+}, \text{ etc.} \quad [29]$$

The electrophilic cleavage in eq 28 is a one-step process which proceeds with a second-order rate constant k_E and in which no intermediates are generated. The transition state depicted in brackets reflects a bond breaking to metal which occurs simultaneously with bond making to electrophile during the transfer of the alkyl group. In contrast, the electron transfer process in eq 29 proceeds by a two-step mechanism in which the initial transfer of an electron from the organometal to the electrophile constitutes the rate-limiting interaction with a second-order rate constant k_{ET}. The radical ion pair shown in brackets is an actual intermediate, but if its collapse to products is more rapid than diffusion from the solvent cage, no paramagnetic species will be observed. Under these circumstances, selectivity studies of the intermediate provide the only suitable alternative for distinguishing these mechanistic pathways. The latter is predicated by the notion that k_E and k_{ET} show much the same characteristics, especially with regard to the structure of the organometal. For example, compare the cleavage of dialkylmercury(II) compounds by acids (electrophilic) and hexachloroiridate (electron transfer), presented above:

The cleavage of dialkylmercury(II) by hexachloroiridate proceeds from a prior rate-limiting electron transfer step,

$$R'HgR + IrCl_6^{2-} \xrightarrow{k_{ET}} R'HgR^{+\bullet} + IrCl_6^{3-}$$

but the selectivity is determined by the fragmentation of the radical-cation $R'HgR^{+\bullet}$ in a fast subsequent reaction,

$$R'HgR^{+\cdot} \longrightarrow \begin{cases} R\cdot + R'Hg^+ \\ R'\cdot + RHg^+ \end{cases}$$

As expected, the rate of this cleavage (log k_{ET}) is linearly related to the ionization potential of the mercurial as shown in Figure 3. A similar correlation is shown by organometals undergoing substitution [Grignard reagent and peroxide (35)] or insertion [tetraalkyllead and TCNE (36)] via electron transfer, and they are also included in Figure 1 for comparison.

The rate constant k_E for the electrophilic protonolysis of an alkyl-mercury bond,

$$R\text{-}HgR' + H^+ \xrightarrow{k_E} R\text{-}H + R'Hg^+$$

can be dissected into two parameters: **C**, which is focused on the cleaved group R, and **L**, which depends only on the leaving group, $R'Hg^+$, as described by eq 8. Figure 4 shows that **L** responds linearly to the ionization potential of the mercurial. Furthermore, **L** also strongly correlates with the Taft polar substituent constant, σ^*, for the alkyl groups listed in Table X. The

Table X. Correlation of **L** and **C** Parameters with Taft Polar (σ^*) and Steric (E_s) Constants.

R	σ^*	E_s
CH_3	0	0
CH_3CH_2	0.10	-0.07
$(CH_3)_2CH$	0.20	-0.47
$(CH_3)_3C$	0.30	-1.54
Leaving Group Effects		
RHg^+	**L** (expt.)	$8.1\ \sigma^* + 0.65\ E_s$ (calc.)
CH_3Hg^+	0	0
$CH_3CH_2Hg^+$	0.76	0.75
$(CH_3)_2CHHg^+$	1.28	1.30
$(CH_3)_3CHg^+$	1.44	1.45
Cleaved Group Effects		
R	**C** (expt.)	$8.1\ \sigma^* + 2.8\ E_s$ (calc.)
CH_3	0	0
CH_3CH_2	0.55	0.61
$(CH_3)_2CH$	0.29	0.30
$(CH_3)_3C$	~-1	-1.9

curvature in Figure 4 for the cleaved group constant, **C**, on proceeding from methyl to t-butyl can be attributed to an increasing steric effect as a result of additional encumbrance by succes-

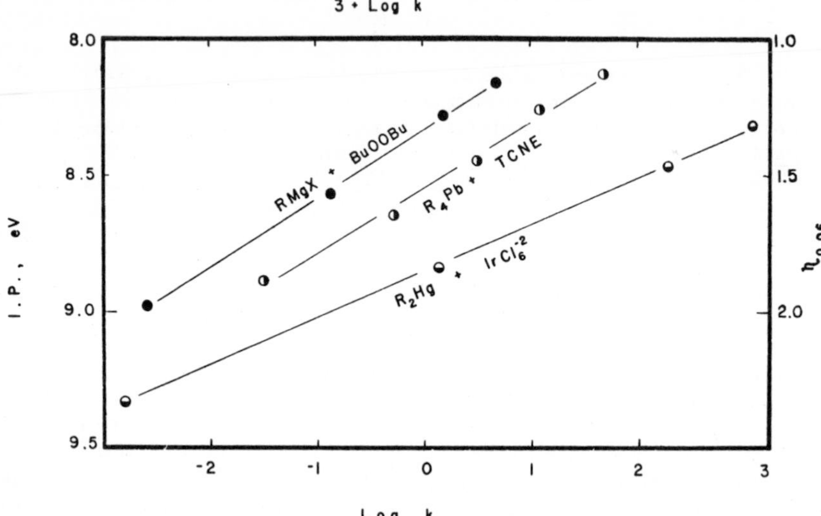

*Figure 3. Electron-transfer processes in electrophilic substitutions. Saturation effects followed by alkyl substituents in the cleavage of organometals during the treatment with various electrophiles: scale left and bottom for (◑) tetraalkyllead with tetracyanoethylene and (⊖) dialkylmercury with hexachloroiridate(IV). Scale right (Tafel potential) and top for Grignard reagents and di-*tert*-butyl peroxide (●).*

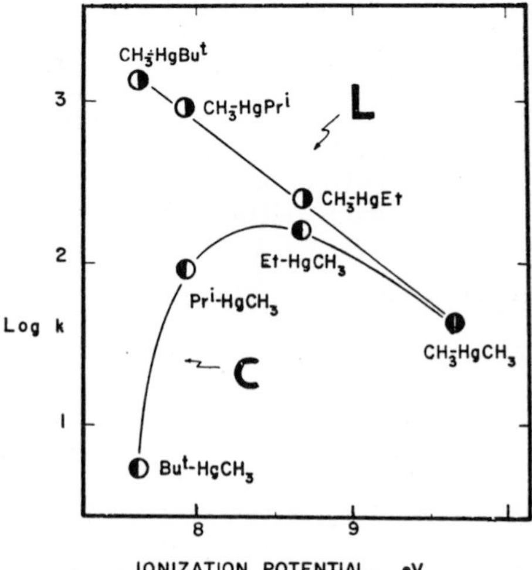

*Figure 4. Correlation of cleaved-group constant **C** and leaving group constant **L** in acetolysis with the ionization potential of the dialkylmercury compound*

sively more α-methyl groups. The latter is also supported by the sizeable contribution of the Taft steric parameter E_s to the correlation with **C**, as shown in Table X. Barring steric effects, both **L** and **C** are thus strongly dependent on the ionization potential of the mercurial; that is, electrophilic attack at an alkyl-metal bond is responsive to electron availability, namely the HOMO energy, in much the same way that the ionization potential is. Described in an alternative way, the transition state for electrophilic cleavage in eq 28 can be considered as a perturbation of the organometal by an electrophile, which is akin to a virtual ionization.

These comparisons show that electron transfer and electrophilic attack both depend heavily on the electron availability in the organometal. It follows that any correlation of the reactivity (i.e., rate constants) with the ionization or oxidation potentials is not sufficient to differentiate the two mechanisms. The difference between the two is largely due to the variations in the steric interactions, which can be large in an inner-sphere, electrophilic process (i.e., for **C** but not **L**) and less important in an outer-sphere, electron transfer process. This conclusion supports the general formulation that electron transfer and electrophilic processes can share a common theme of charge transfer interactions. Such a conclusion also bears on the multiplicity of available mechanisms for the cleavage of organometals. The ready accessibility of such concerted and stepwise processes follows naturally from their basic similarity, and it can make the task of differentiation in individual cases difficult.

These studies also shed light on the effects of polyalkylation of metals on their reactivity to electrophilic and electron transfer cleavages. Thus the rates of cleavage of a single alkyl ligand from Group IVB organometals always decrease in the order: $R_4M > R_3MCl > R_2MCl_2$, and for the mercurials, $R_2Hg \gg RHgCl$, independent of whether an electrophilic or electron transfer process pertains. This reactivity sequence naturally follows from the availability of σ-bonding electrons in the HOMO as listed in Table IV for the mercurials.

V. Homolytic Displacements in Alkyl Transfers

Organocobalt(III) complexes are readily cleaved by chromous ion in aqueous perchloric acid solution (37).

$$RCo^{III}(DMG)_2 + Cr^{2+}_{aq} \xrightarrow[2H^+]{} R\text{-}Cr^{2+}_{aq} + Co^{II}(DMG)_2$$

where DMG = dimethylglyoximate

The alkyl group is transferred to chromium(II) essentially quantitatively. The cleavage formally represents a transfer of an alkyl radical R· to Cr^{2+}, that is an overall reductive cleavage of an alkyl-cobalt(III) bond. The rate of transmetallation follows second-order kinetics,

$$-d[Cr]/dt = k[RCo][Cr] + k'[RCoH^+][Cr^{2+}] \qquad [30]$$

where k and k' relate to the cleavage of the neutral and protonated alkylcobalt(III) species, respectively. The variation in k and k' with the alkyl groups follows the same order, and decreases in the order:

R	=	Me	>	Et	>	n-Pr	>	i-Pr	>	i-Bu
$k'(M^{-1}\ sec^{-1})$	=	$10^{1.4}$		$10^{-1.9}$		$10^{-2.9}$		10^{-4}		$10^{-4.2}$

Alkylcobalamins are also cleaved by chromous ion (38),

$$RCo(corrin) + Cr^{2+} \longrightarrow RCr^{2+} + B_{12r} \qquad [31]$$

with simple second-order kinetics, first-order in each reactant, and independent of pH between 0–2.3. The cleavages of the methyl [$k = 3.6 \times 10^2\ M^{-1}\ sec^{-1}$] and ethyl [$k = 4.4\ M^{-1}\ sec^{-1}$] derivatives proceed with different activation parameters: $\Delta H^{\ddagger} = 3.8$ (Me), 11(Et) kcal mol^{-1}; $\Delta S^{\ddagger} = -34$(Me), -18(Et) eu.

Alkyl transfers from cobalt(III) to chromium(II) as described above are analogous to the reversible exchange between alkylcobalt(III) and cobalt(II) (39,40,41),

$$RCo^{III}(1) + Co^{II}(2) \rightleftharpoons Co^{I}(1) + RCo^{III}(2) \qquad [32]$$

where Co(1) and Co(2) refer to cobalt complexes with slightly different chelating ligands such as dimethylglyoximato and cyclohexanedionedioximato. The overall process in eq 32, which is equivalent to electron transfer, actually occurs by transfer of an alkyl group as a radical as shown by labelling the cobalt atoms with different chelating ligands. The rate of exchange follows second-order kinetics, first-order in cobalt(II) and first-order in alkylcobalt(III). The second-order rate constants decrease in the relative order [$k(Et) = 1.1 \times 10^{-1}\ M^{-1}\ sec^{-1}$]:

R	=	Me	>	Et	>	n-Pr	~	n-Bu	>	i-Pr	>	i-Bu
k_{rel}	=	($\geq 10^{2.6}$)		(10^0)		($10^{-1.2}$)		($10^{-1.4}$)		($10^{-2.5}$)		($10^{-3.3}$)

The transfer of the erythro-PhCHDCHD- group occurs with inversion (42). Coupled with the reactivity trend of alkyl groups, the cleavage is best considered as a homolytic displacement on the carbon center (43). The transition state,

$$[Co^{II} \text{—} C \text{—} Co^{III}]^{\ddagger}$$

is similar to that in electrophilic cleavages occurring with inversion, except the process involves a one-equivalent rather than a two-equivalent change. However, the latter does not appear to be a decisive factor, since alkyl transfer from alkylcobalt(III) to cobalt(I), i.e.,

$$RCo^{III}(1) + Co^{I}(2) \rightleftharpoons Co^{I}(1) + RCo^{III}(2) \qquad [33]$$

occurs with rates and stereochemistry much like that of its cobalt(II) counterpart in eq 32. In particular, inversion of configuration at carbon occurs during alkyl exchange, and the second-order rate constants decrease in the relative order [k(Et) = 10^{-1} M^{-1} sec^{-1}]:

$$\begin{array}{llccccccccc} R & = & Me & > & Et & > & n\text{-}Pr & \sim & n\text{-}Bu & > & i\text{-}Bu \\ k_{rel} & = & (\geq 10^{2.2}) & & (10^{0}) & & (10^{-1.4}) & & (10^{-1.7}) & & (10^{-3.9}) \end{array}$$

Attempts to measure the alkyl exchange of alkylcobalt(III) with cobalt(III) were unfortunately complicated by the slow rates, and thus susceptible to trace catalysis by cobalt(II) species. The low reactivity of cobalt(III) is probably due to its substitution stability, which limits the availability of the active 5-coordinate electrophilic species. Nonetheless, the transition states for alkyl exchanges in all three systems are likely to be similar, effectively involving a linear 3-atom configuration (see above). The similarity in the rates of the cobalt(II) and cobalt(I) reactions suggests that the extra 1 and 2 electrons, respectively, are in a nonbonding orbital centered on both cobalt atoms.

The cleavages of alkyl-metal bonds by each of the three cobalt complexes with oxidation states I, II and III are representative of what is commonly considered to be electrophilic, homolytic and nucleophilic processes, respectively, e.g.,

$$R\text{-}M^{II} \begin{cases} \xrightarrow{Co^{I}} R\text{-}Co^{III} + M^{0} \\ \xrightarrow{Co^{II}} R\text{-}Co^{III} + M^{I} \\ \xrightarrow{Co^{III}} R\text{-}Co^{III} + M^{II} \end{cases}$$

Yet the ready interconversion of each cobalt species by one-equivalent changes,

$$Co^{III} \underset{}{\overset{\epsilon}{\rightleftharpoons}} Co^{II} \underset{}{\overset{\epsilon}{\rightleftharpoons}} Co^{I}$$

raises the issue of whether electron transfer processes are involved in alkyl transfer as discussed in the previous sections. Such processes are especially relevant in view of the ease with which the alkylcobalt complexes themselves undergo oxidation-reduction (<u>44</u>).

$$RCo^{III} \overset{-\epsilon}{\rightleftharpoons} RCo^{IV}$$

For example, the cleavage in eq 32 may involve a two-step process, in which cobalt(II) acts as a nucleophile leading to the initial reduction of alkylcobalt(III), followed by electron transfer.

$$RCo^{III}(1) + Co^{II}(2) \rightleftharpoons Co^{I}(1) + RCo^{IV}(2)$$

$$Co^{I}(1) + RCo^{IV}(2) \longrightarrow Co^{II}(1) + RCo^{III}(2)$$

Such mechanisms are not readily distinguished from a one-step pathway involving direct transfer of an alkyl radical, short of actually detecting the intermediates resulting from electron transfer, and quantitatively relating them to the rate of reaction. There are thermodynamic arguments, however, disfavoring such a formulation (43).

Acknowledgement: I wish to thank Drs. H. Gardner, J. Y. Chen, and W. A. Nugent for their stimulating contributions and the National Science Foundation for financial support.

Literature Cited

1. Wood, J. M., "Biochemical and Biophysical Perspectives in Marine Biology," Vol. 3, D. C. Malins and J. R. Sargent, eds., Academic Press, New York, 1975.
2. Ridley, W. P., Dizikes, L. J., and Wood, J. M., Science (1977), 197, 329.
3. Basolo, F. and Pearson, R. G., "Mechanisms of Inorganic Reactions," 2nd Ed., Wiley-Interscience, New York, 1967.
4. Jenkins, C. L. and Kochi, J. K., J. Am. Chem. Soc. (1972), 94, 842, 845.
5. Ferraudi, G., private communication.
6. Fehlner, T. P., Ulman, J., Nugent, W. A., and Kochi, J. K., Inorg. Chem. (1976), 15, 2544.
7. Taft, R. W., Jr., in "Steric Effects in Organic Chemistry," M. S. Newman, ed., Wiley-Interscience, New York, 1956; cf. also Shorter, J., Quart. Revs. (1970), 24, 433.
8. Holm, T., Acta Chem. Scand. (1974), B28, 809.
9. Hosomi, A. and Traylor, T. G., J. Am. Chem. Soc. (1975), 97, 3682.
10. Lossing, F. P. and Semeluk, G. P., Can. J. Chem. (1970), 48, 955.
11. Streitwieser, A., Jr. and Nair, P. M., Tetrahedron (1959), 5, 149.
12. Gardner, H. C. and Kochi, J. K., J. Am. Chem. Soc. (1975), 97, 1855.
13. Nugent, W. A. and Kochi, J. K., J. Am. Chem. Soc. (1976), 98, 5979.
14. Kreevoy, M. M. and Hansen, R. L., J. Am. Chem. Soc. (1961), 83, 626 ; Kreevoy, M. M., J. Am. Chem. Soc. (1957), 79, 5927.
15. Clinton, N. A., Gardner, H. C., and Kochi, J. K., J. Organomet. Chem. (1973), 56, 227.
16. Gardner, H. C. and Kochi, J. K., J. Am. Chem. Soc. (1974), 96, 1982.

17. Streitwieser, A., Jr. and Nair, P. M., Tetrahedron (1959), 5, 149.
18. Lossing, F. P. and Semeluk, G. P., Can. J. Chem. (1970), 48, 955.
19. Clinton, N. A. and Kochi, J. K., J. Organomet. Chem. (1972), 42, 229.
20. Edwards, J. O. and Pearson, R. G., J. Am. Chem. Soc. (1962), 84, 16.
21. Davis, R. E. and Cohen, A., J. Am. Chem. Soc. (1964), 86, 440.
22. Klopman, G., in Ziegler, K., Kroll, W.-R., Larbig, W., and Steudel, O.-W., J. Liebigs Ann. Chem. (1960), 629, 57.
23. Arbelot, M., Metzger, J., Chanon, M., Guimon, C., and Pfister-Guillouzo, G., J. Am. Chem. Soc. (1974), 96, 6217.
24. Clinton, N. A., Gardner, H. C., and Kochi, J. K., J. Organomet. Chem. (1973), 56, 227.
25. Clinton, N. A. and Kochi, J. K., J. Organomet. Chem. (1972), 42, 241.
26. Chen, J. Y., Gardner, H. C., and Kochi, J. K., J. Am. Chem. Soc. (1976), 98, 6150.
27. Cant, C. S. T., Danby, C. J., and Eland, J. H. D., J. Chem. Soc., Faraday Trans. II (1975), 71, 1015.
28. Gowenlock, B. G., Haynes, R. M., and Majer, J. R., Trans. Faraday Soc. (1962), 58, 1905.
29. Carson, A. S. and Wilmshurst, B. R., J. Chem. Thermodyn. (1971), 3, 251.
30. Jenkins, C. L. and Kochi, J. K., J. Am. Chem. Soc. (1972), 94, 843 ; Kochi, J. K., IUPAC, XXIIIrd Int. Congr. Pure Appl. Chem., Boston, Pure Appl. Chem. Suppl. (1972), 4, 377.
31. Chen, J. Y. and Kochi, J. K., J. Am. Chem. Soc. (1977), 99, 1450.
32. Wright, R. C. and Laurence, G. S., J. Chem. Soc., Chem. Commun. (1972), 132.
33. Adams, G. E., Broszkiewicz, R. B., and Michael, B. D., Trans. Faraday Soc. (1968), 64, 1256.
34. Halpern, J. and Pribanic, M., J. Am. Chem. Soc. (1968), 90, 5942.
35. Nugent, W. A., Bertini, F., and Kochi, J. K., J. Am. Chem. Soc. (1974), 96, 4945.
36. Gardner, H. C. and Kochi, J. K., J. Am. Chem. Soc. (1976), 98, 2460.
37. Espenson, J. H. and Shveima, J. S., J. Am. Chem. Soc. (1973), 95, 4468.
38. Espenson, J. H. and Sellers, T. D., Jr., J. Am. Chem. Soc. (1974), 96, 94.
39. Van Den Bergen, A. and West, B. O., J. Chem. Soc., Chem. Commun. (1971), p. 52; J. Organomet. Chem. (1974), 64, 125.

40. Mestroni, G., Cocevar, C. and Costa, G., Gazz. Chim. Ital. (1973), 103, 273.
41. Dodd, D. and Johnson, M. D., J. Chem. Soc., Chem. Commun. (1971), 1371.
42. Chrzastowski, J. Z., Cooksey, C. J., Johnson, M. D., Lockman, B. L. and Steggles, P. N., J. Am. Chem. Soc. (1975), 97, 932.
43. Dodd, D., Johnson, M. D. and Lockman, B. L., J. Am. Chem. Soc. (1977), 99, 3664.
44. Halpern, J., Chan, M. S., Hanson, J., Roche, T. S., and Topich, J. A., J. Am. Chem. Soc. (1975), 97, 1606.

Discussion

J. H. ESPENSON (Iowa State University): Since both mechanisms, electron transfer and the electrophilic mechanism, show a systematic decrease in rate constants in going from methyl down the series to t-butyl, what is the criterion by which you can use the kinetic data as a means of distinguishing them?

KOCHI: You cannot. The way you must distinguish between these two mechanisms is by focussing on the radical pair that you get in the electron transfer mechanism. The electrophilic mechanism is essentially a one-step mechanism. The rate-limiting step in the electron transfer process leads to an intermediate which then goes on to form product. The way one distinguishes between these two is to focus on the radical pair. One can do that by looking at the selectivity studies, that is, by examining the selectivity in the composition of these organometals. The usual criterion of relating rates to redox potentials is highly questionable.

J. J. ZUCKERMAN (University of Oklahoma): You have a high correlation with ionization potential (ground state, gas phase). Since these are aqueous studies, wouldn't the appropriate parameter be the electrode potential, or is there a high correlation between those two sets of data?

KOCHI: Yes, there is a linear correlation with oxidation potentials of these organometals with the ionization potentials. Solvation effects do not appear to be very strong in this case.

RECEIVED August 22, 1978.

14

Pathways for Formation of Transition Metal-Carbon Bonds in Protic Media

JAMES H. ESPENSON

Ames Laboratory and Department of Chemistry, Iowa State University, Ames, IA 50011

I plan to review work from my group and from the literature concerning the title subject. My intention is to be rather selective, focusing upon only a limited family of compounds and reactions. This main emphasis is on the reactions themselves and the mechanisms by which they occur.

$R-ML_5$ Complexes. The compounds of interest are sigma-bonded organometallic complexes, chiefly of cobalt and chromium. This includes a series of tetradentate macrocyclic complexes of cobalt, RCo(chel)B with $B=H_2O$, py, etc., aquochromium complexes, $[(H_2O)_5Cr-R^{2+}]$, and chromium complexes containing the macrocyclic ligand [15]ane N_4. Representative structures are shown in Figure 1.

The prototype of the organocobalt structures is that of the vitamin B_{12} derivative methylcobalamin. The most successful and abundant family of model compounds are the bis(dimethylglyoximato) complexes developed by Schrauzer (1); these are the compounds $R-Co(dmgH)_2B$. Other cobalt chelate structures are also shown in Figure 1 with semi-systematic names as suggested by Busch (2). Detailed reviews of their preparation and chemistry have been published (3,4).

A large number of compounds with the general formula $(H_2O)_5Cr-R^{2+}$ are now known (5-11). Chromium complexes analogous to those of the organocobalts, although known, are much less abundant. West (12) has prepared some perfluoroalkyls $R_F-Cr(chel)B$, and we have recently prepared an extensive series of the alkyl compounds $[R-Cr([15]ane\ N_4)H_2O]^{2+}(X^-)_2$ (13). Many other classes and examples of organochromium compounds are of course known (14), but their formation will not be reviewed here.

A limited number of related organoiron compounds are also known (15,16,17).

Our consideration of synthetic methods for such sigma organometallics, as well as their reaction chemistry, can be systematized by means of somewhat artificial oxidation state

0-8412-0461-6/78/47-082-235$05.00/0

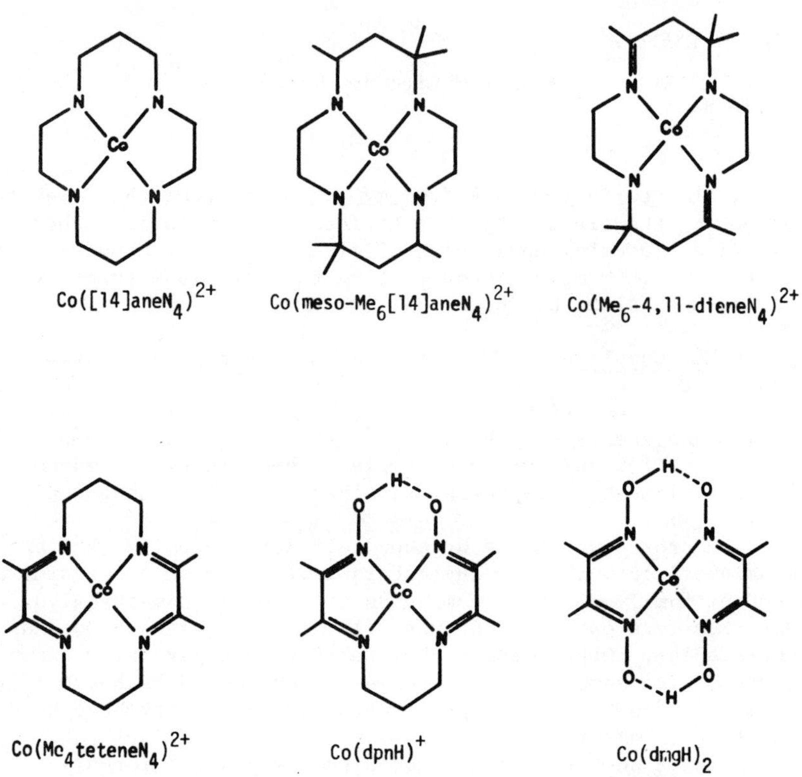

$Co([14]aneN_4)^{2+}$ $Co(meso\text{-}Me_6[14]aneN_4)^{2+}$ $Co(Me_6\text{-}4,11\text{-}dieneN_4)^{2+}$

$Co(Mc_4teteneN_4)^{2+}$ $Co(dpnH)^+$ $Co(dmgH)_2$

Journal of the American Chemical Society

Figure 1. Representative structures of cobalt chelates which form organometallic compounds of the form R–Co(chel)B. Chelate abbreviations are based on the suggestions of Busch (2).

assignments. The organic group R is customarily regarded as a carbanion $R:^-$. Thus the following specific compounds, considering their ionic charges and those on the other ligands (if any), are all derivatives of M(III): CH_3CH_2-Co(dmgH)$_2$py, $[(H_2O)_5Cr\text{-}CH_2CH_3CH_3]^{2+}$, $[CH_3\text{-}Co([14]ane\ N_4)H_2O]^{2+}$, and $[CH_3\text{-}Co(dpnH)H_2O]^+$.

Survey of Preparative Methods. The following reaction scheme summarizes the modes of reactivity of R(Co)B and $(H_2O)_5Cr\text{-}R^{2+}$:

Scheme I. Reactivity patterns:

$$\text{R-(Co}^{III}\text{) or R-Cr}^{2+} \xrightarrow{\text{electrophiles}} \text{R}^{\overline{:}} + (\text{Co}^{III})^+ \text{ or Cr}^{3+} \quad [1]$$

$$\text{R-(Co}^{III}\text{) or R-Cr}^{2+} \xrightarrow{h\nu \text{ or 1-eln reagents}} \text{R}^{\cdot} + (\text{Co}^{II}) \text{ or Cr}^{2+} \quad [2]$$

$$\text{R-(Co}^{III}\text{) or R-Cr}^{2+} \xrightarrow{\text{nucleophiles}} \text{R}^{+} + (\text{Co}^{I})^- \quad [3]$$

The organic groups shown as R^-, $R^{\cdot}$, and R^+ are, of course, combined with other reagents as shown in these specific reactions:

$$R(Co)H_2O + Hg^{2+} \rightarrow RHg^+ + (H_2O)_2(Co^{III})^+ \quad [4]$$

$$(H_2O)_5Cr\text{-}R^{2+} + Br_2 \rightarrow RBr + Cr(H_2O)_6{}^{3+} + Br^- \quad [5]$$

$$R(Co)H_2O + Cr^{2+}_{aq} \rightarrow (H_2O)_5Cr\text{-}R^{2+} + (Co^{II}) \quad [6]$$

It is outside the scope of this review to discuss this reaction chemistry in detail, and it is mentioned here to demonstrate that the reverse of each of the three general reactivity methods can be used to form organometallic derivatives.

Thus a source of a carbanion such as a Grignard reagent or alkyl lithium will react with a M(III) derivative such as a halide to form a metal-carbon bond. Likewise, generation of a carbon-centered free radical from a suitable source in the presence of the M(II) complex gives R-M, provided the radical capture rate is sufficiently rapid (i.e., that M(II) is sufficiently labile to ligand substitution), via the reverse of reaction 2. The nucleophilic process, when permitted by the properties of M(I), often affords the best synthetic method. Thus cobaloximes and related compounds are available from organic halides or tosylates by reactions which are well-characterized as typical S_N2 substitutions (18):

$$(Co^I)^- + RX \rightarrow R(Co^{III}) + X^- \quad [7]$$

Methods Based on M(II) and Free Radicals. The accepted scheme by which many reduced metal complexes react with organic halides to form organometallic products is shown in Scheme II. The earliest examples were reported for $Co(CN)_5{}^{3-}$ (19) and Cr^{2+} (20,21,22).

Scheme II. Two-Step Radical Mechanism:

$$RX + M^{II} \rightarrow XM^{III} + R^{\cdot} \quad [8]$$

$$R^{\cdot} + M^{II} \rightarrow RM^{III} \quad [9]$$

$$\text{Net: } RX + 2M^{II} \rightarrow XM^{III} + RM^{III} \quad [10]$$

The requirements for this to constitute a successful route to $R\text{-}M^{III}$ are (1) the reduced metal complex must be a sufficiently strong and reactive reducing agent that the first step will be feasible, and (2) the rate of combination of M(II) and $R^{\cdot}$ in equation 9 must be sufficiently high to compete with other processes such as radical dimerization and disproportionation.

The first requirement limits the usefulness for both cobalt and chromium. Although the M^{II} complexes $Co(dmgH)_2$ and Cr^{2+}_{aq} react with such activated organic halides as benzyl bromide, they remain unreactive toward saturated alkyl halides such as $\underline{i}$-C_3H_7Br and CH_3I. For systems in which the M^{II} complex is a stronger reducing agent, e.g. $CrSO_4$ in aqueous dimethylformamide (20) or a "Cr^{II}-en" complex in the same solvent (23), the processes do occur as shown.

Recent work by G. J. Samuels (13) has been based on this idea. The complex of Cr^{II} and the tetradentate macrocyclic ligand [15]ane N_4 forms readily, and it is a stronger reducing agent than Cr^{2+}_{aq} (E°=-0.58 V *vs*. -0.41 V for Cr^{2+}_{aq}). It reacts to form organochromium complexes with a wide variety of organic halides according to the stoichiometry of equation 10. The rates of these reactions follow this expression

$$-d[Cr^{II}[15]\text{ane } N_4{}^{2+}]/dt = 2k_8[Cr^{II}[15]\text{ane } N_4{}^{2+}][RX] \quad [11]$$

Values of the second-order rate constant k_8 for a selected group of the halides are summarized in Table I.

Table I
Rate Constants for Reaction of $Cr^{II}[15]$ane $N_4{}^{2+}$ with Alkyl Halides (13)
Conditions: 25°C in 1:1 aq. *tert*-butanol

RX	$k_8/dm^3mol^{-1}s^{-1}$	RX	$k_8/dm^3mol^{-1}s^{-1}$
C_2H_5Br	0.164	c-$C_6H_{11}Br$	0.83
n-C_3H_7Br	0.167	i-C_3H_7I	4.93
n-C_4H_9Br	0.130	C_2H_5I	0.414
6-Br-1-hexene	0.155	$PhCH_2Br$	1.9×10^4
$(CH_3)_3CCH_2Br$	∿0.05	$PhCH_2Cl$	3.2×10^2
i-C_3H_7Br	1.85	t-C_4H_9Br	∿7
c-C_5H_9Br	5.4		

The following findings argue for the two-step mechanism of Scheme II and the involvement of free radical intermediates: (1) The 2:1 stoichiometry of reactants and the production of equimolar concentrations of chromium (III) alkyl and halides; (2) The second-order rate law of equation 11; (3) The reactivity order RI > RBr >> RCl; (4) The reactivity order tertiary > secondary > primary alkyl halide; (5) Production of methylcyclopentane resulting from acid thermolysis of the alkylchromium cation produced in the reaction of 6-bromo-1-hexene, consistent with the known cyclization of the 6-bromo-1-hexenyl radical (24,25).

The second requirement previously mentioned, a high rate of combination of the carbon-centered radical and a second M(II) complex, has been inferred from kinetic evidence by various authors (19,23,26). Direct proof of this and measurements of the rate of this rapid step have come from pulse radiolysis measurements (27,28).

The reaction of the vanadium(II) compound $VCl_2(py)_4$ with benzyl halides produces only the coupling product RR without detection of organovanadium(III) (29). The failure to observe the organovanadium even as an unstable intermediate might be due, as suggested, to the rapidity of subsequent acidolysis or electron transfer (29), but it seems equally plausible to this author that its formation is precluded simply by the low rate of ligand substitution of the d^3 vanadium(II) complex, leaving the organic radicals to undergo coupling reactions.

Among other thermal methods of producing carbon-centered radicals are reactions of organic hydroperoxides, peroxides, peroxo acids, and peroxo esters. This usually uses an electron-transfer step with the M(II) complex to general reactive intermediates which ultimately yield the carbon-centered radical $R^\bullet$. The mechanistic steps shown in Scheme III are illustrative of the reactions (30,31,32,33); the importance of the competitive secondary reactions depends upon the particular peroxide structure and the identity of the M(II) complex.

Scheme III. *t*-Butyl Hydroperoxide and Cr^{2+}_{aq}:

$$(CH_3)_3COOH + Cr^{2+}_{aq} \xrightarrow{H^+} Cr^{3+}_{aq} + (CH_3)_3CO^\bullet \quad [12]$$

$$(CH_3)_3CO^\bullet \longrightarrow (CH_3)_2CO + {}^\bullet CH_3 \quad [13a]$$

$${}^\bullet CH_3 \xrightarrow{Cr^{2+}} Cr\text{-}CH_3^{2+} \quad [13b]$$

$$(CH_3)_3CO^\bullet \xrightarrow{Cr^{2+},\ H^+} (CH_3)_3OH + Cr^{3+} \quad [14]$$

$$(CH_3)_3CO^\bullet \longrightarrow [Cr\text{-}\overset{H}{O}C(CH_3)_3]^{3+} \quad [15]$$

$$(CH_3)_3CO^\bullet \xrightarrow{H\text{-}CH_2OH} (CH_3)_3COH + {}^\bullet CH_2OH \quad [16a]$$

$${}^\bullet CH_2OH \xrightarrow{Cr^{2+}} Cr\text{-}CH_2OH^{2+} \quad [16b]$$

The first step, reaction 12, is the rate-limiting process. Kinetic data for a variety of peroxide structures for Cr^{2+}_{aq} (31) and Co^{II}(chelates) (33) have been published, and will not be reviewed here. The rates for Cr^{2+}, the strongest reducing agent, are considerably higher than those of any of the Co(II) complexes examined.

Organic hydrazines have provided useful routes to organocobalt and organoiron compounds (34,35), although no mechanistic studies have been conducted and the involvement of free radicals is uncertain.

Metal(II) complexes may react with organic halides by other mechanisms as well. With special functional groups electron transfer pathways will be seen (36). An unusual case is found for the cobalt(II) complex Vitamin B_{12r} (37), and for certain iron(II) porphyrins (38) (although XFe(P) and not RFe(P) appears to be formed). In these instances the rate law is

$$-d[M(II)]/dt = k[M(II)]^2[RX] \qquad [17]$$

suggestive perhaps of a dual attack of the separate metal centers at carbon and halogen.

Photochemical Routes *via* Free Radicals. The complexes $(NH_3)_5CoO_2CR^{2+}$ ($R=CH_3$, C_2H_5) undergo clean photoredox decomposition according to

$$(NH_3)_5CoO_2CR^{2+} \xrightarrow[H^+]{h\nu} Co^{2+} + 5NH_4^+ + CO_2 + {}^{\cdot}R \qquad [18]$$

providing a source of a carbon-centered free radical. When the photolysis is carried out in the pressence of an appropriate cobalt(II) complex, it is possible to obtain reasonable yields of $R(Co)H_2O$ product (39).

A photochemical method has also resulted in the successful preparation of organocobalt complexes (40). Methylcobaloxime in solution is quite photosensitive to visible light, the primary photoreaction consisting of homolysis to $^{\cdot}CH_3$ and $Co(dmgH)_2$ (41,42). Although other organocobalt complexes undergo similar photodissociation, often with comparable quantum yields (35), the lower molar absorptivities associated with complexes containing macrocyclic chelate ligands such as Me_6[14]ane N_4 and Me_6[14]4,11-diene N_4 render them, in practice, less photosensitive than methylcobaloxime. A mixture of a small excess of $RCo(dmgH)_2H_2O$ and the desired cobalt(II) complex in deaerated, dilute, aqueous perchloric acid forms the new organocobalt complex upon photolysis with visible light (The acid is useful in decomposing $Co(dmgH)_2$ to Co^{2+}_{aq} and H_2dmg, thereby reducing the extent of recombination of $R^{\cdot}$ and $Co(dmgH)_2$).

Nucleophilic Pathways *via* Cobalt(I). This route is represented by reaction [7], and such reactions are well-characterized S_N2 processes (18,43); the same is true of the reactions of similar rhodium(I) nucleophilies (44). I would call attention to one interesting anomaly in such reactions with dihalides, $Br(CH_2)_nBr$. The cobaloxime(I) nucleophile reacts cleanly in two stages:

$$Br(CH_2)_4Br + (Co^I)^- \rightarrow Br(CH_2)_4(Co) + Br^- \quad [19]$$

$$Br(CH_2)_4(Co) + (Co^I)^- \rightarrow (Co)(CH_2)_4(Co) + Br^- \quad [20]$$

For these processes the rate constants (45) are not too far from statistical: $k_{19}/2=4.8\ dm^3mol^{-1}s^{-1}$ and $k_{20}=2.04\ dm^3mol^{-1}s^{-1}$ (and compare $n\text{-}C_4H_9Br$, $k=2.13\ dm^3mol^{-1}s^{-1}$).

In contrast the cobalt(I) derivative of Vitamin B_{12}, B_{12s}, reacts readily to form "tetramethylene dicobalamin", and only under forcing conditions to produce the intermediate 4-bromobutyl cobalamin (46). This result suggests a substantial "neighboring group effect" for B_{12s} not seen in the cobaloxime system.

Results similar to B_{12s} are found for the reaction of an uncharged rhodium(I) nucleophile with 1,5-dibromopentane (44). The result here is even more dramatic, in that only the dirhodium product was obtained, even under forcing conditions. These puzzling observations are under continuing investigation.

Electrophilic Pathways. Most of the effort in this area has gone into a study of the metal-carbon bond cleavage process rather than formation. Many of the trans-alkylation reactions do also involve formation of a new metal-carbon bond, however, and a brief consideration in the context of the present subject is thus in order. A number of specific reactions are illustrated in Scheme IV.

Scheme IV. Heterolytic Reactions

$$R(Co^{III}) \xrightarrow{Hg^{2+}} (Co^{III})^+ + RHg^+ \quad [22]$$

$$R(Co^{III}) \xrightarrow{CH_3Hg^+} \text{no reaction or very slow reaction} \quad [23]$$

$$R(Co^{III}) \xrightarrow{Br_2} \text{Oxidative cleavage } via\ R(Co^{IV})^+ \quad [24]$$

$$(H_2O)_5Cr^{III}\text{-}R^{2+} \xrightarrow{Hg^{2+}} Cr^{3+}_{aq} + RHg^+ \quad [25]$$

$$(H_2O)_5Cr^{III}\text{-}R^{2+} \xrightarrow{R'Hg^+} Cr^{3+}_{aq} + RHgR' \quad [26]$$

$$(H_2O)_5Cr^{III}\text{-}R^{2+} \xrightarrow{Br_2} RBr + Cr^{3+} + Br^-\ (not\ CrBr^{2+}) \quad [27]$$

$$(H_2O)_5Cr^{III}\text{-}R^{2+} \xrightarrow{IBr} Br^- + Cr^{3+} + RI\ (not\ RBr) \quad [28]$$

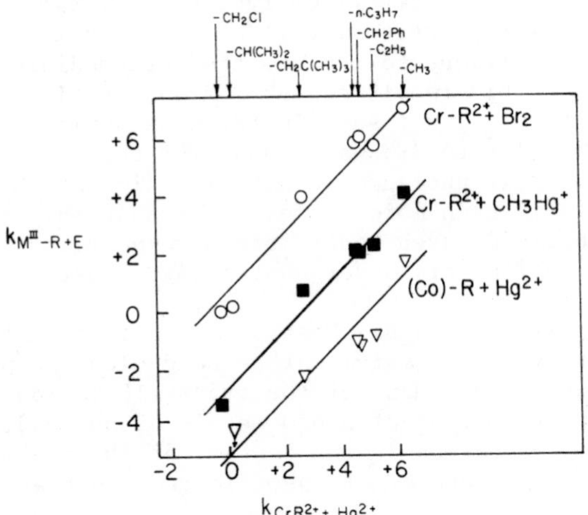

Figure 2. Linear free-energy correlations for electrophilic reactions. The rate constants for different reaction series are plotted on a logarithmic scale against log k for the series $(H_2O)_5Cr$–R^{2+} + Hg^{2+}. Electrophilic reactivity: $K_{M^{III}-R+E}$ vs. $k_{Cr^{III}-R^{2+}+Hg^{2+}}$.

The halogen cleavages of organochromium complexes appear to be straightforward electrophilic processes (47) in contrast to the oxidative processes seen for R(Co) systems with bromine. The mercury (II) reactions have been widely studied and have been summarized for cobalt (48) and chromium (11) systems. Reaction [22] proceeds with inversion at the alpha carbon atom, a probably consequence of steric interactions (48). The stereochemistry of these other electrophilic reactions remain unknown, although the kinetic effects are such (11,47) that inversion seems the probable course in each case.

An interesting linear-free-energy relationship is found for all of these reactions. Figure 2 depicts the rate constant for each reaction plotted (on a logarithmic scale) against that for one series chosen as a standard (Reaction [25] constitutes a good standard since rates were measured for the largest number of R groups). The quality of the correlation is high, suggesting a similar S_E2 mechanism (and stereochemical inversion?) in each case. A thorough review of electrophilic cleavage reactions has recently been published (49).

Literature Cited

1 Schrauzer, G. N., Accounts Chem. Res., (1968) 1, 97.
2 Lovechio, F. V., Gore, E. S., and Busch, D. H., J. Am. Chem. Soc., (1974) 96, 3109.
3 Dodd, D. and Johnson, M. D., J. Organomet. Chem., (1973) 1, 52.
4 Pratt, J. M. and Craig, P. J., Adv. Organomet. Chem., (1977) 99, 5953.
5 Anet, F. A. L., Can. J. Chem., (1959) 37, 58.
6 Dodd, D. and Johnson, M. D., J. Chem. Soc. A., (1968) 34.
7 Schmidt, W., Swinehart, J. H., and Taube, H., J. Am. Chem. Soc., (1971) 93, 1117.
8 Ardon, M., Woolmington, K., and Pernick, A., Inorg. Chem., (1971) 10, 2812.
9 Kochi, J. K. and Buchanan, D. B., J. Am. Chem. Soc., (1965) 87, 859.
10 Espenson, J. H. and Williams, D. A., J. Am. Chem. Soc., (1974) 96, 1008.
11 Leslie, J. P., II and Espenson, J. H., J. Am. Chem. Soc., (1976) 98, 4839.
12 Van Den Bergen, A. M., Murray, K. S., Sheahan, R. M., and West, B. O., J. Organomet. Chem., (1975) 90, 299.
13 Samuels, G. J. and Espenson, J. H., unpublished results.
14 Sneeden, R. P. A., "Organochromium Compounds", Academic Press, New York, 1975.
15 Goedkin, V. L., Peng, S.-M., and Park, Y., J. Am. Chem. Soc., (1974) 96, 284.
16 Taube, R., Drevs, H., and Duc-Hiep, T., Z. für Chem., (1969) 9, 115.

17 Floriani, C. and Calderazzo, F., J. Chem. Soc. A, (1971) 3665.
18 Schrauzer, G. N. and Deutsch, E. A., J. Am. Chem. Soc., (1969) 91, 3341.
19 Halpern, J. and Maher, J. P., J. Am. Chem. Soc., (1965) 87, 5361.
20 Castro, C. E. and Kray, W. C., Jr, J. Am. Chem. Soc., (1963) 85, 2768.
21 Kochi, J. K. and Mocadlo, P. E., J. Am. Chem. Soc., (1966) 88, 4094.
22 Kochi, J. K. and Davis, D. D., J. Am. Chem. Soc., (1964) 86, 5264.
23 Kochi, J. K. and Powers, J. W., J. Am. Chem. Soc., (1970) 92, 137.
24 Walling, C., Cooley, J. H., Ponaras, A. A., and Racah, E. J., J. Am. Chem. Soc., (1966) 88, 5363.
25 Carlsson, D. J. and Ingold, K. U., J. Am. Chem. Soc., (1968) 90, 7047.
26 Espenson, J. H. and Leslie, J. P., II, J. Am. Chem. Soc., (1974) 96, 1954.
27 Cohen, H. and Meyerstein, D., J. Chem. Soc., Chem. Commun., (1972) 320.
28 Cohen, H. and Meyerstein, D., Inorg. Chem., (1974) 13, 2434.
29 Cooper, T. A., J. Am. Chem. Soc., (1973) 95, 4158.
30 Kochi, J. K. and Mocadlo, P. E., J. Org. Chem., (1965) 30, 1134.
31 Hyde, M. R. and Espenson, J. H., J. Am. Chem. Soc., (1976) 98, 4463.
32 Schmidt, W., Swinehart, J. H., and Taube, H., J. Am. Chem. Soc., (1971) 93, 1117.
33 Espenson, J. H. and Martin, A. H., J. Am. Chem. Soc., (1977) 99, 5953.
34 Geodkin, V. L., Peng, S.-M., and Park, Y., J. Am. Chem. Soc., (1974) 96, 284.
35 Mok, C. Y. and Endicott, J. F., J. Am. Chem. Soc., (1978) 100, 123.
36 Marzilli, L. G., Marzilli, P. A., and Halpern, J., J. Am. Chem. Soc., (1970) 92, 5652.
37 Bläser, H. and Halpern, J., unpublished results, as cited by Halpern, J., Ann. N.Y. Acad. Sci., (1974) 239, 2.
38 Castro, C. E., J. Am. Chem. Soc., (1964) 86, 2310.
39 Roche, T. J. and Endicott, J. F., Inorg. Chem., (1974) 13, 1576.
40 Espenson, J. H. and Heckman, R. A., unpublished observations.
41 Schrauzer, G. N., Sibert, J. W., and Windgassen, R. J., J. Am. Chem. Soc., (1968) 90, 6681.
42 Golding, J. T., Kemp, T. J., Sellers, P. J., and Nocchi, T., J. Chem. Soc. Dalton, (1977) 1266.
43 Allen, R. J. and Bunton, C. A., Bioinorg. Chem., (1976) 3, 241.

44 Collman, J. P. and MacLaury, M. R., J. Am. Chem. Soc., (1974) 96, 3019.
45 Espenson, J. H. and Chao, T-H., Inorg. Chem., (1977) 16, 2553.
46 Smith, E. L., Merryon, L., Muggleton, P. W., Johnson, A. W., and Shaw, N., Ann. N.Y. Acad. Sci., (1964) 112, 564.
47 Espenson, J. H. and Samuels, G. J., J. Organomet. Chem., (1976) 113, 143.
48 Fritz, H. L., Espenson, J. H., Williams, D. A., and Molander, G. A., J. Am. Chem. Soc., (1974) 96, 2378.
49 Johnson, M. D., Accounts Chem. Res., (1978) 11, 57.

Discussion

J. M. WOOD (University of Minnesota): Several years ago we synthesized that tetramethylene B_{12} dimer. We wanted to synthesize the bromide intermediate because we wanted to attach it to safrose to use as an affinity column label for purifying enzymes. We were disappointed that we couldn't isolate it, but Professor R. Abeles [Brandeis University] pointed out that with increased concentration of B_{12} in solution, the extinction coefficienct alters; he suggested about 6 years ago that B_{12} has a tendency to dimerize in concentrated solution by forming hydrogen bonds between propionamide groups. That would explain why you principally obtain that product, but if you work in dilute solution you can isolate some of the tetramethylene bromide intermediate.

EPSENSON: Yes, that one can be isolated. Smith did in fact report that [Ann. N. Y. Acad. Sci., (1964) 112, 564] although we have not gone back and worked with that particular one.

W. P. RIDLEY (University of Minnesota): I'm very interested in the reactivity of sulfur [as RS-] with the alkylcobalt complexes. This is extremely important biologically. You listed it as a nucleophilic attack. The evidence with methylcobalamin is that it's a radical mechanism.

ESPENSON: I don't think Schrauzer would agree with that. He gets RSR' products, but not to the extent that you can take as evidence for electrophilic attack.

RIDLEY: No, apparently there is a lag period in the reactivity of these thiols with methylcobalamin, and that's taken to be evidence for the generation of a radical species which is the reacting group.

ESPENSON: That's an interesting result; has that been published?

RIDLEY: Yes, it's in Biochem. Biophys. Acta [Frick *et al.* (1976) 428, 808].

WOOD: There is no change in the rate of that reaction over a pH range from 7 to 14, so as you go to thiolate anion in the system, you don't see any differences in these reactions. We could not repeat Schrauzer's experiments.

ESPENSON: I have no origianl evidence to add to the RS- case cited in the literature. Schrauzer has shown that in low pH or neutral solution you get substitution. At high pH you get the thiolate anion; you get nucleophilic displacement. I don't think he has retracted that so you and he are at odds on this point.

C. P. DUNNE (California State University, Long Beach): I have to support Professor Wood in this controversy. I found that in the system Schrauzer used, a control experiment of the acetate buffer and cobalt complex will generate an enormous amount of so-called radical products. This is one of the major problems that Schrauzer has. He's not really studying the reactions of B_{12} compounds when you eliminate the possibility of radical reactions.

M. KRONSTEIN (Manhattan College, New York): You showed complexes of cobalt and chromium with organics, and at the end of all your reactions you always got separation of R and the metal. How did you identify those, and how does that separated R compare with the R you introduced?

ESPENSON: The R which I showed as R^+, R·, and R^- is not free in that form. R^- would be attached to the attacking reagent and organomercurial and organic bromide, and so on, and it's identified as such by isolating those compounds chemically. Likewise, the metal complex products (chromium(III) or chromium(II), or cobalt(II), cobalt(III), or cobalt(I)) are identified chemically either by their absorption spectra or by separation and chemical identification. Did I correctly understand your question?

KRONSTEIN: I have the opposite experience. If I introduce metal oxides into low organic polymer fractions which continue polymerizing, the metal doesn't go off. Your idea is of importance because there is a big problem that environmentalists face. For example, some antifouling paints or protective antifouling materials in water can give up toxic compounds combining the metal groups and the organic groups. If the metal and the organic group enter the water separately, these would not be as toxic. If metal and R don't separate, the water is made much more toxic. That's why that question would be interesting to study further. [Eds: see chapter by M. L. Good et al.]

RECEIVED August 22, 1978.

15

Unstable Organometallic Intermediates in a Protic Medium

JAMES H. WEBER and MARK W. WITMAN[1]

University of New Hampshire, Parsons Hall, Durham, NH 03824

For many years (1, 2, 3) we have studied the dealkylation reactions of organocobalt compounds (RCo(chel)). The three most-studied processes are nucleophilic attack, electrophilic attack, and homolytic bond cleavage. Typical examples are given in eqn. 1-3. In the nucleophilic attack (eqn. 1) the RH product occurs because of proton extraction by a carb-

$$RCo^{III}(chel) \xrightarrow[\text{e.g. } RS^-,\ CN^-]{Nu^-} NuCo^{III}(chel) + RH \qquad (1)$$

$$RCo^{III}(chel) \xrightarrow[\text{e.g. } Hg^{2+},\ Pd^{2+},\ Tl^{3+}]{E^+} Co^{III}(chel)^+ + RE \qquad (2)$$

$$RCo^{III}(chel) \xrightarrow[\text{anaerobic}]{h\nu,\ R^1SH,\ Cr^{2+},\ Sn^{3+}} R\cdot + Co^{II}(chel) \qquad (3)$$

anionic intermediate. The transient R· product of eqn. 3 is converted to many products depending on the nature of R and the attacking group. Examples are organic products like alkanes, alkenes, dimerized alkanes, and organometallic species such as RCr^{III} and RSn^{IV}.

From the environmental point of view, it is important to note that the electrophilic (eqn. 2) and free radical (eqn. 3) reactions occur with methylcobalamin (4, 5). It is possible that methylcobalamin is involved in microbial methylation of lead (6, 7) by the electrophilic attack mechanism (eqn. 2). For these reasons this paper reports the reactions of the dimethylcobalt complex (8) $(CH_3)_2CoL$ (Structure I) with Zn^{2+}, Cd^{2+} and Pb^{2+} electrophiles. Because we do not know the alkyl donor/alkyl acceptor ratio in the environment, we will discuss a variety of reactant ratios.

Results and Discussion

Although Zn^{2+}, Cd^{2+} and Pb^{2+} are known to be unreactive toward a variety of RCo(chel) (9, 10), they are surprisingly reactive toward the $(CH_3)_2CoL$ carbanion donor. Our data based on these reactions supports

[1]Current Address: Mobay Chemical Corporation, New Martinsville, W. Virginia 26155

0-8412-0461-6/78/47-082-247$05.00/0

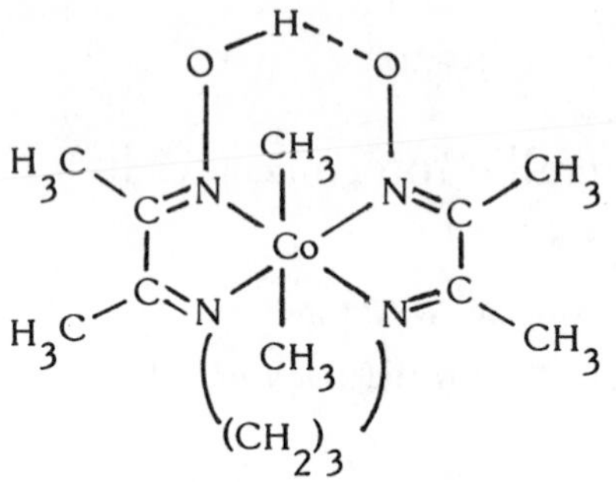

the formation of intermediate organometallic products of the type CH_3M^+ and $(CH_3)_2M$ which are entirely unexpected in view of the protic solvent employed.

In contrast to the reactions of the monoalkyl derivatives RCo(chel) with Hg^{2+} which demonstrate a simple 1:1 stoichiometry (eqn. 4), the

$$RCo(chel) + Hg^{2+} \longrightarrow RHg^+ + Co(chel)^+ \qquad (4)$$

anaerobic reactions of $(CH_3)_2CoL$ with Zn^{2+}, Cd^{2+} and Pb^{2+} are quite unique. Preliminary experiments revealed that two very different reactions occurred depending upon whether the metal ion or complex was in excess. In the presence of excess metal ion an instantaneous reaction was observed. With excess complex, however, the instantaneous reaction was followed by a slower reaction which was complete in a matter of minutes. The biphasic nature of this reaction was confirmed using the stopped-flow kinetics technique. These two classes of reactions will be discussed separately beginning with the experiments using excess metal ion.

Reactions with Excess Zn^{2+}, Cd^{2+} and Pb^{2+}. Upon reaction of $(CH_3)_2CoL$ with an equimolar or an excess of Zn^{2+}, Cd^{2+}, or Pb^{2+} in i-C_3H_7OH the ultraviolet-visible spectrum changes instantaneously. These reactions can be monitored most conveniently by observing the disappearance of the visible maximum at 412 nm (ε = 8380) which is characteristic of $(CH_3)_2CoL$. The complex product ion CH_3CoL^+ was identified by comparison of the spectra of the final solutions with that of a solution made from an analyzed sample of $[CH_3Co(L)H_2O]^+$. From a knowledge of the molar absorptivity of the product ion at 463 nm, (ε = 2120) it is evident that $(CH_3)_2CoL$ is quantitatively converted to CH_3CoL^+ in these reactions. Furthermore, this monomethyl product is indefinitely stable in the presence of excess Zn^{2+}, Cd^{2+}, or Pb^{2+} in accord with the results of Magnuson (10) for the reactions of CH_3CoL^+ with the same metal ions in aqueous solution.

The stoichiometries of these fast reactions were determined by spectral titration. This procedure was complicated by the overlapping of the second slower phase of the reaction which was fairly rapid in the early stages. As a consequence a rapid titration technique was devised to minimize any error incurred due to the competing reaction. The method that was found to be most convenient was to deliver the M^{2+} titrant directly into a spectrophotometric cell, which contained an appropriate amount of complex, using a one microliter syringe. In this manner volume corrections were negligible and the titration could be carried out quite rapidly. The titrations for Zn^{2+}, Cd^{2+} and Pb^{2+} yield end points close to a molar ratio (M^{2+}/complex) equal to 1.0 (Figure 1). The fact that the molar ratios are slightly low reflects the contribution due to a competing reaction (see below).

An interesting feature of this work is the fate of the liberated methyl group. The fast demethylation reactions described above are followed by relatively slow reactions in which methane is evolved. This is in contrast to the demethylation of $(CH_3)_2CoL$ by H_3O^+ in which the methane is liberated instantaneously and quantitatively according to eqn. 5. In fact, for the Cd^{2+} and Pb^{2+} reactions the theoretical amount of

$$(CH_3)_2CoL + H_3O^+ \xrightarrow{\text{fast}} CH_3CoL^+ + CH_4 \tag{5}$$

methane was obtained only after *ca.* 200 hr! The methane could be quantitatively recovered at any *time* during the course of the reaction, however, by the addition of H_3O^+.

On the basis of the slow evolution of methane, the sensitivity of the intermediate species to H_3O^+, and the indicated 1:1 stoichiometry, we postulate that intermediates of the type CH_3Zn^+, CH_3Cd^{2+}, and CH_3Pb^+ are formed in these reactions according to eqn. 6. Further support for

$$(CH_3)_2CoL + M^{2+} \xrightarrow{\text{fast}} CH_3CoL^+ + CH_3M^+ \tag{6}$$

these species comes from a study of the kinetics of their decomposition.

The rate of methane evolution was monitored by GLC for the reactions of $(CH_3)_2CoL$ with the three metal ions. Plots of $\ln(C_\infty - C)$ versus time were linear over at least four half-lives. The rates were independent of excess metal ion concentration, which precudes the possibility of redistribution equilibria of the type shown in eqn. 7. These results are consistent with a process in which the organometallic inter-

$$2\,CH_3M^+ \longrightarrow (CH_3)_2M + M^{2+} \tag{7}$$

mediate is decomposed by $\underline{i}$-C_3H_7OH (eqn. 8). Because the solvent is in

$$CH_3M^+ + \underline{i}\text{-}C_3H_7OH \longrightarrow CH_4 + M(\underline{i}\text{-}C_3H_7O)^+ \tag{8}$$

a large excess over the CH_3M^+ species, a pseudo-first-order rate law is followed.

A summary of the kinetic data for the methane evolution reactions is presented in Table I. The relative stability $CH_3Cd^+ >> CH_3Zn^+$ agrees with the known reactivity of saturated organometallic compounds of the Group IIB metals, Zn > Cd > Hg (11). In addition the reactivity of organometallic compounds of the type RM^+ generally correlates with the ionic character of the carbon-metal bond; both decreasing in the order

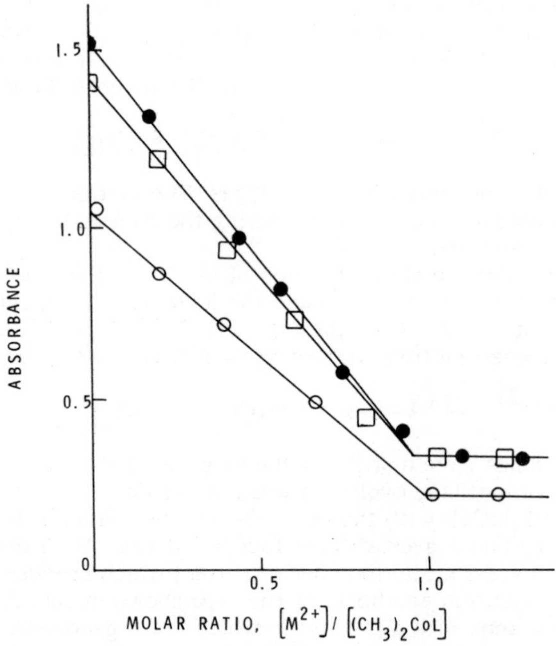

Figure 1. Representative spectral titrations of the fast reaction of $(CH_3)_2CoL$ with Zn^{2+}, ●; Cd^{2+}, □; and Pb^{2+}, ○.

Li > Mg > Zn > Cd > Hg (12). Consequently, the stability of CH_3Pb^+ observed here ($CH_3Pb^+ \sim CH_3Cd^+ > CH_3Zn^+$) agrees with the known covalent character of carbon-lead bonds (13).

Table I. Summary of Kinetic Data for Methane Evolution During the Reactions of $(CH_3)_2CoL$ with Excess Zn^{2+}, Cd^{2+}, and Pb^{2+}.

Electrophile	$10^6 k_{obs}$ (sec^{-1})	$t_{1/2}$ (hr)
Zn^{2+}	128	1.5
Cd^{2+}	3.6	52.6
Pb^{2+}	3.2	60.1

Additional evidence for the formation of intermediates of the type CH_3M^+ comes from the accelerating effect of CH_3OH on the methane evolution reactions. The reactions of $(CH_3)_2CoL$ with excess Zn^{2+}, Cd^{2+}, and Pb^{2+} in solutions containing 20% (v/v) CH_3OH resulted in a considerably faster rate of CH_4 evolution. The fact that the reactions were complete in less than one hour is in agreement with the protonolysis of CH_3M^+ by the more acidic CH_3OH (eqn. 9). A similar reaction in which

$$CH_3M^+ + CH_3OH \longrightarrow CH_4 + M(OCH_3)^+ \qquad (9)$$

$(CH_3)_2CoL$ was reacted with excess $CdCl_2$ in anhydrous methanol resulted in the rapid formation of a white precipitate together with the simultaneous evolution of CH_4. This is consistent with a process (eqn. 10) in which

$$(CH_3)_2CoL + CdCl_2 \longrightarrow CH_3Cd^+ + CH_3CoL^+ \qquad (10a)$$

$$CH_3Cd^+ + Cl^- \xrightarrow{CH_3OH} CH_4 + CdCl(OCH_3) \qquad (10b)$$

unstable CH_3Cd^+ formed in reaction (10a) reacts with CH_3OH in the presence of Cl^- to form solid $CdCl(OCH_3)$ (eqn. 10b). This latter product was filtered from the reaction mixture, washed with CH_3OH, and characterized.

The apparent stability of the CH_3M^+ intermediates is somewhat surprising in view of the protic nature of the solvent. Solvation of the CH_3M^+ species by i-C_3H_7OH presumably contributes to the stabilization of these organometallic intermediates, however, since oxygen donors are known to stabilize alkyl zinc and alkyl cadmium bonds (11). The isolation of stable alkyl zinc (14) and alkyl cadmium alkoxides (15) are good examples of this effect.

Attempts to isolate the CH_3M^+ intermediates using stabilizing ligands such as Br^-, I^-, and bipy were unsuccessful. Despite this failure, however, the demonstrated 1:1 stoichiometry, the methane evolution data, and other evidence (see below) strongly suggest that CH_3M^+ species are indeed intermediates in these reactions.

Reactions with Excess $(CH_3)_2CoL$. The reactions of excess $(CH_3)_2CoL$ with Zn^{2+}, Cd^{2+}, and Pb^{2+} in i-C_3H_7OH are markedly different from the reactions using excess metal ion. Instead of a single instantaneous reaction a two-step reaction was observed in which an initial fast step was followed by a relatively slow second reaction. The stoichiometries of these overall biphasic reactions were determined by spectral titration of $(CH_3)_2CoL$ with standard solutions of appropriate metal ions. The same technique employed for the titration of the fast 1/1 reactions was utilized with the exception that the reactions were allowed to proceed to completion. The results of these titrations for Zn^{2+}, Cd^{2+}, and Pb^{2+} are presented in Figure 2.

It is readily apparent that the end points for all three metal ions are close to a molar ratio (M^{2+}/complex) equal to 0.5; i.e. one mole of added metal ion reacts with two moles of $(CH_3)_2CoL$. This is consistent with a process in which the relatively stable CH_3M^+ intermediates formed in the initial fast reaction (eqn. 11), react with a second mole of $(CH_3)_2CoL$

$$(CH_3)_2CoL + M^{2+} \xrightarrow{\text{fast}} CH_3CoL^+ + CH_3M^+ \quad (11)$$

in a subsequent slower reaction (eqn. 12). The low molar ratio observed

$$(CH_3)_2CoL + CH_3M^+ \xrightarrow{\text{slow}} CH_3CoL^+ + (CH_3)_2M \quad (12)$$

for the Zn^{2+} titration is in agreement with the faster (Table I) protonolysis of CH_3Zn^+ by i-C_3H_7OH (eqn. 8). Thus the reaction is catalytic in the sense that Zn^{2+} is continuously regenerated and reacts further with $(CH_3)_2CoL$ (eqn. 11).

Support for the scheme outlined here comes from experiments conducted by Espenson et al. (16) on the methyl transfer reactions of $(CH_3)_2CoL$ and similar complexes with the analogous CH_3Hg^+ species (eqn. 13). These reactions occur with the indicated 1:1 stoichiometry and

$$(CH_3)_2CoL + CH_3Hg^+ \longrightarrow CH_3CoL^+ + (CH_3)_2Hg \quad (13)$$

proceed to completion. Reaction also occurs with $C_6H_5Hg^+$ (and probably with any other mono-organomercury derivative) as the acceptor, and with $(C_6H_5)_2CoL$ and $(C_7H_7)_2CoL$ as alkyl or aryl donor complexes.

The proposed dimethylzinc, dimethylcadmium and dimethyllead products (eqn. 12) are extremely reactive, and are readily decomposed by compounds (including alcohols) which contain active hydrogen (11, 13). As a consequence, the rapid formation of methane was anticipated in these reactions. In fact, this was observed upon reacting $(CH_3)_2CoL$ with Cd^{2+} or Pb^{2+} in a molar ratio (M^{2+}/complex) equal to 0.5. This observation further corroborates the reaction scheme.

The reaction of $(CH_3)_2CoL$ with Cd^{2+} under these conditions results in the methane evolution profile shown in Figure 3. Two distinct steps are observed. In the first step 50% of the theoretical amount of methane (determined by the acid hydrolysis of $(CH_3)_2CoL$) is evolved in a matter of minutes. This is followed by a slow second step which liberates the remaining methane over a period of 150 hr. The methane evolution pattern

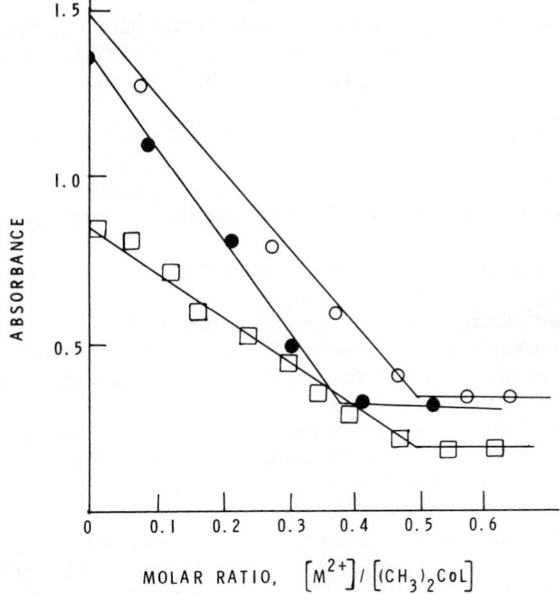

Figure 2. Representative spectral titrations for the overall reaction of $(CH_3)_2CoL$ with Zn^{2+}, ●; Cd^{2+}, □; and Pb^{2+}, ○.

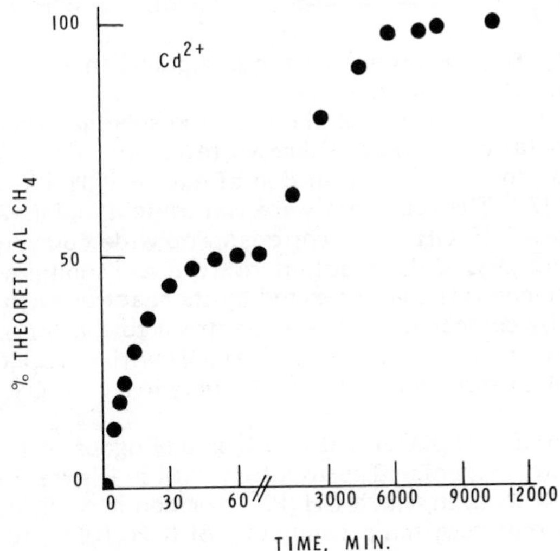

Figure 3. Rate of methane evolution for the reaction of $(CH_3)_2CoL$ with Cd^{2+} ($[Cd^{2+}]/[(CH_3)_2CoL]$ = 0.5)

seen here is consistent with a process in which the $(CH_3)_2Cd$ product formed in eqn. 12 immediately reacts with i-C_3H_7OH to form the well-known tetramer $[CH_3Cd(\underline{i}\text{-}OC_3H_7)]_4$ (15), and CH_4 (eqn. 14). The more

$$(CH_3)_2Cd \xrightarrow[\text{fast}]{\underline{i}\text{-}C_3H_7OH} 1/4\,[CH_3Cd(\underline{i}\text{-}OC_3H_7)]_4 + CH_4 \quad (14)$$

stable tetramer is then slowly decomposed by i-C_3H_7OH to liberate the remaining methyl group (eqn. 15). This interpretation is in agreement with

$$1/4\,[CH_3Cd(\underline{i}\text{-}OC_3H_7)]_4 \xrightarrow[\text{slow}]{\underline{i}\text{-}C_3H_7OH} Cd(\underline{i}\text{-}OC_3H_7)_2 + CH_4 \quad (15)$$

recent kinetic and NMR studies of the reactions of $(CH_3)_2Cd$ and $(C_2H_5)_2Cd$ with a variety of alcohols ROH which show that the $[CH_3Cd(OR)]_4$ products are formed much more rapidly than the $Cd(OR)_2$ products (17).

Additional evidence for the proposed reaction scheme is provided in Figure 4. In this experiment $(CH_3)_2CoL$ was titrated with a standardized Cd^{2+} solution in the usual manner while simultaneously monitoring the quantity of methane evolved. The amount of methane liberated after each addition of titrant was determined by GLC, and compared to the theoretical value which was measured by the acid hydrolysis of $(CH_3)_2CoL$. Figure 4 clearly shows the expected 2/1 (complex/Cd^{2+}) stoichiometry (eqn. 16 and 17). In addition it is evident that 50% of the expected amount

$$(CH_3)_2CoL + Cd^{2+} \longrightarrow CH_3CoL^+ + CH_3Cd^+ \quad (16)$$

$$(CH_3)_2CoL + CH_3Cd^+ \longrightarrow CH_3CoL^+ + (CH_3)_2Cd \quad (17)$$

of methane is liberated upon reaching the end point, thus proving the transient existence of $(CH_3)_2Cd$.

Perhaps the best evidence for the reaction scheme proposed here comes from the isolation of the well-known tetramer $[CH_3Cd(\underline{i}\text{-}OC_3H_7)]_4$ as a product of the reaction of excess $(CH_3)_2CoL$ with Cd^{2+} (eqn. 16 and 17). The reactions were run using a molar ratio $[Cd^{2+}]$/complex = 0.5 in i-C_3H_7OH. The white isopropoxide compound was recovered by freeze-drying the reaction mixture and subliming the residue. The tetramer was characterized by its reaction with 1M $HClO_4$ to yield CH_4 and by comparing its mass spectrum to that of an authentic sample prepared from the reaction of i-C_3H_7OH with $(CH_3)_2Cd$. The molecular ion peak at m/e 744 confirms the polymeric nature of this compound.

The reaction of $(CH_3)_2CoL$ with Pb^{2+} is analogous to the reaction with Cd^{2+} in certain respects. The titration data in Figure 2 is again consistent with a process in which CH_3Pb^+, formed in an initial fast reaction (eqn. 18), reacts with a second mole of $(CH_3)_2CoL$ to

$$(CH_3)_2CoL + Pb^{2+} \xrightarrow{\text{fast}} CH_3CoL^+ + CH_3Pb^+ \quad (18)$$

form $(CH_3)_2Pb$ (eqn. 19). Although divalent organolead compounds are

$$(CH_3)_2)CoL + CH_3Pb^+ \xrightarrow{\text{slow}} CH_3CoL^+ + (CH_3)_2Pb \qquad (19)$$

typically very unstable and frequently disproportionate into organolead(IV) species and lead metal (eqn. 20) (13), this reaction was not observed here.

$$2\,R_2Pb \longrightarrow R_4Pb + Pb^{\circ} \qquad (20)$$

Instead, the reaction of $(CH_3)_2CoL$ with Pb^{2+} (Pb^{2+}/complex = 0.5) resulted in the methane evolution profile shown in Figure 5, and the simultaneous formation of a light-brown precipitate. The fact that only 50% of the methane is ultimately observed suggests that the precipitate is a stable $[CH_3Pb\,(i\text{-}OC_3H_7)]$ compound formed by the protonolysis of $(CH_3)_2Pb$ (eqn. 21). Presumably, the $[CH_3Pb\,(\underline{i}\text{-}OC_3H_7)]$ product is

$$(CH_3)_2Pb + \underline{i}\text{-}C_3H_7OH \longrightarrow CH_3Pb\,(\underline{i}\text{-}OC_3H_7) + CH_4 \qquad (21)$$

polymeric in analogy to $[CH_3Cd\,(\underline{i}\text{-}OC_3H_7)]_4$ since lead alkoxides reportedly exist as polymeric chains linked by oxygen atoms (18). The addition of H_3O^+ to the solution results in immediate evolution of the remaining methane and dissolution of the precipitate (eqn. 22), thus

$$CH_3Pb\,(\underline{i}\text{-}OC_3H) + H_3O^+ \longrightarrow CH_4 + Pb^{2+} + \underline{i}\text{-}C_3H_7OH \qquad (22)$$

lending further support to this interpretation.

Unfortunately, elemental analysis of the isolated precipitate did not confirm the presence of $CH_3Pb\,(i\text{-}OC_3H_7)$. The analysis, which indicated that N was present, suggested that the product was contaiminated with CH_3CoL^+. This was confirmed by dissolving the precipitate in H_2O or H_3O^+ and comparing the spectrum of the resulting orange colored solution to that of CH_3CoL^+. Cold washing with $i\text{-}C_3H_7OH$ did not improve the analysis indicating that CH_3CoL^+ is tightly held. The fact that analogous precipitates were not observed in the reactions of $(CH_3)_2CoL$ with excess Pb^{2+}, or upon mixing solutions of CH_3CoL^+ and Pb^{2+} precludes the presence of Pb^{2+} adducts as have been isolated from the reactions of Hg^{2+} with RCo(salen) complexes (19). Despite these negative results, however, the formation of CH_4 and $i\text{-}C_3H_7OH$ upon dissolution of the precipitate in H_2O^+ suggests that $CH_3Pb\,(\underline{i}\text{-}OC_3H_7)$ is at least a minor product of these reactions.

Kinetics of the Reactions of $(CH_3)_2CoL$ with Excess Zn^{2+} and Cd^{2+}. The kinetic behavior of the 1:1 reaction (eqn. 6) of $(CH_3)_2CoL$ with Zn^{2+} and Cd^{2+} was investigated by monitoring the decrease in absorbance due to the complex at 412 nm. These reactions exhibited half-lives on the order of a few milliseconds, and were studied using the stopped-flow technique. The corresponding reaction with Pb^{2+} was not studied due to the unexplained extreme sensitivity of the reaction to oxygen, which resulted in the formation of brown precipitate.

Plots of the data in the form $\ln(A\text{-}A_\infty)$ versus time were invariably linear over at least four half-lives for both metal ions. Since the metal ions were in excess over the complex, the observed kinetics were pseudo-

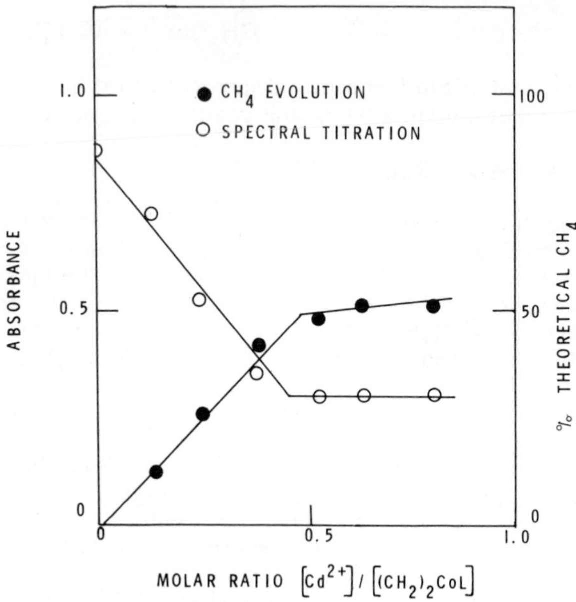

Figure 4. Simultaneous spectral titration and methane evolution study for the reaction of $(CH_3)_2CoL$ with Cd^{2+} ($[Cd^{2+}]/[(CH_3)_2CoL] = 0.5$)

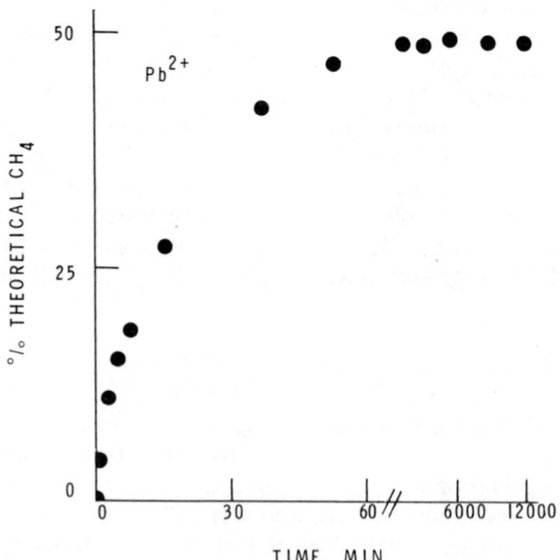

Figure 5. Rate of methane evolution for the reaction of $(CH_3)_2CoL$ with Pb^{2+} ($[Pb^{2+}]/[(CH_3)_2CoL] = 0.5$)

Table II. Summary of Second-Order Rate Constants as a Function of Metal Ion Concentration

Zinc(II)		Cadmium(II)	
10^5 Zn^{2+} (M)	$10^{-5}k_{obs}$/ Zn^{2+} (M^{-1} sec^{-1})	10^5 Cd^{2+} (M)	$10^{-5}k_{obs}$/ Cd^{2+} (M^{-1} sec^{-1})
7.25	11.7	3.75	28.7
10.7	10.7	5.62	25.3
12.9	10.1	7.50	22.6
16.5	9.2	15.0	15.4
22.5	8.0		

first-order. A least squares analysis of the slopes of these lines yielded values for the observed rate constants (k_{obs}) which ranged from 85 to 180 sec^{-1} for Zn^{2+}, and 108 to 231 sec^{-1} for Cd^{2+}. The metal ion dependence of the reactions was determined by varying $[Zn^{2+}]$ and $[Cd^{2+}]$ from 7.25 x 10^{-5} to 2.25 x 10^{-4} M while maintaining the ionic strength constant at 1.8 x 10^{-3} M. The apparent second-order rate constants decreased markedly with increasing metal ion concentration throughout the range studied as can be seen in Table II.

A reasonable explanation of this phenomenon is that there is a rapid pre-equilibrium between the reactants and a substitutionally labile 1:1 adduct that is being saturated as the concentration of the metal ion is increased (eqn. 23). This process is followed by rate-determining

$$R_2CoL + M^{2+} \underset{k_b}{\overset{k_f}{\rightleftharpoons}} M^{2+} \cdot R_2CoL \qquad (23)$$

decomposition of the adduct to products (eqn. 24), where R = CH_3 in

$$M^{2+} \cdot R_2CoL \xrightarrow[\text{solvent}]{k_1} RM^+ + RCoLsolv^+ \qquad (24)$$

this scheme. If $K = k_f/k_b$ and $k_f, k_b >> k_1$, the rate law (eqn. 25) is derived (20). Plots of the reciprocal of the observed pseudo-first-order rate

$$\frac{-d[R_2CoL]}{dt} = \frac{k_1K[M^{2+}][R_2CoL]}{1 + K[M^{2+}]} \qquad (25)$$

constant versus the reciprocal of the metal ion concentration are linear in agreement with this scheme. The data indicating this linear relationship are presented in Figure 6. The kinetic and equilibrium parameters were extracted from the slopes ($1/k_1K$) and intercepts ($1/k_1$) of these lines and are summarized in Table III.

Table III. Summary of Kinetic and Equilibrium Data[a] for the Reaction of $(CH_3)_2CoL$ with Zn^{2+} and Cd^{2+}.

Electrophile	$10^{-3}K(M^{-1})$	$10^{-2}k_1(sec^{-1})$	$10^{-6}k_1K(M^{-1}sec^{-1})$
Zn^{2+}	3.8	3.8	1.4
Cd^{2+}	10.6	4.0	4.2

[a] At 22.2 C, μ = 1.8 x 10^{-3} ($LiClO_4 \cdot 3H_2O$) in $\underline{i}$-C_3H_7OH.

The proposed first-order decomposition of the adduct (eqn. 24) is not easily distinguished from a second-order reaction of metal ion with free R_2CoL (eqn. 26) since both processes lead to the same kinetics (20).

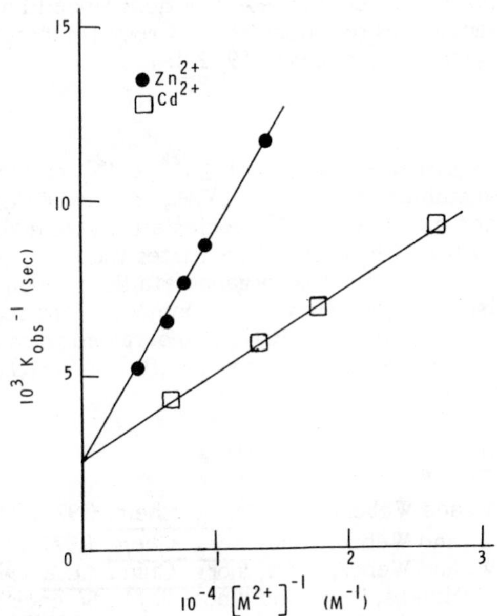

Figure 6. Variation of k_{obs}^{-1} with $[M^{2+}]^{-1}$ for the reactions of $(CH_3)_2CoL$ with excess Zn^{2+} and Cd^{2+}

$$R_2CoL + M^{2+} \longrightarrow RM^+ + RCoL^+ \quad (26)$$

However, we favor the former reaction (eqn. 24) in view of the nearly identical k_1 values observed for both the Zn^{2+} and Cd^{2+} reactions. If the leaving group RM^+ is loosely held in the adduct similar k_1 values are expected. The slight difference in $k_1[Cd^{2+}] > k_1[Zn^{2+}]$ may reflect the size difference and thus the lability of the leaving group in the order $RCd^+ > RZn^+$. Presumably, adduct decomposition occurs by way of a five-coordinate intermediate since CH_3 is known to promote the dissociation of trans ligands in octahedral cobalt complexes (20). Furthermore, five-coordinate alkyl cobalt intermediates are well known in the anation reactions of $[RCo(L)H_2O]^+$ and alkylcobaloximes (21). It is not surprising that adduct formation occurs in these reactions since 1:1 adducts were detected in the analogous reactions of the Group IIB metal ion Hg^{2+} with several monoalkylcobalt complexes (19, 22).

Conclusions

The reactions of $(CH_3)_2CoL$ with Zn^{2+}, Cd^{2+}, and Pb^{2+} in i-C_3H_7OH show unexpected stability for CH_3Zn^+, CH_3Cd^+, and CH_3Pb^+ in a protic medium. The fact that the CH_3M^+ species are stable enough to react with a second mole of $(CH_3)_2CoL$ accentuates their unexpected behavior. This chemistry demonstrates that organometallic compounds thought incapable of existing in the aquatic environment might occur and eventually be found. For example the $(CH_3)_4Pb$ obtained from Pb^{2+} by microbial processes (6, 7) might arise via the CH_3Pb^+ or $(CH_3)_2Pb$ intermediates observed in this work.

Literature Cited

1. Witman, M. W. and Weber, J. H., Inorg. Chem. (1976), 15, 2375.
2. Witman, M. W. and Weber, J. H., Inorg. Chem. (1977), 16, 2512.
3. Witman, M. W. and Weber, J. H., Inorg. Chim. Acta (1977), 23, 263.
4. Ridley, W. P., Dizikes, L. J., and Wood, J. M., Science (1977), 197, 329.
5. Dizikes, L. J., Ridley, W. P., and Wood, J. M., J. Am. Chem. Soc. (1978), 100, 1010.
6. Schmidt, U. and Huber, F., Nature (1976), 259, 157.
7. Wong, P. T. S., Chau, Y. K., and Luxon P. L., Nature (1975), 253, 263.
8. Witman, M. W. and Weber, J. H., Synth. React. Inorg. Met.-Org. Chem. (1977), 7, 143.
9. Agnes, G., Bendle, S., Hill, H. A. O., Williams, F. R., and Williams, R. J. P., Chem. Commun. (1971), 850.
10. Magnuson, V. E., Ph.D.Dissertation, University of New Hampshire, 1974.
11. Coates, G. E., Green, M. L., and Wade, K., "Organometallic Compounds", Vol. I, pp. 121-147, Methuen and Co. Ltd., London, 1967.
12. Swan, J. M. and Black, D. St. C., "Organometallics in Organic Syntheses", p. 39, Chapman and Hall, London, 1974.
13. Shapiro, H. and Frey, F. W., "The Organic Compounds of Lead", Interscience Publishers, New York, 1968.

14. Herold, R. J., Aggarwal, S. L., and Neff, V., Can. J. Chem. (1963), 41, 1368.
15. Coates, G. E. and Lauder, A., J. Chem. Soc. (A) (1966), 264.
16. Espenson, J. H., Fritz, H. L., Heckman, R. A., and Nicolini, C., Inorg. Chem. (1976), 15, 906.
17. Emptoz, G. and Huet, F., J. Organometal. Chem. (1974), 82, 139.
18. Amberger, E. and Grossich, R. H., Chem. Ber. (1965), 98, 3795.
19. Tauzher, G., Dreos, R., Costa, G., and Green, M., J. Organometal Chem. (1974), 81, 107.
20. Wilkins, R. G., "The Study of Kinetics and Mechanisms of Reactions of Transition Metal Complexes", pp. 26-31, Allyn and Bacon, Inc., Boston, 1974.
21. Costa, G., Mestroni, G., Tauzher, G., Goodall, D. M., Green, M., and Hill, H. A. O., Chem. Commun. (1970), 34.
22. Adin, A., and Espenson, J. H., Chem. Commun. (1971), 653.

Discussion

D. W. MARGERUM (Purdue University): When you had the metal precursor complex with the dimethylmetal system, did you have direct evidence of that complex? There is a third kinetic mechanism which could explain the data; that is, a steady-state, rate-limiting phenomenon rather than a pre-equilibrium.

WEBER: Did we get direct evidence for the dimethylcadmium adduct? We didn't. If that mechanism has the same rate law, I'm not sure we could differentiate without direct evidence which mechanism is correct. Would the mechanism you mention give the same kind of rate constant with different electrophiles?

MARGERUM: Yes, it could give exactly the same type of interpretation, but you would never be able to spot the intermediate, because it would never be there in appreciable concentrations. Your equilibrium constant suggests that you ought to be able to see some direct evidence of the complex.

WEBER: We do see slight spectral changes during the reaction; shifts in the absorption which might be indicative of that kind of adduct.

MARGERUM: That might be sufficient, and that ought to be adaptable to quantitatively check your proposal.

J. J. ZUCKERMAN (University of Oklahoma): I wasn't quite sure of the medium in which those reactions were carried out. Was this isopropanol or was that wet isopropanol?

WEBER: That was dry isopropyl alcohol.

ZUCKERMAN: I was wondering about the implications for protic media. Would the methylcobalt compound simply act as a hydrolysis catalyst for the metal ions to produce oxy-hydroxides or hydroxides which would then precipitate with the evolution of methane from the cobalt compound?

WEBER: You mean if we were doing the reactions in aquatic media.

ZUCKERMAN: Yes. Would the products be the hydroxide and methane from the cobalt?

WEBER: The dimethyl complex does react with water. The rate constant was about 80 $M^{-1}s^{-1}$ if you add water to isopropyl alcohol. I'm pretty sure it would react much more quickly with these metal ion electrophiles, and I predict that some kind of metal hydroxide would precipitate.

ZUCKERMAN: In the case of your alkyllead compound, is that a homogeneous system at that point or has that alkyllead alkoxide complex removed itself from the medium? Is there a wettability problem in that case?

WEBER: In both cases, with the cadmium and with the lead, after you get 50 percent of methane you get a white or off-white precipitate. In the case of cadmium, it's that methylcadmium isopropoxide tetramer. In the case of lead, there is also precipitate.

F. HUBER (University of Dortmund): Did you find out what happens on the oxidation of your monomethyllead product? I understood that there is a high sensitivity for oxidation of the product.

WEBER: We had no success in studying the reactions if we allowed oxygen into the reaction mixture. We got solid products of some kind. These might have been lead peroxides or lead oxides. We could not identify any of the lead precipitates by using x-ray powder patterns; there were no lead metal or lead oxides, for example, as a product.

HUBER: Did you make a normal elemental analysis of the product, _e.g._, C to Pb ratio?

WEBER: It did not analyze to anything that we could figure out.

HUBER: A second question. Is this a specific reaction of your cobalt compound, or did you try other methylcobalt species and react them with Pb^{2+}?

WEBER: No, we reacted only this particular dimethylcobalt compound. I'm convinced that with a couple of other ligands Dr. Espenson cited, you would get the same kind of reactions.

HUBER: Did you find the last 50% of your methane from the lead compound if you add acid?

WEBER: In the case of the cadmium reaction, you just wait about 100 or 150 hours to get the second half of the methane. In the lead reaction, if you wait even 200 hours, you do not get the second half at all. However, any time you put in acid you instantaneously get the second half.

HUBER: Isn't it possible that you have a disproportionation, and when you add acid you have a normal acidolysis of tetramethyllead?

WEBER: We looked for various products in this reaction. We considered the possibility of disproportionation; we found no tetramethyllead; we found no hexamethyldilead, which is another logical product; we found no lead metal. So we could find no sign of disproportionation in this reaction.

RECEIVED August 22, 1978.

16

Chloramine Equilibria and the Kinetics of Disproportionation in Aqueous Solution

EDWARD T. GRAY, JR.,[1] DALE W. MARGERUM, and RONALD P. HUFFMAN[2]

Department of Chemistry, Purdue University, West Lafayette, IN 47907

Introduction

The chlorination of water as a means of disinfection continues to enjoy popularity because of the low cost, simplicity, and great effectiveness of the technique. Although the process of disinfection is not well understood, it is known that as chlorine is introduced into waste water, the amines present react rapidly with the chlorine to form chloramines. Chloramines have been shown to be toxic to fish when present at lower than ppb levels (1). However, quantitative analytical procedures sensitive at this level are not available. The high toxicity to aquatic life has prompted investigations in this laboratory aimed at extending existing analytical techniques to lower levels of sensitivity.

One question which must immediately be addressed in this work is the effect of changing pH upon the distribution, or the very existence, of the chloramines present in a sample taken for analysis. In the case of ammonia, for example, the equilibria of the chloramine species can be expressed by eq 1-3. Equation 1

$$NH_3 + HOCl \xrightleftharpoons{K_{HYD}} NH_2Cl + H_2O \qquad (1)$$

$$2NH_2Cl + H^+ \xrightleftharpoons{K_{MD}} NHCl_2 + NH_4^+ \qquad (2)$$

[1]Current address: Dept. of Chemistry, U. of Hartford, W. Hartford, CT 06117

[2]Current address: Upjohn Co., P.O.B. 685, Laporte, TX 77571

0-8412-0461-6/78/47-082-264$05.00/0

$$3NHCl_2 + H^+ \xrightleftharpoons{K_{DT}} 2NCl_3 + NH_4^+ \quad (3)$$

might also be expressed, in basic media, as shown in eq 4.

$$NH_3 + OCl^- \rightleftharpoons NH_2Cl + OH^- \quad (4)$$

Values of K_{HYD}, K_{MD}, and K_{DT} have historically been difficult to obtain. Corbett, Metcalf, and Soper (2) reported a value for K_{HYD} of $3.6 \times 10^9\ M^{-1}$ (15°C, $K_w = 4.7 \times 10^{-14}\ M^2$, K_a of HOCl = 2.7×10^{-8} M) while Anbar and Yagil (3) offer evidence that if the number can be measured at equilibrium, the value is greater than $10^{15}\ M^{-1}$ (27.3°C, $\mu = 1.0$ M, calculated using a K_a for HOCl = 4×10^{-8} M). Bridging this large gap, Granstrom (4) calculated the value for K_{HYD} from kinetic data to be $2.4 \times 10^{11}\ M^{-1}$ (25°C, $\mu = 0.45$ M, K_a of HOCl = 2.9×10^{-8} M).

The equilibria shown in eq 2 and 3 have not been evaluated experimentally because of the reported lack of stability of the polychlorinated species (2, 5, 6, 7). However, Jolly (8) was able to estimate values of $10^6\ M^{-1}$ and $10^4\ M^{-1}$ for the equilibrium constants K_{MD} and K_{DT}, respectively. This was done using data of Chapin (5) and Corbett, Metcalf, and Soper (2) even though the latter authors conclude that their data do not represent an equilibrium condition. Using the same data of Chapin (5), Morris (9) calculated K_{MD} to be $2.3 \times 10^7\ M^{-1}$. Using the hydrolysis constant of Corbett, Metcalf, and Soper (2), and his estimates of K_{MD} and K_{DT}, Jolly (8) was also able to estimate electrochemical potentials for the ammoniacal chloramines.

In the present work, all chloramine solutions are prepared with excess amine present and are manipulated with the use of a double-two-jet tangential mixing system, similar to those used in stopped-flow spectrophotometers. This assures rapid and efficient achievement of homogeneity (less than 100 μs). Chloramine solutions produced in this manner are significantly more stable than those produced by hand mixing; they have exhibited no light sensitivity; and experimental results can be consistently duplicated. Monochloramine solutions prepared in this manner are stable at room temperature, undergo disproportionation to $NHCl_2$ without apparent side reactions at pH <5, and readily form NCl_3 in more acidic solutions. Monochloramine and dichloramine are stable in the presence of each other in acidic solution for over 24 h but in the presence of NCl_3 the chloramines slowly decompose. These reproducible properties have allowed the direct experimental determination of K_{MD}, K_{DT}, and K_{HYD} at equilibrium.

Values of K_{MD} for chloramines prepared from amines other than ammonia also have been determined in this work. In order to determine K_{HYD} for some of these other chloramine species, equilibria of the type shown in eq 5 have been established.

$$NH_2Cl + RNH_2 \rightleftharpoons RNHCl + NH_3 \qquad (5)$$

These K_{MD} and K_{HYD} values allow the calculation of electrochemical potentials for these chloramines.

Results and Discussion

I. Equilibria

Determination of K_{MD}. Monochloramines of ammonia, methylamine, β-alanine, and glycylglycine are prepared in this work by rapidly mixing solutions of these amines with a solution of hypochlorite at pH ≃ 11. The amine concentration is kept in excess and the transfer of the chlorine is observed to be rapid and complete. Dichloramines are formed upon acidification. Figure 1 is an example of repetitive spectra of a solution of NH_2Cl after it is adjusted to pH 3.8. The λ_{max} of NH_2Cl at 243 nm can be seen in the first spectrum. In subsequent spectra, over a one-hour period, the absorbance at 294 nm (λ_{max} of $NHCl_2$) increases and the absorbance at 243 nm decreases accordingly. Two clear isosbestic points at 277 nm and 231 nm indicate that the total available chlorine is constant during the reaction. Equilibrium mixtures of monochloramine and dichloramine are observed to be quite stable for at least 24 h and have shown no sensitivity to room light when prepared in this manner.

Values of the molar absorptivity of NH_2Cl at various wavelengths are easily obtained because of the total conversion from hypochlorite and NH_3 in basic media with excess ammonia. However, similar values of $NHCl_2$ presented in the literature vary from 265 M^{-1} cm^{-1} (6) to 300 M^{-1} cm^{-1} (10) and our present work shows that $NHCl_2$ is not fully formed in solution at equilibrium. If the molar absorptivity of $NHCl_2$ at 294 nm ($\varepsilon_{294}^{NHCl_2}$) were known, values of $[NH_2Cl]$ and $[NHCl_2]$ at equilibrium could be determined from eq 6 and 7, where A_{294} is the absorbance of the mixture at

$$A_{294} = b(\varepsilon_{294}^{NH_2Cl}[NH_2Cl] + \varepsilon_{294}^{NHCl_2}[NHCl_2]) \tag{6}$$

$$[HOCl]_T = [NH_2Cl] + 2[NHCl_2] \tag{7}$$

294 nm, b is the pathlength, $\varepsilon_{294}^{NH_2Cl}$ and $\varepsilon_{294}^{NHCl_2}$ are the molar absorptivities of monochloramine and dichloramine at 294, respectively, $[HOCl]_T$ refers to the total available chlorine in terms of the concentration of HOCl.

Solutions of NH_2Cl at constant $[NH_3]_T$ were brought to various values of pH in the range 3.5-6.0. Since $\varepsilon_{294}^{NHCl_2}$ was not well known, equilibrium concentrations of NH_2Cl and $NHCl_2$ and values of K_{MD} were calculated while varying $\varepsilon_{294}^{NHCl_2}$ by iteration. The relative standard deviation of K_{MD} reached a minimum when $\varepsilon_{294}^{NHCl_2} = 267\ M^{-1}\ cm^{-1}$. The value of K_{MD} for this condition was $(4.3 \pm 0.6) \times 10^6\ M^{-1}$. The molar absorptivity is in agreement with that reported by Morris (6) and K_{MD} is in remarkable agreement with the estimate of Jolly (8).

Similar treatment of methylamine, β-alanine, and glycylglycine give the K_{MD} values in Table I. Figure 2 shows the absorbance at 304 nm of equilibrium mixtures of N-chloroglycyl-

Table I

Summary of Equilibria (μ = 0.50 M ($NaClO_4$), 25.0°C).

Equilibrium	K
$NH_3 + HOCl \rightleftharpoons NH_2Cl + H_2O$	$1.5 \times 10^{11} M^{-1}$
$NH_2Cl + HOCl \rightleftharpoons NHCl_2 + H_2O$	$2.3 \times 10^{8} M^{-1}$
$2NH_2Cl + H^+ \rightleftharpoons NHCl_2 + NH_4^+$	$4.3 \times 10^{6} M^{-1}$
$2CH_3NHCl + H^+ \rightleftharpoons CH_3NCl_2 + CH_3NH_3^+$	$3.5 \times 10^{6} M^{-1}$
2β-alanine-Cl + $H^+ \rightleftharpoons$ β-alanine-Cl_2 + β-alanineH^+	$5.2 \times 10^{7} M^{-1}$
2glycylglycine-Cl + $H^+ \rightleftharpoons$ glycylglycine-Cl_2 + glycylglycineH^+	$1.42 \times 10^{5} M^{-1}$
β-alanine + $NH_2Cl \rightleftharpoons NH_3$ + β-alanine-Cl	250
$CH_3NH_2 + NH_2Cl \rightleftharpoons NH_3 + CH_3NHCl$	830
β-alanine + HOCl $\rightleftharpoons$ β-alanine-Cl + H_2O	$3.8 \times 10^{13} M^{-1}$
β-alanine-Cl + HOCl $\rightleftharpoons$ β-alanine-Cl_2 + H_2O	$1.4 \times 10^{11} M^{-1}$
$CH_3NH_2 + HOCl \rightleftharpoons CH_3NHCl + H_2O$	$1.2 \times 10^{14} M^{-1}$
$CH_3NHCl + HOCl \rightleftharpoons CH_3NCl_2 + H_2O$	$7.5 \times 10^{9} M^{-1}$

glycine and N,N-dichloroglycylglycine vs. -log $[H^+]$. The solid line is the calculated fit of the resolved equilibrium constant.

Determination of K_{DT}. Monochloramine ($\sim 10^{-3}$M) is converted to a mixture of $NHCl_2$ and NCl_3 within 5 minutes by raising the hydrogen ion concentration to 0.01 M or higher using efficient mixing techniques. Figure 3 gives spectra of a solution of 2.5×10^{-3} M NH_2Cl in 1 M $HClO_4$ which first forms a maximum at 294 nm ($NHCl_2$). This peak subsequently decreases as the absorbance at 336 grows, indicating the formation of NCl_3. The two clean isosbestic points at 277 nm and 321 nm again indicate no available chlorine is being reduced in the transition from $NHCl_2$ to NCl_3. The dotted line spectrum, taken after one hour, shows a slight decay of the entire spectrum. Therefore, the presence of NCl_3 does introduce some instability to the equilibrium mixture.

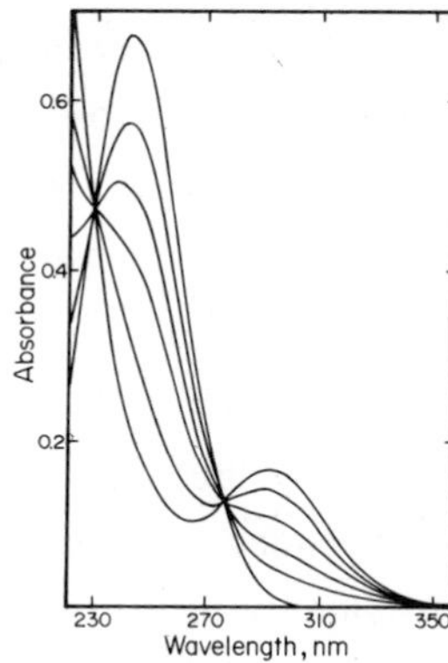

Figure 1. Spectral changes following a pH jump, from pH 11 to 3.8, of a monochloramine solution (λ_{max} = 243) showing the formation of dichloramine (λ_{max} = 294) as equilibrium is approached.

Isosbestic points are 277 and 231 nm. (Scans taken over a period of 30 min, $[NH_3]_T = 9 \times 10^{-3}$M, $[HOCl]_T = 1.5 \times 10^{-3}$M, 1.00-cm path length, 0.50M ($NaClO_4$), 25.05°C)

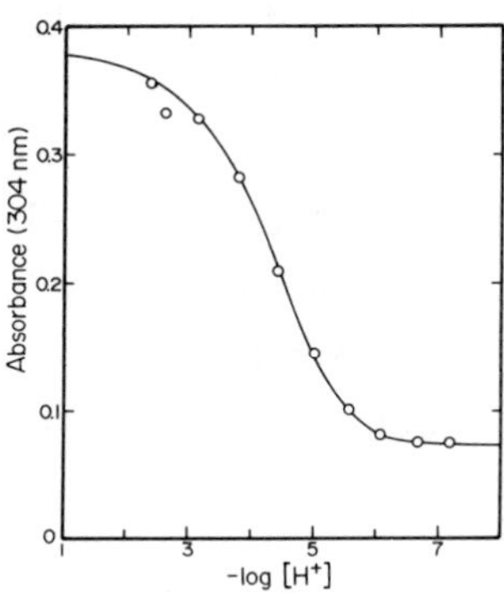

Figure 2. Determination of the disproportionation equilibrium constant of N-chloroglycylglycine.

The solid line is the calculated fit using $K_{DIS} = 1.42 \times 10^5$ for $2RNHCl + H^+ = RNCl_2 + RNH_3^+$. ([Glycylglycine = 1.13×10^{-3}M, $[HOCl]_T = 1.40 \times 10^{-3}$M, 2.00-cm path length, $\mu = 0.50$M ($NaClO_4$), 25.0°C).

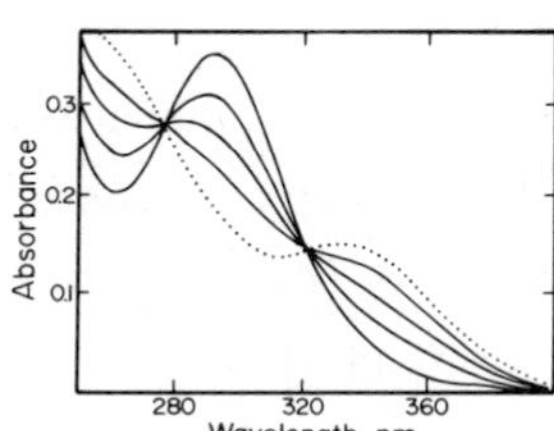

Figure 3. Spectral changes following a pH jump, from pH 11 to 1 M $HClO_4$, of a monochloramine solution showing the loss of $NHCl_2$ (λ_{max} = 294 nm) and the formation of NCl_3 (λ_{max} = 336 nm).

Isobestic points are at 277 and 321 nm. The dotted line, taken 1 hr after pH jump, shows a slight decay. (Scans taken over a period of 1 hr $[NH_3]_T = 10^{-2}$M, $[HOCl]_T = 2.5 \times 10^{-3}$M, 0.50M ($NaClO_4$), 25.0°C.)

Since the total available chlorine in a $NHCl_2/NCl_3$ mixture decays slowly, this value must be determined for each equilibrium to be studied. Therefore, the molar absorptivity of the $NHCl_2/NCl_3$ mixture at 321 nm in terms of total available chlorine was established before significant decay had occurred. This molar absorptivity was determined to be 54 M^{-1} cm^{-1}.

Solutions of various hydrogen ion concentrations were prepared and K_{DT} was calculated in the same manner as previously described for K_{MD}, except that the absorbance was followed at 336 nm and $\varepsilon_{336}^{NCl_3}$ was iteratively varied. The value of K_{DT} exhibited a minimum relative standard deviation when $\varepsilon_{336}^{NCl_3}$ = 272 M^{-1} cm^{-1} and K_{DT} = 130 ± 40 M. This molar absorptivity is slightly higher than the values of 260 M^{-1} cm^{-1} (12) and 265 M^{-1} cm^{-1} (6) previously reported. The value of K_{DT} is significantly lower than Jolly's (8) estimate of 10^4M.

Confirmation of K_{HYD}. The disparity of reported values of K_{HYD} necessitated a reevaluation of the problem. The chlorine atom can be removed from monochloramine *via* two possible pathways, both of which have been observed. Equation 8 is the hydrolysis

$$NH_2Cl + H_2O \longrightarrow NH_3 + HOCl \tag{8}$$

$$NH_2Cl + OH^- \longrightarrow NH_2OH + Cl^- \tag{9}$$

of NH_2Cl with the chlorine atom bound to the hydroxide as the product. Equation 9 involves hydroxide attack on nitrogen to give hydroxylamine and release chloride ion. It is eq 8 which is involved in K_{HYD} and eq 9 which is the reaction Anbar and Yagil (3) conclude interferes to the extent that K_{HYD} cannot be measured at equilibrium. Note that high base concentrations will shift the equilibria in both eq 9 and eq 8 (due to $HOCl + OH^- \longrightarrow OCl^- + H_2O$) to the right. However, we find that in 0.1 M NaOH the rate of the reaction in eq 8 is approximately 100 times that of eq 9 and that conditions can be chosen to permit the equilibrium in eq 8 to be measured without serious interference from eq 9. Our experimental value of K_{HYD} is 1.5 x 10^{11} M^{-1}, which is in approximate agreement with Granstrom's value (4).

The result can be expanded to calculate hydrolysis constants for other monochloramines. Methylchloramine and N-chloro-β-alanine do not decompose readily by either eq 8 or eq 9, even in 1 M NaOH. However, when β-alanine is mixed with NH_2Cl at pH 8 and the absorbance spectrum is scanned repetitively, a set of spectra similar to that shown in Fig. 4 is obtained. The single, sharp isosbestic point indicates the establishment of an equilibrium mixture of NH_2Cl and β-alanine. The same method was used with CH_3NH_2 in place of β-alanine and equilibrium constants for the transfer of chlorine, K_T (eq 5) were determined as shown in Table I. Since eq 5 is the summation of the hydrolysis constants of two chloramines, K_{HYD} values were calculated for N-chloro-β-alanine and methylchloramine and are given in Table I.

If the electrochemical potentials of Latimer (11) for Cl_2, HOCl, and OCl^- are used as a point of reference, the appropriate potentials for chloramine species of NH_3, β-alanine, and CH_3NH_2 can be calculated. These are given in Table II. These potentials are consistent with the previous reports of the strong oxidizing power of NH_3 chloramines (8). The values for β-alanine and CH_3NH_2 are slightly less but still indicate high oxidizing power.

II. Kinetics of Disproportionation

General Behavior. When a monochlorinated primary amine prepared in basic media is jumped to high acid, a second-order reaction occurs and a dichloramine is formed (in certain cases other decomposition reactions may occur and, in the case of NH_3, NCl_3 may form, but these are not part of the present discussion). As the initially basic monochloramine solutions are jumped to increasingly higher acid, more and more of the initial absorbance is lost prior to the formation of the $RNCl_2$ species. This absorbance loss occurs within the mixing time of the stopped-flow instrument

Table II

Standard Reduction Potentials (vs. NHE, 25.0°C).[a]

Reaction Type	E^{o}'s (volts) for the chloramine of: Ammonia	β-alanine	Methylamine
$RNHCl + H_2O + 2e^- \rightarrow Cl^- + RNH_2 + OH^-$ [b]	0.75	0.68	0.66
$RNH_2Cl^+ + H^+ + 2e^- \longrightarrow Cl^- + RNH_3^+$ [c]	1.40	1.37	1.35
$RNHCl + 2H^+ + 2e^- \longrightarrow Cl^- + RNH_3^+$ [c]	1.44	1.39	1.39
$RNH_2Cl^+ + H^+ + e^- \rightarrow 1/2\ Cl_2 + RNH_3^+$ [c]	1.45	1.39	1.35
$RNHCl + 2H^+ + e^- \longrightarrow 1/2\ Cl_2 + RNH_3^+$ [c]	1.53	1.42	1.43
$RNCl_2 + 2H_2O + 4e^- \longrightarrow 2Cl^- + RNH_2 + 2OH^-$ [b]	0.79	0.71	0.72
$RNCl_2 + 3H^+ + 4e^- \longrightarrow 2Cl^- + RNH_3^+$ [c]	1.34	1.27	1.30
$RNCl_2 + 3H^+ + 2e^- \longrightarrow Cl_2 + RNH_3^+$ [c]	1.33	1.19	1.24
$RNCl_2 + H^+ + 2e^- \longrightarrow Cl^- + RNHCl$ [d]	1.24	1.16	1.20
$RNCl_2 + 2H^+ + 2e^- \longrightarrow Cl^- + RNH_2Cl^+$ [d]	1.29	1.17	1.24

[a]Referenced to the potentials for chlorine species in ref. (11). [b]1 M amine and 1 M OH^-. [c]1 M amine H^+ and 1 M H^+. [d]1 M chloramine and 1 M H^+.

and is due to the protonation of the monochloramine. The amount of $RNCl_2$ formed is directly dependent on the amount of RNHCl initially present, but is independent of the magnitude of the initial absorbance jump.

The rate of $RNCl_2$ formation, however, is dependent on the amount of the initial absorbance jump. Figure 5 is a plot of the observed-second-order rate constant of the formation of $NHCl_2$, N,N-dichloro-β-alanine, and N,N,-dichloroglycylglycine as a function of acidity. It is noteworthy that the kinetic maximum occurs when 50% of the initial absorbance of NH_2Cl is lost during the pH jump. This implies strongly that the reaction involved is (using NH_3 as an example):

$$NH_2Cl + NH_3Cl^+ \xrightarrow{k_{DIS}} NHCl_2 + NH_4^+ \quad (10)$$

The rate expression for this microscopic reaction is

$$\frac{d[RNCl_2]}{dt} = k_{DIS}[NH_2Cl][NH_3Cl^+] = 1/2\ (\frac{-d[RNHCl]}{dt}) \quad (11)$$

In terms of $[NH_2Cl]_T (=[NH_2Cl] + [NH_3Cl^+])$

$$\frac{-d[RNHCl]}{dt} = \frac{2k_{DIS}K_H[H^+][NH_2Cl]_T^2}{(1 + K_H[H^+])^2} \quad (12)$$

where K_H is the protonation constant for monochloramine (eq 13).

$$NH_2Cl + H^+ \xrightarrow{K_H} NH_3Cl^+ \quad (13)$$

Values of K_H and k_{DIS} were resolved for chloramines of NH_3, CH_3-NH_2, β-alanine, glycine, and glycylglycine as compared in Table III. The values for glycylglycine are the result of a considerable extrapolation but are certainly as accurate as they would be if the 0.5 M ionic strength restriction had been broken in order to achieve the 2-5 M acid necessary to obtain data in the region of the kinetic maximum.

It should be noted that these values of k_{DIS} are rather small and the K_H values are higher than expected (12). Hence protonated monochloramines will exist at low pH for a significant time period at low total concentrations of RNHCl. Accordingly, equilibrium constants and electrochemical potentials are included for the protonated species in Tables I and II.

III. Distribution of Ammoniacal Chloramines

Figure 6 displays the calculated distribution of millimolar chlorine in a five-fold excess of ammonia as a function of pH. This is analogous to the conditions used for the preparation of stock monochloramine in this work. Besides the complete formation of monochloramine above pH 9.5, it is interesting to note

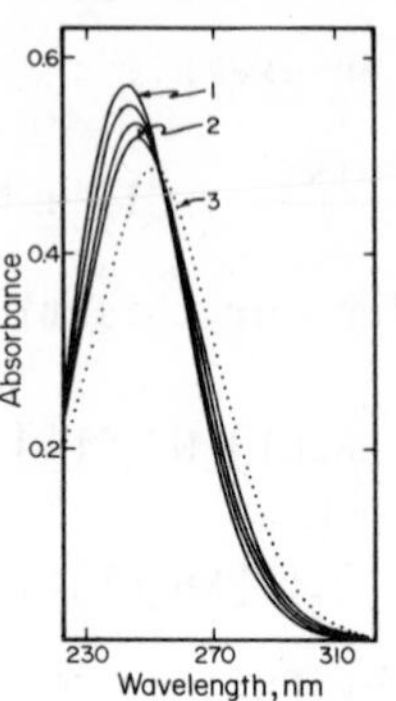

Figure 4. Approach of a mixture of NH_3, β-alanine and their monochloramines toward equilibrium.

(1) initial spectrum; (2) equilibrium spectrum, [HOCl] = 1.31×10^{-3}M, $[NH_3]_T$ = 0.059M, $[\beta\text{-alanine}]_T$ = 0.0022M, $-\log[H^+]$ = 9.08, 1-cm path length, μ = 0.50M ($NaClO_4$) 25.0°C; (3) . . . fully formed N-chloro-β-alanine.

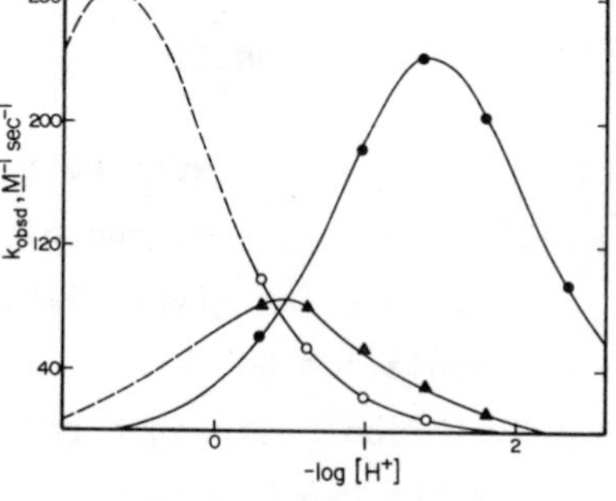

Figure 5. Dependence of k_{obs} (M^{-1} sec^{-1}) on [H^+] for the disproportionation $\left(\frac{-d[RNHCl]}{dt} = 2k_{obs}[RNHCl]_T^2\right)$ of monochloramine (●), N-chloro-β-alanine (△), and N-chloroglycylglycine (○). ([Chloramine] = 1.06×10^{-3}M, μ = 0.50M ($NaClO_4$), 25.0°C).

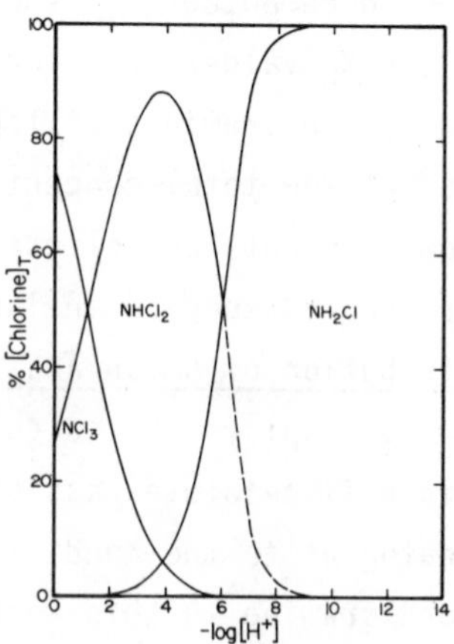

Figure 6. Distribution of chloramines in aqueous solution ($[NH_3]_T = 10^{-2}$M, $[HOCl]_T$ = 2.5×10^{-3}M, 0.50M ($NaClO_4$), 25.0°C.) The decomposition of $NHCl_2$ above pH 6 is neglected.

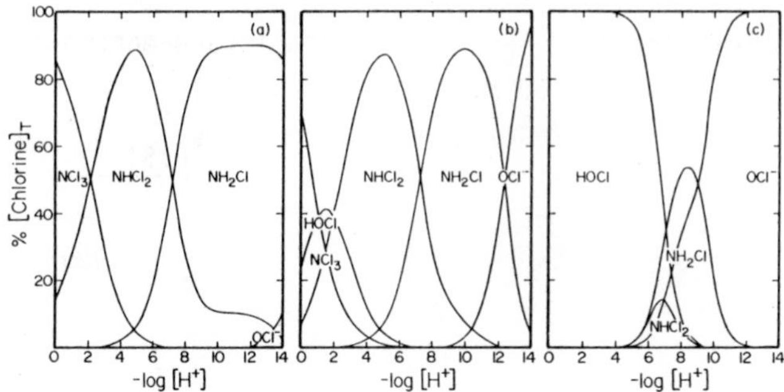

Figure 7. Distributions of chloramines in aqueous solutions with $[NH_3]_T = [HOCl]_T$, 25.0°C, 0.50M ($NaClO_4$). The decomposition of $NHCl_2$ above pH 6 is neglected: (a) 10^{-3}M, (b) 10^{-6}M, (c) 10^{-9}M.

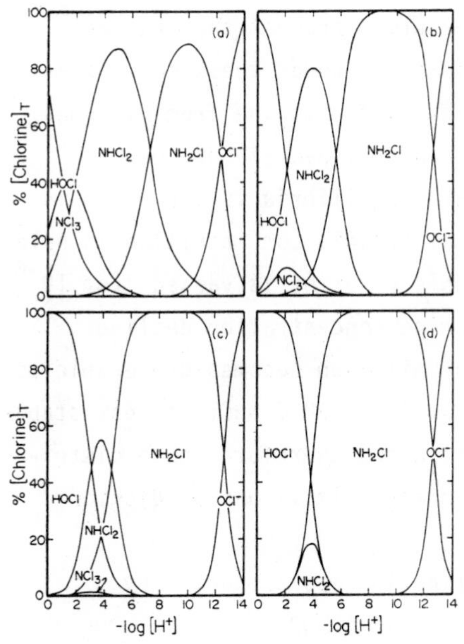

Figure 8. Distributions of chloramines in aqueous solution with a constant ammonia concentration and varying $[HOCl]_T$ ($[NH_3]_T = 10^{-6}$, 25.0°C, $\mu = 0.50$M $NaClO_4$).

The decomposition of $NHCl_2$ above pH 6 is neglected: (a) $[HOCl]_T = 10^{-6}$M, (b) $[HOCl]_T = 10^{-7}$M, (c) $[HOCl]_T = 10^{-8}$M, (d) $[HOCl]_T = 10^{-9}$M.

Table III

Protonation Constants of Monochloramines and the Rate Constants for Disproportionation from

$$-d[\text{monochloramine}]/dt = k_{DIS}[\text{monochloramine}][\text{H-monochloramine}^+]$$

(μ = 0.50 M ($NaClO_4$), 25.0°C)

Amine	K_H, M^{-1}	k_{DIS}, M^{-1} sec^{-1}
Methylamine	36	60
Ammonia	28	980
β-alanine	2.4	330
Glycine	0.37	1200
Glycylglycine	0.21	1100

that the equilibria do not allow full formation of dichloramine or trichloramine under these conditions. Figure 7 illustrates the calculated distributions of chloramines at equilibrium after mixing equimolar hypochlorous acid and ammonia at various concentration levels. These three plots present the changes which occur upon dilution of chloramines under these concentration and pH conditions. The transfer of chlorine from nitrogen to oxygen (hydrolysis) is the predominant overall process.

Another facet of the chloramine equilibria is presented in Figure 8. In this figure the total ammonia concentration is held constant at 10^{-6} M and the available chlorine is varied from 10^{-6} to 10^{-9} M. As the available chlorine concentration declines, chlorine singly bound to oxygen or nitrogen becomes preferable to polychlorinated nitrogen moieties. Above pH 6 $NHCl_2$ is not stable relative to self oxidation-reduction and therefore these distributions represent only the maximum possible values of dichloramine.

Acknowledgment. This investigation was supported by National Science Foundation Grant CHE75-15500.

References and Notes

1. Merkins, J. C., Water Waste Treatment J., (1958), 7, 150.
2. Corbett, R. E., Metcalf, W. S., and Soper, F. G., J. Chem. Soc., (1953), 1927.
3. Anbar, M. and Yagil, G., J. Am. Chem. Soc., (1962), 84, 1790.
4. Granstrom, M. L., Ph.D. Thesis, Harvard University, 1954.
5. Chapin, R. M., J. Am. Chem. Soc., (1929), 51, 2113.
6. Morris, J. C., "Kinetics of Reactions Between Aqueous Chlorine and Nitrogen Compounds" in "Principles and Applications of Water Chemistry," Faust, S. D. and Hunter, J. D., eds., Wiley, N. Y., 1964, pp 23-53.
7. Metcalf, W. S., J. Chem. Soc. (1942), 148.
8. Jolly, W. L., J. Phys. Chem., (1956), 60, 507.
9. Weil, I. and Morris, J. C., 116th National Meeting of the American Chemical Society, Atlantic City, N. J., September, 1949.
10. Galal-Gorchev, H. and Morris, J. C., Inorg. Chem., (1965), 4, 899.
11. Latimer, W. M., "The Oxidation States of the Elements and Their Potentials in Aqueous Solutions," 2nd ed., Prentice Hall, Inc., New York, 1952, pp 53-55.
12. Weil, I. and Morris, J. C., J. Am. Chem. Soc., (1949), 71, 3123.

RECEIVED August 22, 1978.

17

Chlorination and the Formation of *N*-Chloro Compounds in Water Treatment

DALE W. MARGERUM, EDWARD T. GRAY, JR.,[1] and RONALD P. HUFFMAN[2]
Department of Chemistry, Purdue University, West Lafayette, IN 47907

Chlorine is widely used for the disinfection of potable water and wastewater. It is estimated that 0.3 to 0.4 million tons are used annually for sanitary purposes (1). Chlorine and hypochlorous acid (HOCl) are much more effective than hypochlorite ion (OCl^-). Thus, for disinfection four times as much as total chlorine is needed at pH 9 than at pH 7. Below pH 7, a 2×10^{-6} M concentration of HOCl is sufficient for 99.6% virus inactivation, 99.999% coliform kill, and 99.999% cyst disinfection (2). Ammonia, which is present in waste water and even in many drinking and river waters, will react rapidly with HOCl to form monochloramine (NH_2Cl). Hence NH_2Cl becomes the principal chlorine disinfectant in any water containing as much as 0.1 mg/L of NH_3-N and less than 1.0 mg/L of Cl_2. Additional chlorine will give dichloramine ($NHCl_2$) and trichloramine (NCl_3). Figure 1, the decomposition of residual chlorine is accomplished by the destruction of the chloramines (presumably to N_2 and NO_3^-) with excess chlorine. A typical breakpoint occurs at 8 mg Cl_2/L per 1 mg N/L. Beyond the breakpoint additional chlorine dosage introduces more HOCl and often N-chloro-organic compounds as well (2).

[1] Current address: Dept. of Chemistry, U. of Hartford, W. Hartford, CT 06117
[2] Current address: Upjohn Co., P.O.B. 685, Laporte, TX 77571

0-8412-0461-6/78/47-082-278$05.00/0

Chloramines are toxic to fish. Thus, 0.08 ppm residual chlorine causes fish kills and long exposure to ppb levels of NH_2Cl are lethal to rainbow trout (3). The toxicities of free and combined chlorine (i.e. chloramines) are considered to be the same order of magnitude (3, 4). Fresh water fishes suffer anoxia due to chlorine or chloramine oxidation of hemoglobin to methemoglobin (5). It is important to know the ease of formation, the stability, and reactivity of chloramines in order to understand their environmental effects.

The present paper is concerned with some of the solution chemistry of chloramines, especially the kinetics of formation and dissociation of various N-chloro compounds. Earlier work by Morris and his associates (6, 7, 8, 9, 10) has provided important basic information about some of the properties of chloramines. It has been possible to expand on this base through the use of stopped-flow spectrophotometry and by taking advantage of the ability to form relatively stable solutions of N-chloro species. In the preceding paper we have discussed the determination of accurate equilibrium constants for the chloramines and the kinetics of disproportionation of monochloramines. We now consider the kinetics of transfer of active chlorine to oxygen and nitrogen as well as its transfer between oxygen and nitrogen compounds.

Reactions of Cl_2. The transfer of Cl^+ to oxygen and to nitrogen from Cl_2 is rapid but measurable for several different reactions. The kinetics of hydrolysis of Cl_2 (eq 1) have been

$$Cl_2 + H_2O \underset{k_{-1}}{\overset{k_1}{\rightleftharpoons}} HOCl + Cl^- + H^+ \qquad (1)$$

measured by Eigen and Kustin (11) and by Lifshitz and Perlmutter-Hayman (12). In the present work we needed to know the equilibrium constant for eq 1 in 0.50 M $NaClO_4$ at 25.0°C and in the process of this determination obtained the rate constants which are in substantial agreement with the earlier studies. Our values are: $k_1 = 28.6\ s^{-1}$, $k_{-1} = 2.8 \times 10^4\ M^{-2}\ s^{-1}$, and $k_1/k_{-1} = 1.0 \times 10^{-3}\ M^2$. It can be seen from these rate constants that the equilibration of Cl_2 with HOCl and Cl^- occurs within seconds.

The rates of Cl_2 reaction with ammonia, alkylamines and amino acids (eq 2) are very fast and, in fact, are nearly at the

$$Cl_2 + RNH_2 \xrightarrow{k_{Cl_2}^{RNH_2}} RNHCl + H^+ + Cl^- \qquad (2)$$

diffusion-controlled limit. The protonated amine species, RNH_3^+, are not reactive with Cl_2 or with HOCl. We have resolved rate constants (Table I) for the reaction of these active chlorine species with the free amines by appropriate variation of the H^+ and Cl^- concentrations. Once the monochloramines are formed they also react rapidly with Cl_2 (eq 3). Since protonated monochlor-

$$Cl_2 + RNHCl \xrightarrow{k_{Cl_2}^{RNHCl}} RNCl_2 + H^+ + Cl^- \qquad (3)$$

amine, RNH_2Cl^+, forms only at very low pH, the rate of formation of $RNCl_2$ from Cl_2 is not as easily suppressed in acidic solution as is the rate of formation of RNHCl. As a result dichloramines may be observed as the products of the chlorination of amines and the rate-determining step is eq 2 rather than eq 3.

Table I

Second-Order Rate Constants ($M^{-1}s^{-1}$) for the Chlorination of Amines and Amino Acids, 25.0°C

	$k_{Cl_2}^{RNH_2}$	$k_{Cl_2}^{RNH_2{}^o}$	$k_{HOCl}^{RNH_2}$
Ammonia	4.0×10^9	---	2.9×10^6
Methylamine	2.8×10^9	---	1.9×10^8
Glycine	---	1.5×10^9	5.0×10^7
Glycylglycine	---	2.1×10^9	5.3×10^6
α-alanine	---	1.0×10^9	5.4×10^7
β-alanine	---	1.3×10^9	8.9×10^7

The doubly protonated amino acids, $^{+}NH_3RCOOH$, are not reactive with Cl_2 and the concentration of the completely unprotonated form, NH_2RCOO^-, is negligible under the acidic conditions used. Although a monoprotonated form does react with Cl_2, it is proposed that this is not the zwitterion species, $^{+}NH_3RCOO^-$, but rather is the neutral species, NH_2RCOOH, which is present as a minor component. Rate constants calculated in terms of this RNH_2^{0} species in Table I agree with those found for the free amines and also are close to the diffusion-limiting values.

Reactions of HOCl. The transfer of Cl^+ from oxygen to nitrogen in the reaction of hypochlorous acid with ammonia (eq 4)

$$HOCl + NH_3 \xrightarrow{k^{NH_3}_{HOCl}} NH_2Cl + H_2O \qquad (4)$$

is rapid and this is the main pathway for monochloramine formation in chlorination reactions. Weil and Morris (6) used data taken below pH 6.5 (where NH_4^+ formation suppresses the rate) and data taken above pH 12 (where OCl^- formation suppresses the rate) to give a rate constant of $6.2 \times 10^6\ M^{-1}s^{-1}$ for reaction 4. In our stopped-flow studies the reaction was measured over the intermediate pH range as shown in Figure 2 where k_{obsd} is a pseudo-first-order rate constant (eq 5) using excess ammonia.

$$\frac{d[NH_2Cl]}{dt} = k_{obsd}[HOCl]_T \qquad (5)$$

The solid line in Figure 2 is calculated using a value of $2.8 \times 10^6\ M^{-1}s^{-1}$ (0.1 M $NaClO_4$), 25.0°C) for the second-order rate constant, $k^{NH_3}_{HOCl}$. This value is based on protonation constants of $10^{9.39}M^{-1}$ for NH_3 and $10^{7.44}\ M^{-1}$ for OCl^- also measured in 0.1 M $NaClO_4$ at 25.0°C. The results confirm the postulate that only HOCl and NH_3 are reactants.

The rate constants for HOCl with some alkyl amines, amino acids and peptides are summarized in Table I. In contrast with the behavior of Cl_2, the rate constants for the HOCl reactions

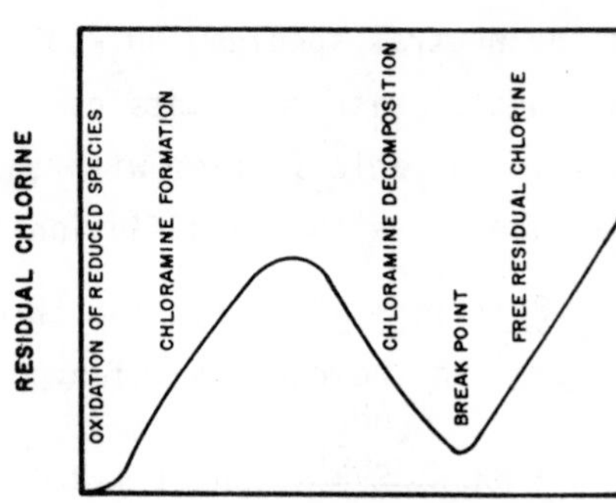

Water Chlorination

Figure 1. Relationship between chlorine dosage and residual chlorine for breakpoint chlorination (2)

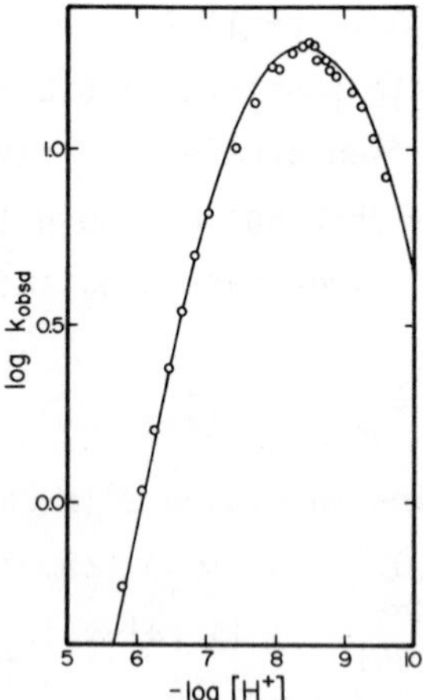

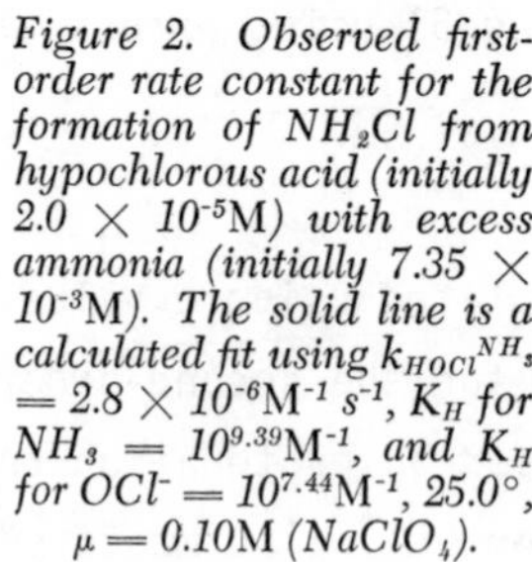

Figure 2. Observed first-order rate constant for the formation of NH_2Cl from hypochlorous acid (initially 2.0×10^{-5}M) with excess ammonia (initially 7.35×10^{-3}M). The solid line is a calculated fit using $k_{HOCl}^{NH_3} = 2.8 \times 10^{-6} M^{-1} s^{-1}$, K_H for $NH_3 = 10^{9.39} M^{-1}$, and K_H for $OCl^- = 10^{7.44} M^{-1}$, 25.0°, $\mu = 0.10$M ($NaClO_4$).

increase as the basicity of the amines increase as seen in Figure 3. All the amines except ammonia fall along a line correlating log $k_{HOCl}^{RNH_2}$ with log $K_H^{RNH_2}$. This suggests a nucleophilic attack by the amine on the Cl atom in HOCl as the rate-determining step in the reaction mechanism.

The rate constants for the reaction of monochloramines with HOCl (eq 6) are given in Table II. The smaller rate constants

$$RNHCl + HOCl \longrightarrow RNCl_2 + H_2O \qquad (6)$$

agree with the diminished nucleophilicity of the monochloramines. Figure 4 shows the correlation for both RNH_2 and RNHCl. The slope of the log-log plot is 0.61.

Table II

Protonation Constants and Rate Constants for HOCl Reaction with Monochloramines, 0.50 M ($NaClO_4$), 25.0°C

	log K_H	k_{HOCl}^{RNHCl}, M^{-1} s^{-1}
monochloramine	1.44	150
methylchloramine	1.55	352
N-chloro-β-alanine	0.41	278
N-chloroglycylglycine	-0.67	8.7

Below pH 4-5 the addition of HOCl to RNH_2 solutions results in $RNCl_2$ as the observed product rather than RNHCl despite the fact that the $k_{HOCl}^{RNH_2}$ rate constants are much larger than the k_{HOCl}^{RNHCl} rate constants. The reason is that the ratio of RNH_2/RNH_3^+ is very small, while the ratio of $RNHCl/RNH_2Cl^+$ is large and hence $k_{HOCl}^{RNHCl}[RNHCl] \gg k_{HOCl}^{RNH_2}[RNH_2]$.

Reactions of Monochloramines. The rate of hydrolysis of NH_2Cl (eq 7) is relatively slow. A rate constant of 1.9×10^{-5}

$$NH_2Cl + H_2O \xrightarrow{k_{H_2O}^{NH_2Cl}} NH_3 + HOCl \qquad (7)$$

s^{-1} for $k_{H_2O}^{NH_2Cl}$ can be calculated from the formation rate constant

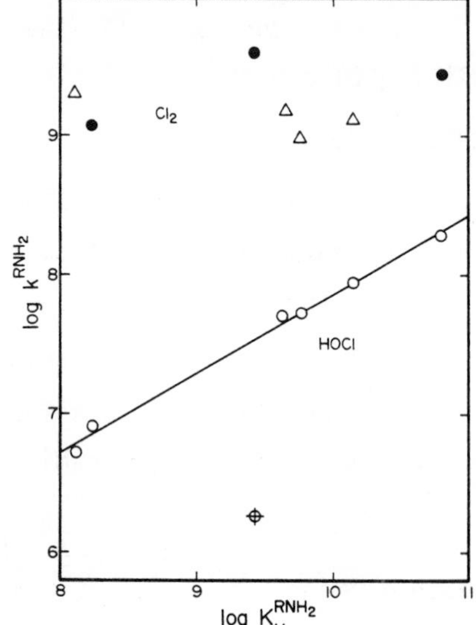

Figure 3. Comparison of second-order rate constants (25.0°C) for the chlorination of amines as a function of their base strength. (●) RNH_2 and Cl_2, (△) amino acids (NH_2RCOOH form) and Cl_2, (○) RNH_2 and HOCl, and (⊕) NH_3 and HOCl.

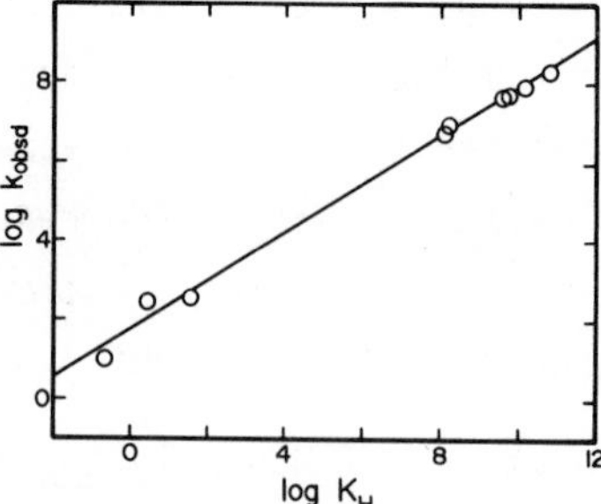

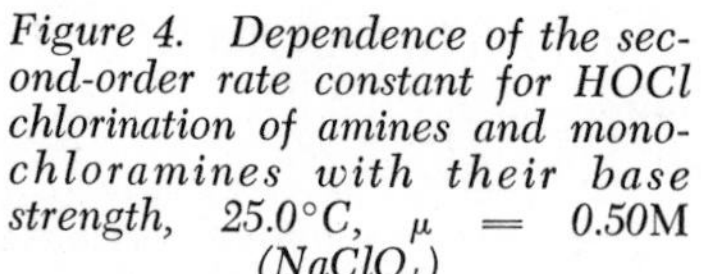

Figure 4. Dependence of the second-order rate constant for HOCl chlorination of amines and monochloramines with their base strength, 25.0°C, μ = 0.50M ($NaClO_4$)

and the equilibrium constant given in the preceding paper. Once NH_2Cl forms it is slow to release active chlorine in the form of HOCl. Another hydrolysis reaction to give hydroxylamine (eq 8) is

$$NH_2Cl + OH^- \xrightarrow{k_{OH}^{NH_2Cl}} NH_2OH + Cl^- \qquad (8)$$

a significant pathway only in strong base. Anbar and Yagil (13) report a value of 6.3 x 10^{-5} $M^{-1}s^{-1}$ for $k_{OH}^{NH_2Cl}$. Rate constants for the hydrolysis of monochloramines and dichloramines (releasing HOCl) are summarized in Table III.

Table III

Rate Constants for the Hydrolysis of Chloramines, 25.0°C, 0.5 M ($NaClO_4$)

	k_{H_2O}, s^{-1}
NH_2Cl	1.9 x 10^{-5}
N-Cl-β-alanine	2.1 x 10^{-6}
CH_3NHCl	1.6 x 10^{-6}
$NHCl_2$	6.5 x 10^{-7}
N,N-diCl-β-alanine	1.6 x 10^{-9}
CH_3NCl_2	4.7 x 10^{-8}

As shown in the preceding paper monochloramines will disproportionate in acid in accord with eq 9. It is interesting that

$$RNHCl + RNH_2Cl^+ \xrightarrow{k_{DIS}} RNCl_2 + RNH_3^+ \qquad (9)$$

the value of k_{DIS} decreases as the base strength of RNH_2 decreases. The contrast with the HOCl chlorination dependence is striking as shown in Figure 5. This suggests a shift in the rate-determining step to the breaking of the N-Cl bond after nucleophilic attack by RNHCl on the chlorine of RNH_2Cl^+.

Is the direct transfer of Cl^+ from one amine nitrogen to another possible? Yes, recent work by Snyder and Margerum (14) shows that the rate of chlorine transfer in eq 10 is faster than

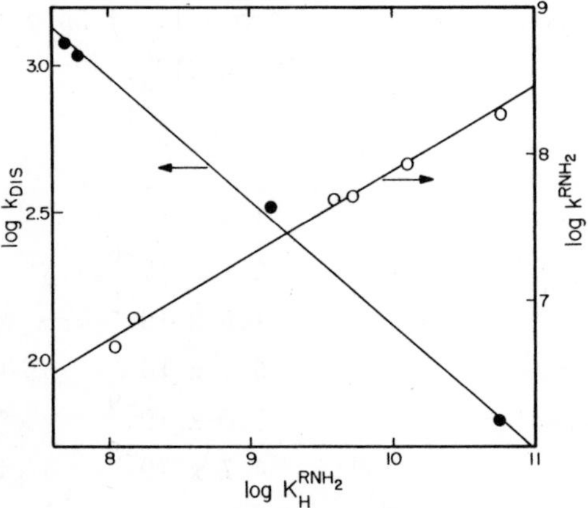

Figure 5. Comparison of the rate constants for monochloramine disproportionation (k_{DIS}) and the rate constants for the reaction of amines with HOCl (k^{RNH_2}) with the basicity of the amines, 25.0°C, $\mu = 0.50$M ($NaClO_4$).

$$NH_2Cl + NH_2CH_2COO^- \longrightarrow NH_3 + ClNHCH_2COO^- \quad (10)$$

the rate of hydrolysis. The observed rate constant is 1.5 $M^{-1}s^{-1}$ from pH 4 to 9. The observed rate constant is invariant over this pH range because the reactive species are NH_3Cl^+ and NH_2CH_2-COO^- whereas the predominant species in solution are NH_2Cl and $NH_3{}^+CH_2COO^-$. Protonated monochloramine, NH_3Cl^+, is a very effective chlorinating agent as seen from the data in Table IV. The

Table IV

Rate Constants for the Direct Transfer of Chlorine from NH_3Cl^+ to Amino Acids and Peptides (14), 25.0°C

	k, M^{-1} s^{-1}
glycylglycine$^-$	1.3 x 10^7
glycine$^-$	2.5 x 10^8
β-alanine$^-$	2.6 x 10^8

NH_3Cl^+ ion reacts considerably faster than HOCl and is only about an order of magnitude slower than Cl_2. On the other hand, the protonation constant to form NH_3Cl^+ from NH_2Cl is only 28 M^{-1}, so the fraction of protonated species present in neutral pH is small and the observed second-order rate constants are not large (*e.g.* 1.5 M^{-1} s^{-1} for glycine).

Reactions of Dichloramines. The hydrolysis of $NHCl_2$ (eq 11)

$$NHCl_2 + H_2O \longrightarrow NH_2Cl + HOCl \quad (11)$$

is slower than the hydrolysis of NH_2Cl. Similarly, other dichloramines (Table III) have small rate constants for their hydrolysis reactions. Obviously the instability of $NHCl_2$ is not due to simple hydrolysis reactions.

The rate constant and equilibrium constant for the disproportionation reaction of monochloramine given in the preceding paper can be used to calculate a rate constant of 6.4 x 10^{-3} M^{-1} s^{-1} for the reaction in eq 12. Therefore greater than 10^{-3} M

$$NHCl_2 + NH_4^+ \longrightarrow 2NH_2Cl + H^+ \qquad (12)$$

concentrations of ammonia or other amines could speed the conversion of dichloramines to monochloramines. The equilibrium position of the reaction in eq 12 also shifts to the right as the pH becomes greater than 6.

Other reactions of $NHCl_2$ must take place in neutral or basic solutions in order to account for the instability reported for this species, but as yet these reactions are not well defined.

Discussion. The kinetics data for the transfer of Cl^+ to and from nitrogen atoms in amines show that RNHCl and $RNCl_2$ are formed rapidly and are slow to release Cl^+ in acid. Although $NHCl_2$ has often been considered an extremely unstable compound in regard to self oxidation-reduction, this is not the case in acidic solutions in the absence of additional HOCl. The reactions of $NHCl_2$ and other dichloramines in neutral and basic solutions need to be studied. Monochloramines are relatively stable in slightly basic and slightly acidic solutions and act as a reservoir of active chlorine. The persistence of residual chlorine in river water has been attributed to the slow decay of monochloramine (15). The half life of this residual chlorine is comparable with the half life for the hydrolysis of NH_2Cl.

Table V summarizes some possible reactions of NH_2Cl after it forms in the chlorination of water. Reaction (a) is thermodynamically unfavorable (the equilibrium constant is 6.7×10^{-12} M^{-1}) and would occur only if other processes were removing the reaction products. Reaction (b) is much too slow to be of concern except in strong base. Reaction (c) is thermodynamically favorable only below pH 6 and it is second-order process in monochloramine so this decay would not be a favorable pathway. The self redox decomposition of $NHCl_2$ at higher pH could help to drive the reaction but the rate of reaction (c) is slower at high pH. Reaction (d) depends on the concentration of other amines,

Table V

Possible Cl^+ Transfer Reactions of NH_2Cl

(a) $NH_2Cl + H_2O \rightleftharpoons HOCl + NH_3$

$t_{1/2}$ = 10 hrs

(b) $NH_2Cl + OH^- \longrightarrow NH_2OH + Cl^-$

at pH 8 $t_{1/2}$ = 350 yr

(c) $2NH_2Cl + H^+ \longrightarrow NHCl_2 + NH_4^+$

at pH 6 and 10^{-3} M NH_2Cl the first $t_{1/2}$ = 10 hrs

(d) NH_2Cl + gly $\longrightarrow$ Cl-gly + NH_3

for 10^{-3} M gly, $t_{1/2}$ = 0.13 hrs

amino acids, peptides or proteins and is one of the more favorable pathways if these concentrations are appreciable (*i.e.* more than 10^{-5} M). The N-chloroamino acid products are even slower to hydrolyze than is NH_2Cl. Relatively little is known about the oxidation of substrates by chloramines or about species which might catalyze the self oxidation-reduction of chloramines. The transfer of Cl^+ from one N-compound to another seems a likely path to permit the toxic chloramines to form in marine life. It is not yet clear what redox reactions are actually responsible for the chloramine toxicity.

Acknowledgment. This investigation was supported by National Science Foundation Grant CHE 75-15500.

Literature Cited

1. White, G. C. in "Water Chlorination" Vol. 1, R. L. Jolley, ed., Ann Arbor Science Publ., Ann Arbor, Mich., p 2.
2. Johnson, J. D., *ibid*., p 41.
3. Merkens, J. C., *Water Waste Treat. J.*, (1958), *7*, 150.
4. Brungs, W. A., *J. Water Poll. Control Fed.*, (1973), *45*, 2180.
5. Grothe, D. R. and Eaton, J. W., *Trans. Am. Fish.*, (1975), *104*, 800.
6. Weil, I. and Morris, J. C., *J. Am. Chem. Soc.*, (1949), *71*, 1664.
7. Granstrom, M. L., Ph.D. Thesis, Harvard University, 1954.
8. Friend, A. G., Ph.D. Thesis, Harvard University, 1956.
9. Samples, W. R., Ph.D. Thesis, Harvard University, 1959.
10. Morris, J. C., in "Principles and Applications of Water Chemistry," S. D. Faust and J. D. Gunter, eds., Wiley, New York, N.Y., 1967, p 23.
11. Eigen, M. and Kustin, K., *J. Am. Chem. Soc.*, (1962), *84*, 1355.
12. Lifshitz, A. and Perlmutter-Hayman, B., *J. Phys. Chem.*, (1960), *64*, 1663.
13. Anbar, M., and Yagil, G., *J. Am. Chem. Soc.*, (1962), *84*, 1790.
14. Snyder, M. P. and Margerum, D. W. unpublished data.
15. Ref. 2, p 49.

Discussion

F. E. BRINCKMAN (National Bureau of Standards): Referring to Table V, you have a large excess of solvent (water) which is not reactive with chloramine species. The glycines show a short half-life in terms of reaction with chloramine. What about organometals? I raise this because much faith is being placed in chlorination and/or ozonation processes for waste water treatment involving organometals. Have you done any studies in this direction?

MARGERUM: No, I did not. There is a vast number of experiments, not only with organometals, but also with unsaturated carbon compounds, etc., that we have not gotten to. Even for simple oxidizable metal ions, the information is not available now.

BRINCKMAN: I think we are saying where the action is. I was quite impressed with the enormous rate differences for these substrates, particularly involving organonitrogen, or the glycine in this case.

J. J. ZUCKERMAN (University of Oklahoma): I was facinated by the chlorinated amino acids. Are they isolable species and do they exist under environmental conditions in aqueous media as zwitterions?

MARGERUM: First, you can form these chloroamino acids and keep them in solution, either as the monochloro- or dichloro-amino acids, and then they are not zwitterions. Both of these species, as Professor Gray and I have shown, can be kept for a long time in solution.

ZUCKERMAN: Can they be taken out of aqueous solution into lipids or organic media?

MARGERUM: The water solubility changes dramatically as you go from monochloramine to trichloramine. Trichloramine is quite fat-soluble, and the same thing is true as you to to the alkylamines. The N-chloroalkylamines can be quite fat soluble depending on what's on the rest of the chain. This is one of the reasons we were asking the question of whether we could have direct transfer from one N-chloro to another N-chloro species because it's quite possible that these could be concentrated in fatty areas. The monochloramine is hydrophilic; it has a low solubility in organic solvents.

R. L. JOLLEY (Oak Ridge National Laboratory): You showed the protonated monochloramine reacting with the glycyl-glycine. What was the pH where you determined your $10^7\ M^{-1}s^{-1}$?

MARGERUM: We varied the pH for these studies from 3 to 12. From pH 4 to 9, the observed rate was constant. At those pH levels, the amount of protonated monochloramine is very low, so we have to assign the proton to one of the two species, either to the amine or the monochloramine. We assign it to the monochloramine as the more logical reactant. This is how we get that rate constant that I gave.

JOLLEY: So the half-life of 1.3 hours was over a pH range of 5 to 9 which you would expect in natural waters?

MARGERUM: Right. By the way, I would like to mention I used Dr. Jolley's figures with several of these slides.

G. GORDON (Miami University, Ohio): Water treatment is moving in the direction of combined chlorine-chlorine dioxide because of intermediates and the use of some of these intermediates. Have you looked at any of the oxidation reactions of chloramines using ClO_2 to see what the products might be and what the rates are?

MARGERUM: We haven't. There is a tremendous amount of chemistry here which needs to be done. Much of the earlier work ran into all these problems (instabilities, etc.). I'm sure that as we investigate reasons for some of the instabilities, we will see some of the reactions you mention.

RECEIVED August 22, 1978.

18

Ozone and Hydroxyl Radical-Initiated Oxidations of Organic and Organometallic Trace Impurities in Water

JÜRG HOIGNÉ and HEINZ BADER

Swiss Federal Institute of Technology Zurich, Institute for Aquatic Sciences and Water Pollution Control, EAWAG, CH-8600 Dubendorf, Switzerland

Ozonation is sometimes thought of when impurities present in drinking water or wastewater have to be oxidized. For instance in Switzerland, drinking water for a considerable number of communities is supplied from lakes. Most of these waters are ozonized in order to improve taste, colour and disinfection. The lakes which are used as water resources for these drinking waters are situated in densely populated areas. They are bordered by many industries. Accidents might occur at any time through which the reservoir might be polluted by toxic or taste forming chemicals. We therefore are interested to learn more about the kinetics by which ozonation can oxidize organic pollutants even in cases when the pollutants are only present as trace impurities.

Experience has shown that only very extended ozonation can lead to full mineralization of the solutes - i.e., oxidation to carbonate, nitrate, sulphate and oxidized metal ions. However, because of technical and economic reasons, ozonation processes, applied for water treatment, generally have to be stopped after a reaction time of 10 to 30 minutes and only an ozone concentration below 10^{-4} M can be maintained during the ozonation process. The amount of ozone added is generally less than 10^{-3} mole/L. We therefore are confronted with the question: Which solutes are oxidized to what extent during the time available in a practical ozonation process?

The Reaction Model

Most of the results we found so far on ozonation processes in water fit well into the following model (see Fig. 1):

On the one hand, part of the ozone (O_3) dissolved in water reacts directly with the solute M. Such direct reactions are highly selective and for many solutes rather slow (time scale of minutes). On the other hand, part of the ozone added to the water decomposes before it reacts with solutes or before it is stripped off. This

0-8412-0461-6/78/47-082-292$05.00/0

decomposition leads to free radicals. Among these the OH˙ radicals belong to the most reactive oxidants known to occur in water. Up to 0.5 mole of these OH˙ radicals are formed per mole of ozone decomposed (1). OH˙radicals can easily oxidize organic, organo-metallic and inorganic solutes. They exhibit little substrate selectivity. Even bicarbonate and carbonate ions are oxidized by OH˙ radicals; thus OH˙ radicals are consumed quickly even in non-polluted waters (microsecond time scale).

The decomposition of ozone is catalysed by hydroxide ions and proceeds faster the higher the pH. The decomposition of ozone is additionally accelerated by a radical-type chain reaction in which OH˙radicals can act as chain carriers. Some types of solutes, for instance aromatic hydrocarbons, react with OH˙ radicals and form secondary radicals (R˙) which also act as radical chain-carriers. Other solutes, for example bicarbonate ions (2) or methylmercury hydroxide only reduce the primary radicals to species which do not act as chain carriers. Such solutes quench the radical-type chain reaction. Therefore the lifetime of O_3 in water depends on the pH as well as on the solutes present. Examples of lifetimes are given in Fig. 2.

The overall oxidation initiated by ozonation is a superposition of oxidations by the "direct reaction" and by the "OH˙ radical reaction". The two processes are optimized by different process parameters and lead to different daughter products. For many systems encountered in practice these two processes are of comparable importance. Data on the direct reactions can be obtained by studying ozonations performed at low pH values, where the ozone decomposition becomes slow, and by adding free radical scavengers which inhibit the effect of the "radical reaction". Data on the "radical reaction" can be obtained by following the depletion of solutes which only react very slowly in "direct reactions" and (or) by limiting the lifetime of the ozone for instance by increasing the pH.

Experimental

Experiments performed to determine rate constants with which ammonia or organic solutes react with ozone are described in references (3) and (4). A description of the measurements of relative rates with which solutes are eliminated is given in references (1) and (3).

Methylmercury hydroxide was prepared from CH_3HgCl (Merck; synthesis grade) by methods described in ref. (13). It was stored in lightprotected bottles as 0.45 M solutions. Its oxidation by ozonation was determined from the amount of methyl mercury mineralized to inorganic species found after all ozone added had reacted in closed bottles. Minor modifications of the Hatch and Ott method

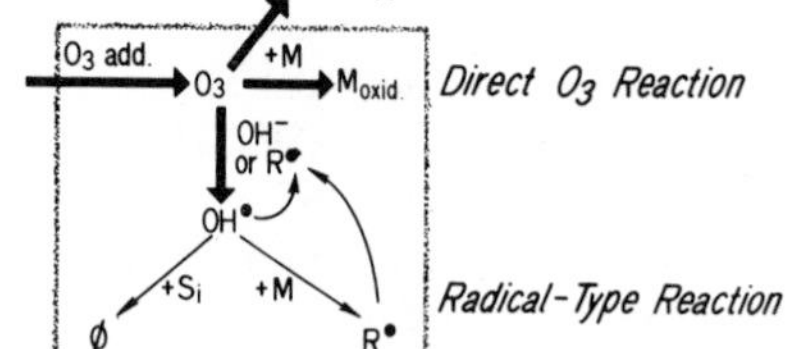

Figure 1. Scheme of reactions initiated by ozone in water

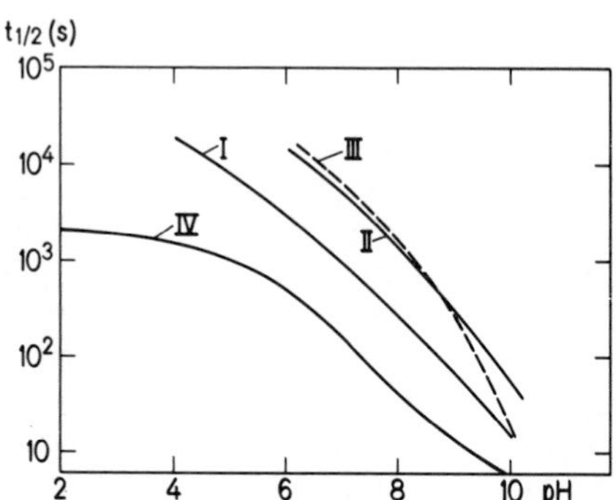

Figure 2. Half-life of ozone in different types of water vs. pH. (I) distilled water; (II) distilled water + 2mM carbonate ($\Sigma(CO_2)$); (III) distilled water + 0.3mM CH_3HgOH; (IV) wastewater (70% secondary effluent; DOC = 7 g/m^3).

based on atomic absorption spectroscopy (14) were applied for the determination of the amount of methylmercury mineralized. The concentration of the remaining methylmercury was calculated from the difference between the amount of methylmercury applied and the amount found as mineralized. For each series the 100% value was determined from samples in which all the methylmercury was oxidized at pH 10.5 by adding an excess of ozone. This 100% value was in good agreement with values found from direct calibrations with inorganic mercury compounds. All absorption signals were corrected for the signal found on non-oxidized samples (about 2% of 100% value).

Organotin compounds were laboratory samples supplied from the National Bureau of Standards. The concentrations of the aqueous solutions were measured by atomic absorption (flame).

Hydrogen peroxide was taken from a 30% Merck p.a. product which contained about 30 mg/l total organic carbon (measured after decomposition of H_2O_2 by MnO_2). This value could not be changed by a simple distillation procedure. Hg compounds were not present in significant amounts. The concentration of the H_2O_2 was determined by iodometric titration .

Results and Discussion

The direct reaction of ozone with solutes. Some solutes (M) become oxidized directly by molecular ozone:

$$1/\eta \cdot O_3 + M \xrightarrow{k} M_{oxid.}$$

Experiments show that the rates of these reactions are 1st order with respect to ozone and, as a rule, nearly 1st order with respect to solute concentration (3,4). Therefore, the rate by which a solute is oxidized, and the rate by which ozone is depleted becomes:

$$- d[M]/dt = - \eta \cdot d[O_3]/dt = \eta \cdot k \cdot [O_3][M] \qquad (1)$$

The change of the relative concentration of M with time in a batch- or plug-type reactor can be formulated by integrating equation (1) over reaction time t:

$$- \ln ([M]_e/[M]_o) = \eta \cdot k \cdot \overline{[O_3]} \cdot t \qquad (2)$$

where, $[M]_o$, $[M]_e$: initial and end concentration of M

η: yieldfactor of M eliminated per ozone used (mole/mole)

k rate constant of overall consumption of ozone by M and instable daughter products.

$[\overline{O_3}]$: mean O_3 conc. during reaction period t.

The relative solute elimination by this "direct reaction" depends only on the mean concentration of ozone (mean over reaction time), the time the ozonation lasts, the rate constant k and the yield factor η (3). The yield factor is about 1.0 in cases where the solute is an olefinic compound. It is also about 1.0 when organomercurials are ozonized in chloromethylene (8). It is somewhat lower for many other types of solutes and assumes a value of about 0.25 in the case of ammonia oxidation in water (4).

We determined k values for many solutes in aqueous solutions by measuring the rate of ozone consumption in presence of a solute concentration which was high enough not to become changed significantly by the ozone reactions and observing reaction conditions, such as for pH, to avoid interferences by free radical reactions (3, 4). For such cases we may write from equation (1) for a batch-type reactor:

$$-\ln([O_3]_t / [O_3]_o) = k \cdot [M]_o \cdot t$$

or: $$1/\tau_{O_3} = k \cdot [M]_o$$

where, $[O_3]_o$, $[O_3]_t$: initial concentration of O_3 and conc. at time t

τ_{O_3} : lifetime of O_3.

The rate constants can be calculated from the lifetime values of ozone measured for different concentrations of a specified solute M. The main problem for all these measurements in water was to avoid disturbances by radical reactions (compare Fig. 1). Criteria for the occurrence of pure direct ozone reactions were: (i) the kinetics stayed first order in ozone and solute concentrations over the entire concentration range measured; (ii) in the pH range employed, small changes of pH (one pH unit) had no effect on the rate of the ozone depletion; (iii) the addition of radical scavengers (bicarbonate or propanol) had no effect on the results.

In Fig. 3 are presented a few of about 60 rate constants we have measured so far. (3). The rate constants may be read on the left-hand scale. Also "elimination lines" are drawn into this figure indicating eliminations down to 37% (by a factor of e), to 1% or to 0.01 %, respectively. The required ozonation time leading to such eliminations may be read off the horizontal time scale. The scale in Fig. 3 is calculated for a plug- or batch-type reactor and for a mean ozone concentration of 10^{-4} M (about $5 g/m^3$). This ozone concentration can be considered as a limiting value achieved when commercial ozonators are operated with dry air. The scale would increase inversely to the ozone concentration.

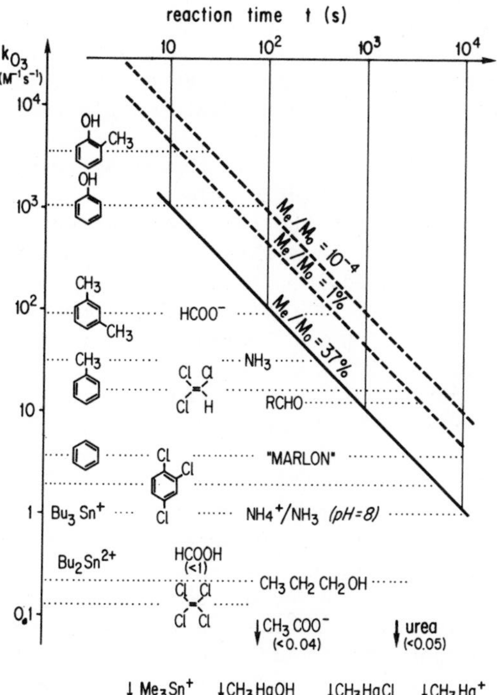

Figure 3. Examples of direct reactions of ozone with solutes in water. Ordinate values: reaction rate constants (3). Abscissa values: calculated reaction time needed to reduce the solute concentration to the relative value indicated on the elimination lines. Assumptions: $\eta = 1.0$; $[O_3] = 10^{-4}$M ($\sim 5\ g/m^3$).

The relation between required ozonation time and reaction-rate constant given in this figure shows that in a drinking water or in a secondary effluent of a domestic wastewater (at pH below 9)solutes may react significantly by direct reactions with ozone only if their rate constant is beyond about 50 $M^{-1}s^{-1}$. For solutes of lower reactivity and in case of lower ozone concentrations, oxidations may rather follow a preliminary decomposition of ozone molecules to $OH^{\cdot}$ radicals (to be discussed). Fig. 3 does not include data on solutes which react so fast with ozone that the rate of the chemical reaction is beyond the rate-determining step of the overall process. For instance nitrite and compounds characterized by unsubstituted olefinic double bonds are above scale.

The compounds listed show that the direct reactions of ozone molecules with solutes are highly selective. Even small modifications of a chemical structure have large effects on the rate constants. Examples: aldehydes, which occur as an ozonation product in lakewater (5), need 1000 sec of exposure to 10^{-4}M ozone for an elimination to 37% of their initial concentration. A comparable reaction time would be required for toluene. However, benzene would require 5000 sec.

Examination of the examples quoted in Fig. 3 indicates that ozone also reacts in water as an electrophile. Substituents which decrease the electronegativity of the reaction centers of the molecules render them nonreactive. Extended studies on substituted benzenes showed that the reaction rates of these compounds increase with the electronegativity as expected from data known for other electrophilic reactions. When applying the Stock and Brown linear energy relation (6) a reaction constant $\rho = -2.9$ was found when σ-values were based on σ_{ρ}^{+}-values (3).

The simple types of organotin cations we tested did not interact with ozone within a reaction time of practical interest. The reaction rate constants determined were approximately: Me_3Sn^+ (pH 3.3) 0.006 $M^{-1}s^{-1}$: Bu_3Sn^+ (pH 2.3 - 4.0) 1.7; Bu_2Sn^{+2} (pH 1.9) 1.5 (1.2 mM solution of the chloride compound, no buffer added). Approximate rate constants are included in Fig. 3. These rate constants are within the range of values expected for reactions of ozone with corresponding alkyl groups such as present in alkyl alcohols (3).(The corresponding organotin hydroxides did not show enough solubility for performing this type of measurements).

No direct reaction of methylmercury cation, methylmercury chloride or methylmercury hydroxide with ozone could be measured. Such solutes even stabilized the ozone concentration to some extend by quenching the radical induced chain reaction which otherwise would accelerate the decomposition of ozone (see Fig.2). Their rate constants for the direct interaction with ozone must therefore be lower than the values indicated in Fig.3, i.e.k < 10^{-2} $M^{-1}s^{-1}$.

TABLE 1

Direct Reactions of Ozone with Organomercurials

M	k ($M^{-1} \cdot s^{-1}$)
CH_3HgCl	0.007
CH_3CH_2HgCl	0.003
$(CH_3)_2CHHgCl$	0.02
$(CH_3)_2CHHgI$	0.03
$(CH_3)_2CHOH$	0.04

*) measured in CH_2Cl_2, 0°C

Examples from Waters et al. (8)

The organomercurials for which we measured the rate constants seem not to represent exceptions. This can be illustrated by comparing constants Waters et al. measured on these and other types of organomercurials in methylene chloride (8) (see examples in Table 1). Even though rate constants in water may differ by orders of magnitudes from those measured in non-aqueous systems (3), such comparisons can give some informations on the general trends. Waters et al. also showed that n-alkylmercury halides react with ozone to produce all possible acids and ketones upon chain breakdown. That means, that the rate of cleavage of carbon - carbon bonds competes with the rate of cleavage of the carbon - mercury bond and that also for this reason the scission of the carbon - mercury bond must be assumed to be a rather slow process.

Based on these examples we must expect that organomercurials will generally have a small chance to get directly oxidized by molecular ozone in a water treatment process except in cases where the reactivity of such compounds is due to other functional groups present in the organic part of the molecule.

Oxidations initiated by OH˙ radical reactions. The amount of the secondary oxidants produced in an ozonation process is proportional to the amount of ozone decomposed in water. The oxidations due to these secondary processes are initiated mainly by OH˙ radicals. (Decomposition of ozone can also lead to further secondary oxidants. From kinetic studies on systems containing different types of radical scavengers we conclude that only the OH˙ radicals determine the kinetics of the radical type reactions). The fraction of these OH˙ radicals available for oxidation of a specified solute M is proportional to the rate with which these oxidants react with M, divided by the sum of all the rates by which these are consumed by all other solutes present in the water:

	reaction:	rate of $OH^{\cdot}$ consumption:	
$O_{3,\Delta} + \eta' OH^{\cdot}$	$\xrightarrow{+ M}$	$k'_M[M] \cdot [OH^{\cdot}]$	(4)
	$\xrightarrow{+ \Sigma S_i}$	$\Sigma(k'_i \cdot [S_i]) \cdot [OH^{\cdot}]$	

Where, $O_{3,\Delta}$: O_3 decomposed during process

η': yield for $OH^{\cdot}$ formation from $O_{3,\Delta}$

k'_M: 2nd order reaction-rate constant for $OH^{\cdot}$ with M

$\Sigma(k'_i\ [S_i])$: rate of $OH^{\cdot}$ scavenging by all solutes present including by O_3 and M.

The reactivity of $OH^{\cdot}$ radicals towards most organic solutes present in a water is high. Even carbonate ions, free ammonia, hydrogen peroxide and ozone itself may react with them (for selected k' values see Fig. 4). Therefore, in natural water or in wastewater many solutes contribute to the consumption of this secondary oxidant.

We can characterize the oxidation of a specified trace pollutant M in a given water by the following expressions which can be derived from formulations given above in equation 4 (9):

If: $$k'_M\ [M] << k'_i\ [S_i] \quad (5)$$

we may write: $$\ln ([M]_e/[M]_o = -O_{3,\Delta}/\Omega_M$$

or: $$[M]_e/[M]_o = e^{-O_{3,\Delta}/\Omega_M} \quad (6)$$

where: $$\Omega_M = \frac{1}{\eta'.\eta''} \cdot \frac{\Sigma(k_i'[S_i])}{k'_M} \quad (7)$$

thereby: Ω_M = oxidation competition value= amount of ozone which must be decomposed per volume to reduce the solute concentration [M] by a factor of e(i.e. to 37% of its initial value $[M]_e/[M]_o = 0{,}37$ for $O_{3,\Delta} = \Omega_M$).

thereby: η'' = number of oxidations initiated per $OH^{\cdot}$ radical consumed by solute M.

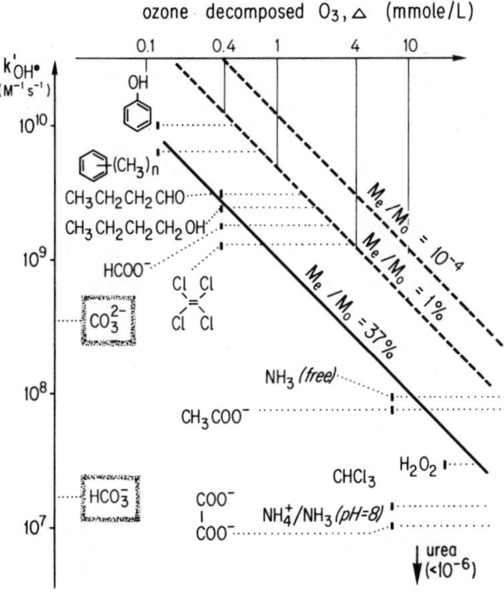

Figure 4. Examples of oxidations initiated by OH˙ radical reactions. Ordinate values: reaction rate constants of OH˙ radical reactions (see *Ref.* 1). *Abscissa: calculated amount of ozone to be decomposed to OH˙ radicals to achieve a solute oxidation leading to a relative solute concentration indicated on the elimination lines. Assumptions:* $\eta' \cdot \eta'' = 0.5$; *rate of radical scavenging* $= \Sigma (k_i[S_i]) = 5 \cdot 10^5 s^{-1}$. *(Data from Ref.* 9 *for secondary effluents.)*

The oxidation competition-values, Ω_M values, can easily be determined in a water for solutes M when conditions prevail under which these solutes M react only with OH$^\bullet$ radicals and not with ozone or other oxidants.

As long as M represents a trace pollutant (condition eq. 5 fulfilled) Ω_M values do not depend on solute concentration. Experimental examples measured on such types of waters which do not change their competition value,(that means their scavenging efficiency) during the ozonation, are given in Fig. 5,6,9,11. Fig. 5 represents the efficiency with which ozone added to a model solution, containing a small load of octanol, mineralizes methylmercury hydroxyde (resp. phosphate). The linearity of the depletion curve, when plotted in a semi-log scale, corresponds formally with equation 6. The efficiency of the oxidation is not changed, even when the solute concentration of the methylmercury is changed by two orders of magnitudes. In water from the lake of Zurich comparable depletion rates (Ω values) for toluene were found in case where the toluene concentration was increased from the background level of the lakewater (25 $\mu g/m^3$) by spiking to 250 mg/m^3, i.e. by a factor of 10^4 (5).However, the Ω_M values are characterized by the composition of the aqueous solution (compare eq.7) Results on experiments in which the Ω values for methylmercury or benzene were measured vs. increased loads of model scavenger (n-octanol) are shown in Fig. 7. Comparable relationships are found when other impurities are spiked to the aqueous solutions.

Fig. 8 shows in case of benzene how the amount of ozone used per solute elimination ($\Delta O_3/\Delta M$) changes with pH when carbonate is present. (The expression $\Delta O_3/\Delta M$ was used in our earlier experiments where we applied solute concentrations which were large and therefore did not fulfil condition of equation 5. However, these values give a semi-quantitative information on the Ω values).Further examples for different types of solutions containing carbonates as a hydroxyl radical scavenger are given in reference (2). All the examples quoted on carbonate containing solutions demonstrate that at higher pH values a higher amount of ozone is required to achieve a certain elimination factor: this is due to the dissociation of bicarbonate ions (HCO_3^-) to carbonate ions (CO_3^{2-}) which scavenge OH$^\bullet$radicals much faster (see k' values given in Fig.4). Also in natural water part of the pH effect encountered is due to the increased amount of carbonate ions present at the higher pH value (2). Such an increase of Ω values with pH determined for a lake water is exemplified in Fig. 9.

In a secondary effluent of a domestic wastewater the Ω values are about ten times higher than in lakewater (pH 8). This means, that about 10 times more ozone is required to achieve a

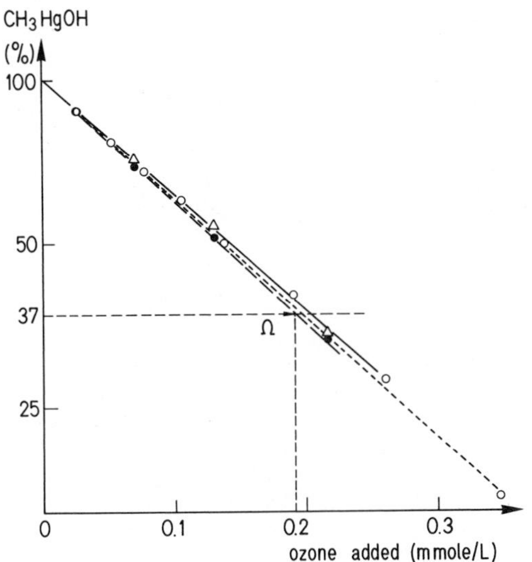

Figure 5. Mineralization of methylmercury hydroxide by the action of ozone decomposed to OH˙ radicals. The solutions contained 5 g/m³ octanol added as a model-type impurity; pH 10.5; 0.05M sodium phosphate buffer. [M]₀: (●), 5 · 10⁻⁹M CH₃HgOH; (○), 5 · 10⁻⁸M CH₃HgOH; (△), 5 · 10⁻⁷M CH₃HgOH.

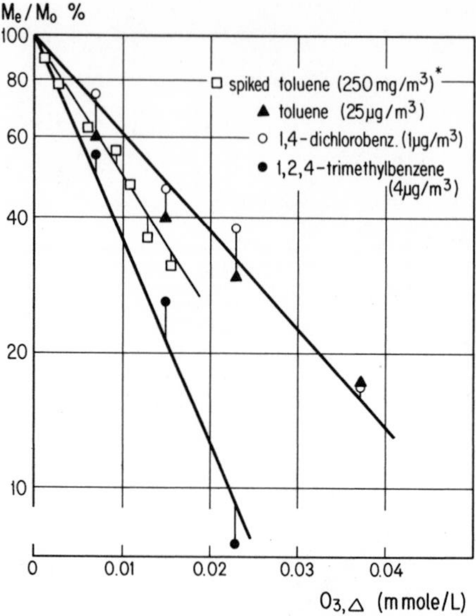

International Symposium on Ozon and Wasser

Figure 6. Elimination of trace impurities in Lake Zurich water upon decomposition of different amounts of ozone. In one experiment 250 mg/m³ of toluene was spiked into the water before ozonation. (Measurements performed on samples M from different dates) (5).

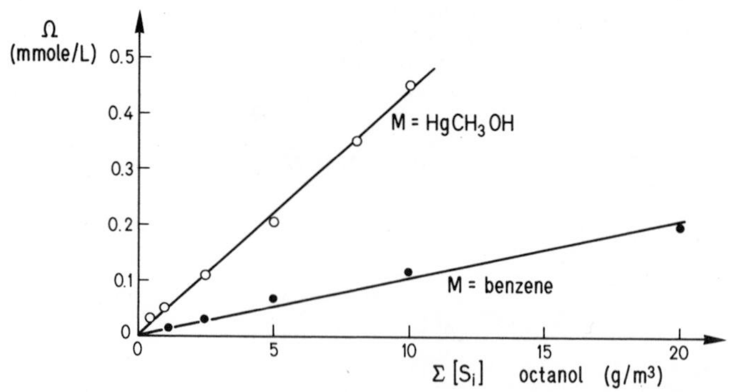

Figure 7. Oxidation competition values (Ω) for the elimination of micropollutants ($HgCH_3OH$ or benzene) in water vs. the concentration of octanol which was used as a model type of scavenger: pH 10.5; 0.05M sodium phosphate buffer.

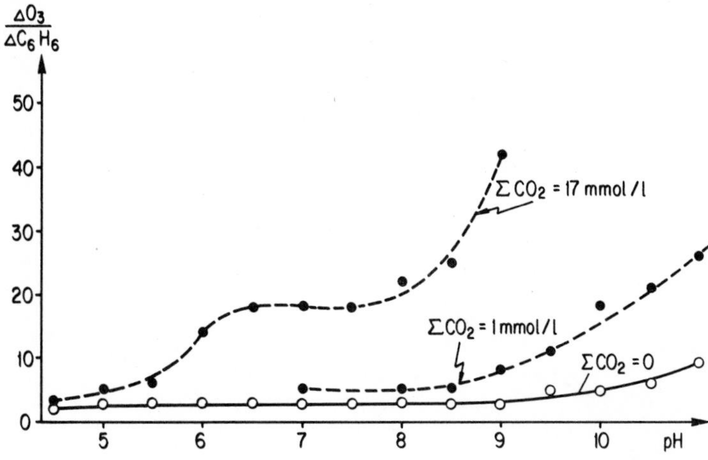

Vom Wasser

Figure 8. Ozone to be decomposed per oxidation of benzene (mol/mol) in waters of specified carbonate concentrations at different pH values. The carbonate ions are more reactive scavengers for OH˙ radicals than the bicarbonate ions. Therefore the effect of a given amount of total carboate increases with pH (2).

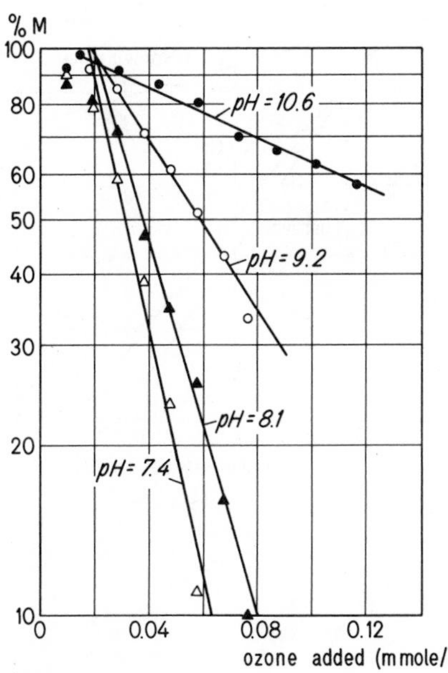

Figure 9. Elimination of benzene spiked to water of Lac de Bret by decomposition of ozone. DOC = 4 g/m³; Σ [CO_2] = 1.6mM. The pH was varied by 0.05M sodium phosphate buffer.

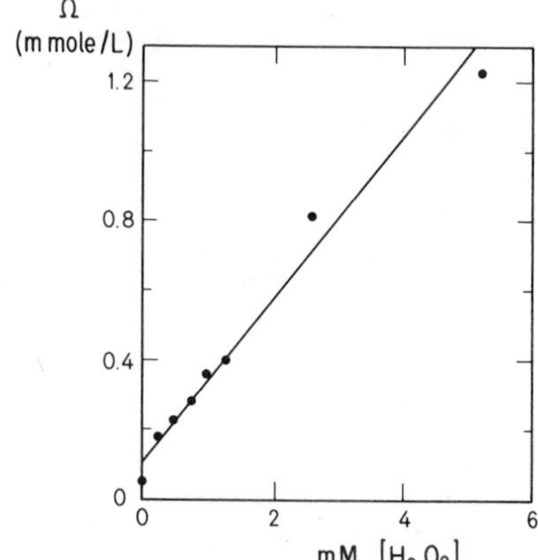

Figure 10. Effect of added hydrogen peroxide on the oxidation competition value, Ω of a water. Data for the mineralization of methylmercury hydroxide. $[CH_3HgOH] = 5.10^{-8}M$; pH = 8; 0.05M sodium borate buffer.

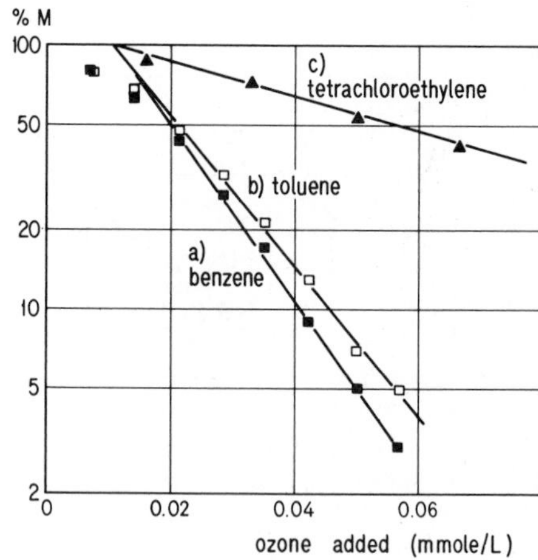

Figure 11. Elimination of different types of solutes spiked into water of Lac de Bret (DOC = 4 g/m³ Σ(CO_2) ~ 1.6mM; ph 8.3; benzene 80 mg/m³; toluene 103 mg/m³, tetrachloroethylene 500 mg/m³.

comparable elimination factor (9). In addition, some of the initial amount of ozone added to a wastewater is used up in a spontaneous reaction and only beyond this the added ozone becomes available for other oxidations. However, the efficiency with which further ozone additions oxidize solutes is not changed appreciably by further pre-ozonation treatment (see ref. 9).

Fig. 10 shows the scavenging effect of hydrogen peroxide on the oxidizing species formed when ozone decomposes: The Ω value for methylmercury hydroxide increases by about 0.25 mM ozone per mM H_2O_2 added. An effect of this order of magnitude is expected from eq. 7 and the relative rate constants with which $OH^{\cdot}$ radicals select between H_2O_2 and methylmercury hydroxide.

$OH^{\cdot}$ radicals reacting with H_2O_2 become reduced by H_2O_2. Thereby $HO_2^{\cdot}$ (O_2^-) is formed by:

$$OH^{\cdot} + H_2O_2 \rightarrow H_2O + HO_2^{\cdot}$$

$$HO_2^{\cdot} \rightleftarrows O_2^- + H^+$$ (for lit.see ref. (1,4))

Now the experiments using H_2O_2 show that the secondary radicals formed do not contribute significantly to the oxidation of methylmercury hydroxide. That means that O_2^-, which is expected to be also produced during the docomposition of ozone, does not contribute to the oxidation of methylmercury species.

In a given water the amount of ozone required to eliminate a solute B by a factor e, compared with that required for elimination of impurity A results from equation 7:

$$\Omega_A/\Omega_B = \eta''_B \cdot k'_B \;/\; \eta''_A \cdot k'_A \;\cong\; k'_B \;/\; k'_A$$

The approximation on the right hand side can be tolerated for many systems, because the yield factor η'' for many solutes are of comparable magnitudes. That means that the slope of the logarithmic expression of the relative solute concentration vs. the amount of ozone decomposed decreases with the relative reaction-rate constant with which $OH^{\cdot}$ radicals react with a solute. An example is given in Fig. 11. Because of the importance of $OH^{\cdot}$ radical reactions in radiation chemistry of aqueous systems, lists of hundreds of rate constants for $OH^{\cdot}$ radicals are published (for literature see reference (1)). Fig. 4 represents a few of such values. It shows that most of the organic solutes have constants in the order of magnitude of 10^9 $M^{-1}s^{-1}$. (The aspect of this figure is biased somewhat because we also wanted to emphasize those solutes which exhibit rather low constants). Fig. 4 also shows elimination lines. Projecting from these lines vertically to the abscissa, the amount of decomposed ozone required to yield a certain amount of solute oxidation by this "radical-type reaction" may be read. The scale is based on eq. 6, assuming a yield factor $\eta'.\eta'' = 0.5$ and a sum of the rates of radical consumption, $\Sigma(k'_i.[S_i])$, of $5\cdot10^5$ s^{-1}. These numerical assumptions give an upper limit for the order of magnitude found for secondary effluents at pH 8.

If we compare the oxidation competition values Ω, measured for different solutes, we find that the Ω value for methylmercury hydroxide is smaller than that determined for benzene (measured in a 0.05 molar solution of borate/boric acid of pH 8). Because benzene and methylmercury hydroxide exhibit comparable yield factors when oxidized in pure water ($\eta'.\eta'' \sim 0.3$) this ratio in Ω values reflects the relative rate constant for reactions of the two solutes with $OH^{\cdot}$ radicals. In case of methylmercury compounds these ratios of Ω values, however, are strongly influenced by the types of anions present in the solution even when the characteristics of the solution itself are not changed in respect to the rate of scavenging of $OH^{\cdot}$ radicals. Fig. 12 shows how the Ω values change at given pH with chloride ion concentration. From this it seems that the methylmercury chloride has a higher rate constant for reactions with $OH^{\cdot}$ radicals than the methylmercury hydroxide: In presence of high concentrations of chloride, methylmercury is oxidized nearly as fast as benzene. At present we, however, cannot explain why Ω changes when the pH is increased from pH 7.4 to 8.4 and why the chloride effects the Ω values already at about 3 to 10 times lower concentrations than those necessary to convert CH_3HgOH significantly to CH_3HgCl. The range of reactivity however occurs in the chloride-concentration region in which the final product is not $Hg(OH)_2$ but rather Hg(OH)Cl. (For equilibrium-constants see e.g. reference 10). This enhanced oxidation in presence of small chloride concentrations is not due to the formation of OCl^- by the ozonation: Neither a direct addition of OCl^- nor a preozonation of the chloride containing water lead to a comparable mineralization of methylmercury.

A concentration of only 0.05 M sodiumphosphate/phosphoric acid buffer at pH 8 also increased the measured rate of the methylmercury oxidation by a factor of about 2 when compared with the rate found in a corresponding borate solution. This would mean that the phosphate complex is also more easily oxidized by $OH^{\cdot}$ radicals than the methylmercury hydroxide. However, at pH 10.5 the rate was the same, whether the pH was adjusted by phosphate (0.05 M) or sodium hydroxide. We assume that this difference is due to the higher equilibrium concenctration of the methylmercury hydroxide at this higher pH value.

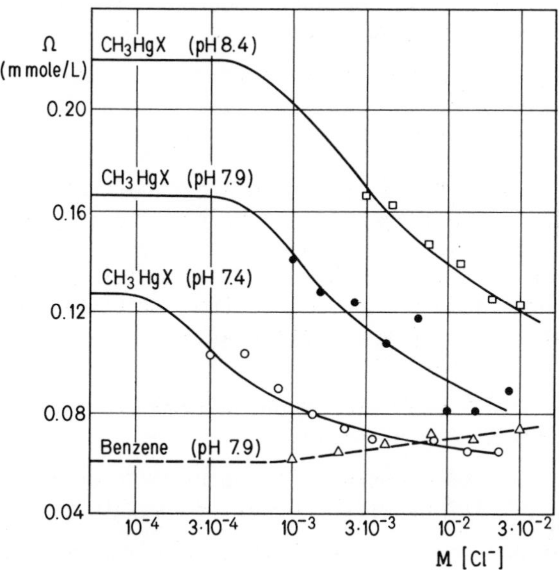

Figure 12. Oxidation competition value (Ω) measured for water containing 2.5 g/m³ octanol as a model impurity. Ω is measured for the methylmercury mineralization or benzene oxidation in the presence of different chloride ion concentrations. $CH_3HgX = 5.10^{-8}M$; 0.05M sodium borate buffer.

Conclusions

For quantitative assessments of oxidations initiated by ozone in water the direct reaction of ozone with the specified solutes and the OH·radical type mechanism have to be accounted for. The extent of the direct reactions can be calculated from

- mean ozone concentration during reaction time, $[\bar{O}_3]$
- reaction time, t
- second order reaction-rate constant for the direct reactions of molecular ozone, k

k values can be determined by experiments on model solutions when process parameters are chosen in a way such to eliminate all interactions by the free radical mechanism initiated by decomposed ozone.

However, many of the organic and organometallic solutes react too slowly and are hardly oxidized by the direct reaction with molecular ozone. But oxidations of all types of organic and organometallic solutes can be initiated by hydoxyl radicals which are formed when ozone decomposes to free radicals. The amount of oxidations of specified trace pollutants initiated by this mechanisms can be estimated from:

- amount of OH·radicals produced from ozone decomposed during process
- reaction-rate constant with which OH·radicals initiate oxidations of the specified solute M
- sum of all the rates with which all radical scavengers present in the water consume the OH·radicals in competition to the specified solutes.

A water can be characterized considering the extent of OH· radical consumption by using a substance such as benzene as a reference solute. For a few hundred solutes the relative rate-constants for reactions with OH radicals, can be deduced from literature data. Therefore, kinetic standardisations based on a kinetically well known solute such as benzene are meaningful.

From the described ozonation experiments we may conclude that the oxidation of methylmercury compounds is initiated only by the OH·radical type mechanism. No data on the rate of reactions of organometallics with OH·radicals are however quoted in the literature. Our experiments in which we initiated the OH·radicals by decomposing ozone show that methylmercury chloride or phosphate react with a rate constant comparable to that of benzene. However, the methylmercury hydroxide reacts in absence of chloride four times slower, and in absence of phosphate two times slower. Therefore four or two times more ozone has to be decomposed to OH˙ radicals to achieve a comparable elimination factor for these micropollutants in a given water.

More research will be needed to explain the phenomena that even such low concentrations of chloride ions, which are too small to convert a significant part of the methylmercury hydroxide to methylmercury chloride, significantly changed the apparent rates of oxidations of the methylmercury.

Acknowledgements

We thank Dr. F.E. Brinckman of the National Bureau of Standards for the encouragement to include organometallic compounds in our research program; he and his coworkers provided us with the tinorganic compounds and a description of methylmercury speciation as a function of anion concentration. Their comments on our results are highly appreciated. We also thank Prof. W. Stumm for comments and continous interest and to Dr. H. Hohl for communicating experiences on aquous methylmercury systems.

Literature Cited

1. Hoigné J. and Bader H., Water Res., 10, 377 (1976).
2. Hoigné J. and Bader H., Vom Wasser, 48, 283-304 (1977).
3. Hoigné J. and Bader H., in prep. (1978).
4. Hoigné J. and Bader H., Environm. Sci. Technol., 12, 99 (1978).
5. Hoigné J., Bader H. and Zürcher F., Internat.Symp.Ozon and Wasser, Berlin, Mai 1977, (Proceedings in Press).
6. Exner O., in "Advances in Linear Free Energy Relationships", J. Shorter ed. (Plenum Press 1972), p.32.
7. Spialter L., LeRoy P., Bernstein St., Swansinger W.A., Buell G.R. and Michael E., in "Advances in Chemistry Series 112", P.S. Baily, ed. (ACS 1972), pp. 65-77.
8. Waters W.L., Pike P.E. and Rivera J.G.,:ibid. pp.78-99.
9. Hoigné J. and Bader H., Intern. Assoc. on Water Pollution Res., 96th Intern.Conf., Stockholm (1978) in: Progr. in Watertechnology, 10, 657-671 (1978).
10. Erni I.W., "Relaxationskinetische Untersuchungen von Methylquecksilberübertragungsreaktionen", Dissert., Swiss Fed.Inst. Techn. Zurich (1977).
11. Pike P.E., Marsh P.G., Erickson R.E., and Waters W.L. Tetrahedron Lett. 31, 2679-2682 (1970).
12. Weber P. and Waters W.L., Proc. Montana Academy of Sciences, 32, 66-69 (1972).
13. Geier G., Erni I. and Steiner R., Helv. Chim. Acta 60, 1, 9-18 (1977).
14. Hatch W.R. and Ott W.L., Anal.Chem. 40, 14, 2085-2087 (1968).

Discussion

J. J. ZUCKERMAN (University of Oklahoma): The ultimate oxidation products of the organic compounds would be carbon dioxide and water, and of the organometallic compounds would be metal oxides, carbon dioxide and water. Do you know what the products are of the ozonolysis of natural waters containing these materials?

HOIGNÉ: Where we studied methylmercury we measured inorganic mercury by atomic absorption. In water, we measured ozonation products formed from different alkylmercury compounds and found different carbonic acids, aldehydes, and so forth. That is, C-C bond scission competed with the alkylmercury scission. That means that the alkylmercury bond reacts so slowly that even the alkyl group scissions can compete with this reaction.

ZUCKERMAN: Did you isolate any compounds in the case of larger alkyl groups which were hydroxylated on the alkyl chain?

HOIGNÉ: No. Based on the literature, and our analyses, we find mostly acids, ketones, or aldehydes. For instance, when we ozonize water from the lake of Zurich, we form a tremendous amount of aliphatic aldehydes which we see by gas chromatography.

ZUCKERMAN: You are using a chemical oxidant. In the Soviet literature there is a hint that electrolysis of waste water might be an effective way to deal with organometallic pollutants. Are you aware of this and do you have any comment on that?

HOIGNÉ: Yes. However, I do not know where to put it in relation to my work. You have to be very careful to define your system, considering pH and all other factors. Otherwise it is really difficult to use this information.

W. A. PRYOR (Louisiana State University): Is there evidence for epoxides from olefins or arenes?

HOIGNÉ: I'm not aware that epoxides can be formed in water, except in one case. Where you oxidize aldrin to get dieldrin. There have been many studies on epoxide formation in non-aqueous solutions. From these results we may assume you get epoxides, but only when the double bond which is oxidized is in a very rigid framework. That is also the case for the aldrin-dieldrin oxidation I mentioned.

F. GRAY (Colgate-Palmolive): Do you oxidize the chloride to hypochlorite or chlorine in these reactions?

HOIGNÉ: I don't know. They disappear when I look at the gas chromatography after ozonation. The yield factor is about 1.0

when compared with ozone utilized, and it is about 0.5 when compared with a free radical type (·OH) of reaction.

R. L. JOLLEY (Oak Ridge National Laboratory): This statement is relative to the epoxide question. The only epoxide to have been reported in aqueous ozonation was by Dr. Carlson, University of Minnesota. He reports that an epoxide was formed on aqueous ozonation of oleic acid. What his concentrations and yields were, I don't know.

RECEIVED August 22, 1978.

19

Partition of Organoelements in Octanol/Water/Air Systems

STANLEY P. WASIK

Chemical Thermodynamics Division, National Bureau of Standards, Washington, DC 20234

Introduction

The ability of organic compounds to bioconcentrate in the marine environment was found to be dependent upon the partition behavior of molecules between lipid and aqueous phases (1,2). The organic/water partition coefficient is defined as the ratio of the concentration of a chemical in the organic solvent to its concentration in water under equilibrium conditions in an organic solvent/ water system. Neely, Branson, and Blau (1) demonstrated that bioconcentration factors for chlorobenzenes and chlorophenols in trout muscles from water containing low levels of these compounds could be successfully correlated with their partition coefficients in the n-octanol/water system. Dunn and Hansch (3) compiled hydrophobic interactions of a large number of organic compounds and showed that these interactions quantitatively correlated with partition coefficients of organic/water systems. Leo (4) suggested that hydrophobicity, as measured by the n-octanol/water partition coefficient, is the most important parameter in bioaccumulation and biotransport. The role of hydrophobicity in non-biological transport may not be dominant but should, nonetheless, be important. Although values for the n-octanol/ water partition coefficient have been reported in the literature for over 10,000 compounds, there is a large number of environmentally significant molecules that are absent. One group of molecules for which the n-octanol/water partition coefficient have not been reported is the organometal compounds.

Wasik and coworkers (5,6) reported a method for determining the water/air partition coefficient from measurements of the concentration of the solute in the head-space in equilibrium with the liquid phase. In this paper, are presented, by a similiar method, our measurements of the n-octanol/water partition coefficients of dimethylmercury in distilled and sea water over a temperature range of 0-25°C. The apparatus is described and the various factors effecting the precision and accuracy of the partition coefficients are discussed.

Experimental Methods and Calculations

Apparatus. A schematic diagram of the apparatus and the equilibration cell used to measure the water/air partition coefficient is shown in Figure 1. The glass equilibration cell of volume V_t which contains a volume, V_L, of the aqueous solution and a small volume of the gas phase, $V_c = V_t - V_L$, is contained in a water bath where the temperature is controlled to within 0.01°C. The gas circulating pump was constructed from a stainless steel bellows (7.6 x 3 cm) by the NBS shop. The bellows was enclosed in a stainless steel jacket and was driven by pulsating compressed air with a duty cycle of 3 seconds. The air-solute mixture was pumped into the apparatus through the 1/16-inch tube connecting the check valve with the 4-port gas valve. The two glass check valves were made by the NBS glass shop. The pump circulated the air-solute vapor through a 6-port gas sampling valve. By switching this valve, a small sample volume of the gas phase (V_s = 0.1 cm^3) can be sent to the gas chromatograph for analysis without interrupting the circulation of the air-solute mixture. The gas mixture is then directed to a 4-port gas valve which in the closed-position passes the gas phase back to the pump, and in the open position directs the gas mixture through the equilibration cell and then back to the pump. The components of the apparatus are connected together with 1/16-inch stainless steel tubing and commercial tubing connectors. The total volume of the pump, valves, and tubing, V_R, is 85.25 cm^3. This part of the apparatus is contained in an air bath at 100 $\pm$ 0.05°C.

The equilibration cell used to measure the octanol/air partition coefficient was constructed in the form of a U; the bottom part of the U was a glass rectangular tube (4 x 1 x 1 cm) and the two side arms were glass tubes (1/4-inch OD x 3 cm). The cell was connected to the 4-port valve with 1/16-inch tubing and commercial reducing fittings. The n-octanol (saturated with water) was weighed in the cell and the volume of the solution was calculated from its density. The n-octanol was deposited on the bottom of the cell and the air-solute mixture circulated above it. The volume of the cell was 4.821 cm^3.

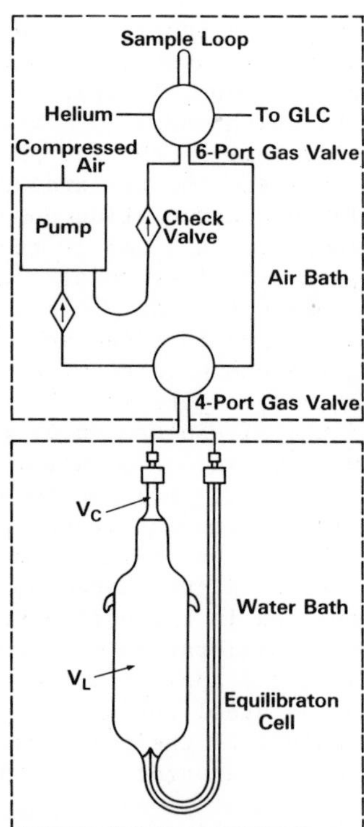

Figure 1. Schematic showing equilibration cell and apparatus for determination of water/air partition coefficients

Before assembly, the stainless steel portions of the apparatus were cleaned with trichloroethylene; the glass parts were washed with 10 percent (by weight) aqueous HF and rinsed several times with distilled water.

The chromatgraphic column was a Scot column (15 m x 0.5 mm ID) prepared with finely ground diatomaceous earth on a fused support and coated with a mixture of m-bis-(m-phenoxylphenoxyl) benzene and Apiezon L. The effluent gas was monitored by a hydrogen flame detector and an electronic integrator was used to measure the peak areas.

Commercial n-octanol was purified by successive washings with HCl, with dilute sodium bicarbonate solution, and finally with distilled water. The n-octanol, dried with sodium sulfate, was distilled under reduced pressure. The distilled sample was found by gas chromatography to be 99.97 mole percent n-octanol. The water used in the preparation of solutions was doubly distilled over potassium permanganate. The purity of the commercial diethylmercury and dimethylmercury was determined by gas chromatography to be 99.4 mole percent. The compound was used without further purification. The artificial sea water was prepared according to a recipe given by Sverdrup et al (7) and had a chloridity value of 19.0 percent. The chloridity is the total amount of chlorine, bromine, and iodine (in grams) contained in 1 kg of sea water where bromine and iodine are replaced by chlorine.

n-Octanol/Water Partition Coefficient. Consider a system composed of two immiscible liquid phases, n-octanol and water, and a gas phase, air, into which volatile solute vapor is introduced. Let the equilibrium concentration of the solute in the air be C_A, in the octanol (saturated with water) be C_o, and in the water (saturated with octanol) be C_w. The octanol/water partition coefficient, K(o,w), is defined as the ratio C_o/C_w, the water/air partition coefficient, K(w/a), as C_w/C_a, and the octanol/air partition coefficient, K(o/a) as C_o/C_a. At equilibrium

$$K(o/w) = C_o/C_w = K(o/a)/K(w/a) \qquad (1)$$

providing the solute concentration is sufficiently low in each phase so that there are no interactions between the solute molecules. For volatile solutes K(o/w) is difficult to determine accurately from individual measurements of C_o and C_w due to loss of solute by evaporation during sample manipulation. K(o/w) may be determined more conveniently and accurately by measuring K(o/a) and K(w/a) in separate experiments and then calculating K(o/w) from their ratio.

Water/Air Partition Coefficient. Distilled or sea water was weighed in the equilibration cell and its volume was determined from its density. The air-solute vapor mixture was introduced into the apparatus with the 4-port valve in the closed position. The gas mixture was then circulated through the apparatus for approximately 10 minutes and a small volume of the gas mixture was injected into the gas chromatograph via the 6-port gas sampling valve. The sampling was repeated until the solute peak area reached a constant value. The 4-port valve was then switched to the open position which allowed the gas mixture to bubble through the equilibration cell. After approximately 30 minutes successive periodic analyses of the gaseous mixture were made until the solute peak area reached a constant value.

From the ideal gas law the total moles of solute, n, in the apparatus is given by

$$n_t = PV_R/RT_R \tag{2}$$

where P is the partial pressure of the solute before equilibrium with the liquid phase, T_R is the temperature of the air bath in kelvins, and R is the gas constant. After equilibration with the liquid phase

$$n_t = \frac{P'V_R}{RT_R} + \frac{P'V_c}{RT_c} + \frac{P'K(w/a)V_L}{RT_c} \tag{3}$$

where P' is the partial pressure of the solute after equilibration with the liquid phase, T_c is the temperature of the water bath in kelvins and $K(w/a) = C_w/C_a = \frac{n_L RT_c}{V_L P'}$ where n_L is the moles of solute in the liquid phase. Combining equations (3) and (2), substituting A/A' = P/P' (where A and A' are the solute peak areas before and after equilibration, respectively), and solving for K(w/a)

$$K(w/a) = \frac{T_c(A/A' - 1)V_R}{T_R V_L} - V_c/V_L \tag{4}$$

This method for determining K(w/a) is the "solute absorption method."

K(w/a) may be determined by a second method. At equilibrium, with the 4-port valve in the open position, n_t may be expressed as

$$n_t = \frac{P_1V_R}{RT_R} + \frac{P_1V_c}{RT_c} + \frac{P_1K(w/a)V_L}{RT_c} \tag{5}$$

where P_1 is the partial pressure of the solute at the first equilibrium. The 4-port valve is switched to the closed position and the solute in V_R is purged and replaced with clean air. The 4-port valve is then switched to the open position and the air is circulated through the equilibration cell until equilibrium is established. Then

$$n_t = \frac{P_2V_R}{RT_R} + \frac{P_2V_c}{RT_c} + \frac{P_2K(w/a)V_L}{RT_c} + \frac{P_1V_R}{RT_R} \quad (6)$$

where P_2 is the partial pressure of the solute at the second equilibrium. Combining equations (5) and (6), substituting A_1/A_2 for P_1/P_2, and solving for K(w/a) gives

$$K(w/a) = \frac{V_RT_c}{T_R(A_1/A_2 - 1)V_2} - V_c/V_L \quad (7)$$

The procedure may be repeated i number of times giving the expression

$$K(w/a) = \frac{V_RT_c}{T_R(A_i/A_{i+1}) - 1)V_L} - V_c/V_L \quad (8)$$

when A_i is the solute peak area at the ith equilibrium. Similiarly it can be shown that

$$A_1/A_i = [\frac{V_RT_c}{T_R(K(w/a)V_L + V_c} + 1]^{i-1} \quad (9)$$

or

$$\log(A_1/A_i) = (i - 1)\log [\frac{V_RT_c}{T_R(K(w/a)V_L + V_c} + 1] \quad (10)$$

This method for determining K(w/a) is the "solute extraction method."

Partition Coefficients at Different Temperature. It is possible to obtain K(w/a) at various temperatures in terms of the values at some reference temperature, T_o. This can be done simply by measuring the relative solute partial pressure in the gas phase at different temperatures of the equilibration cell. Consider the system in equilibrium with the liquid phase at T_o

$$n_t = \frac{P_{T_o}V_R}{RT_R} + \frac{P_{T_o}V_c}{RT_o} + \frac{P_{T_o}K(w/a)V_L}{RT_o} \quad (11)$$

where P_{T_o} is the partial pressure of the solute and $K(w/a)_{T_o}$ is known from either a solute absorption or a solute extraction experiment. When the temperature of the liquid phase is changed to T

$$n_t = \frac{P_TV_R}{T_RR} + \frac{P_TV_c}{RT} + \frac{P_TK(w/a)_TV_L}{RT} \quad (12)$$

where P_T is the partial pressure of the solute at temperature T and $K(w/a)_T$ is the water/air partition coefficient. Combining equations (11) and (12), substituting A_{T_o}/A_T for P_{T_o}/P_T and solving for $K(w/a)_T$

$$K(w/a)_T = \frac{V_R T}{V_L T_R}(A_{T_o}/A_T - 1) + \frac{V_c}{V_L}(A_{T_o}/A_T - 1) + \frac{A_{T_o} T\, K(w/a)_{T_o}}{A_T T_o} \quad (13)$$

In determining $K(w/a)_T$ by this method, a small correction should be made for V_c and V_L due to the change in density of the liquid phase. This method for determining $K(w/a)_T$ is the "temperature variation method." In deriving equations (4), (7), and (13), the example was for a single solute system. In practice many solutes may be used in the system as long as they are at very low concentrations and can be separated by gas chromatography. Neither of these conditions presents a serious experimental problem.

n-Octanol/Air Partition Coefficient. The octanol/air partition coefficient, K(o/a), was measured by either the solute absorption, solute extraction, or the temperature variation method described above except that the U-shaped equilibration cell was used in place of the bubble cell. The U-shaped cell was designed to keep V_c small and to be able to weigh octanol conveniently.

Results

Values for K(w/a), K(o/a), and K(o/w) for dimethylmercury in distilled and sea water are given in tables I and II, respectively. Absolute values of K(w/a) and K(o/a) in distilled water were determined using the solute absorption method at two temperatures, 15.42 and 5.38°C. Values of K(w/a) and K(o/a) at other temperatures were obtained using the temperature variation method. Absolute values of K(w/a) and K(o/a) in sea water were determined at two temperatures 18.42 and 6.42°C. The remaining K(w/a) and K(o/a) were obtained using the temperature variation method. The numbers in parentheses are the standard deviations of the partition coefficients.

No K(o/w) values for dimethylmercury are reported in the literature. In order to compare K(o/w) values measured in this work with literature values, we determined K(o/w) for benzene (132), toluene (482), and ethylbenzene (1420) in distilled water at 25°C. The values may be compared with those values measured Hansch et al (8), e.g., benzene (135), toluene (490), and ethylbenzene (1413).

Discussion

The main cause of the spread in the K(w/a) and K(o/a) values arise from dispersions in the peak area measurements by the electronic integrator. This spread was reduced by turning the circulating pump off before sample injection into the gas

TABLE I

Temp. (°C)	Number of Observations	Dimethylmercury Partition Coefficients in Distilled Water K(w/A)	K(o/A)	K(o/W)
5.38	6	12.41(.050)	1800(6.2)	145.0(.76)
8.02	6	10.51(.041)	1590(5.4)	151.4(.78)
10.40	8	9.11(0.41)	1435(5.4)	157.1(.91)
13.10	7	7.82(.038)	1272(4.8)	162.6(.99)
15.42	6	6.80(.049)	1160(3.9)	170.6(1.08)
18.62	6	5.71(.031)	1004(3.9)	175.8(1.1)
20.00	9	5.25(.032)	950(3.6)	180.9(1.2)
24.48	6	4.13(.028)	790(3.4)	191.3(1.3)

TABLE II

Temp. (°C)	Number of Observations	Dimethylmercury Partition Coefficients in Sea Water K(w/A)	K(o/A)	K(o/W)
6.42	7	8.52(.041)	1720(9.1)	201.9(1.4)
10.02	6	6.82(.034)	1466(6.2)	215.0(1.4)
13.08	7	5.68(.030)	1282(5.2)	225.7(1.5)
15.64	7	4.90(.022)	1151(5.1)	234.9(1.5)
18.42	8	4.15(.021)	1010(5.0)	243.4(1.7)
20.00	6	3.82(.017)	950(4.4)	248.7(1.6)
25.00	6	2.88(.011)	760(3.6)	263.9(1.6)

chromatograph. A group of 30 observations of dimethylmercury peak areas was found to have a coefficient of variation of 0.75 percent with the pump on while the same number of observations gave a variation of 0.15 percent with the pump off. The pump was conveniently turned off by stopping the supply of compressed air to the pump.

In the methods described above for measuring K(o/a) and K(w/a), the term (A/A_1 - 1) appears in equations (4), (7), and (13). The error, particularly in the K(w/a) calculations, may be large for small values of A/A_1. This error may be reduced by choosing the volume of the liquid phase such that A/A_1 is large. In practice, several equilibration cells of different volumes should be available for this use.

This error may also be reduced by the proper choice of the method used for measuring K(w/a). For instance, the volume of the equilibration cell used in this study was 114.2 cm^3 with the liquid volume, V_L, equal to 112.2 cm^3. Using the solute adsorption method for dimethylmercury in distilled water at 16.02°C it was found that A/A_1 = 10.217 (K(w/a) = 6.00). An error of 0.50 percent in A/A_1 resulted in a 0.549 percent error in K(w/A). Under the same experimental conditions the solute extraction method gave A/A_1 = 1.098, resulting in an error of 5.31 percent in K(w/a). If, under the same experimental conditions, a solute with K(w/a) = 0.10 was measured using the solute extraction method; A/A_1 would equal 6.00 resulting in an error of 0.7% in K(w/a) whereas the solute absorption method would give A/A_1 = 1.20 and an error of 3.5 percent. In general, the solute absorption method is preferred for K(w/a) values greater than unity, while the solute extraction method is preferred for K(w/a) values less than unity.

A possible systematic error is that a significant fraction of the solute vapor could be adsorbed on the surface of the apparatus rather than in the gas phase. To evaluate this effect, the extraction factor (equation 10, where $V_L = 0$, $V_c = V_t$) for the dry apparatus was examined for dimethylmercury over a 1000-fold concentration range by repeatedly extracting the gas in V_R. The slope $\log[\frac{V_R T_c}{V_T T_R} + 1]$ of the plot $\log(A_1/A_i)$ versus i-1 was linear and equal to the calculated value. Two such experiments at air bath temperature of 100°C and 50°C gave the same results. These experiments indicate that surface adsorption was a negligible factor in the K(w/a) and K(o/a) measurements and that the hydrogen flame detector was linear over the solute concentration range measured. The same experiments were repeated with methane as the solute and gave similiar results.

In an identical experiment with diethylmercury as the solute it was observed that the solute peak area did not obtain a constant value but decreased with time. Two impurity peaks; one identified by gas chromatography as n-butane and one as possibly ethane increased in time. Lowering the air-bath temperature from 100° to 50°C slowed the thermal decomposition of diethylmercury but did not stop it. These experiments were done in the absence of light in order to rule out well-known photodecomposition processes (9).

The methods described in this paper for measuring K(w/a) and K(o/a) are applicable to compounds having a vapor pressure greater than 0.10 Torr. For compounds with vapor pressures less than 0.1 Torr adsorption of the solute on the walls of the apparatus would probably lead to erroneous values of K(w/a) and K(o/a). In any case, for compunds having low vapor pressures or compounds that adsorb strongly on surfaces, the procedure described above should be used to determine the extent of surface adsorption.

Using Neely et al (1) linear regression equation to correlate the octanol partition coefficient with bioconcentration of chlorohydrocarbons in trout muscle we calculated a bioconcentration factor of twenty for dimethylmercury in fresh water. This presumes that the lipophilicities of dimethylmercury and the chlorohydrocarbons either occur by similiar processes or that the molecular differences are not that important in rate-determining uptake steps.

The K(o/w) values for dimethylmercury are approximately 37 percent greater in sea water than fresh water which would indicate that the bioconcentration factor is greater in sea water than in fresh water.

Literature Cited

1. Neeley, W. B., Branson, D. R., and Blau, G. E., Environ. Sci. Technol. 8, 1113 (1974).

2. Branson, D. R., et al., Proceedings of Symposium Structure Activity Correlations in Studies of Toxicity and Bioconcentration with Aquatic Organisms, Canada Center for Inland Waters, Burlington, Ontario, Canada (1975).

3. Dunn, W., and Hansch, C., J. Pharm. Sci. 61, 1 (1972).

4. Leo, A. J., Symposium on Nonbiological Transport and Transformation, National Bureau of Standards, Gaithersburg, MD (1976).

5. Wasik, S. P., Brown, R. L., and Minor, J. I., J. Environ. Sci. Health A11(1), 99 (1976).

6. Brown, R. L., and Wasik, S. P., J. of Res. National Bureau of Standards 78A, 453 (1974).

7. Sverdrup, H. V., Johnson, M. W., and Fleming, R. H., The Oceans Prentice-Hall, Inc., Englewood Cliffs, New Jersey (1942).

8. Hansch, C., Quinlan, J. E., and Lawrence, G. L., J. Organic Chem. 33, 347 (1968).

9. Rebbert, R. E., and Steacic, E. W. R., Can. J. Chem. 31, 631 (1953).

Discussion

F. E. BRINCKMAN (National Bureau of Standards): By doing the linear free energy relationship as a function of temperature one can generate a thermodynamic property, in this case, the heat of solution. This is a very important concern for organometal chemists who do not have such information for these kinds of species. Experimentally, what kind of volatility do you need if you want to do a system as a function of salinity and temperature; for example, one of these organotin species which has ionic properties and strongly solvates in water?

WASIK: We have other methods of determining these partition coefficients. This is what we call our volatile solute method. Roughly, it's good for solutes of a boiling point less than 200°. Above 200° we have to use liquid chromatography as a detector.

G. E. PARRIS (Food and Drug Administration): People who do partition coefficients, such as octanol-water, where they put the two phases in direct contact, are usually concerned about there being some water solubility in octanol and some octanol solubility in water. How much difference does it make doing it indirectly where you actually measure (or can get a calculated value) for water-octanol, as opposed to water-saturated octanol versus octanol-saturated water?

WASIK: We did the experiment where we determined the octanol-air partition coefficient in octanol saturated with water and in octanol; there was about 2% difference. The octanol-air partition coefficient can also be measured by gas chromatography by measuring the retention time of the solute.

PARRIS: Is this figure of 2% typical in view of the heats of solvation for different types of molecules or do you think it might vary widely?

WASIK: Of course it would vary widely according to the solute.

J. S. THAYER (University of Cincinnati): What are the properties of octanol that make it a desirable solvent for use in this particular study?

WASIK: From the physical chemical standpoint, it's a poor choice. We would have been better off with something less volatile and less soluble. Water is fairly soluble in octanol. But I think people just kept measuring it and measuring it, and its conventional use grew. Once you measure the octanol-water partition coefficient of a parent compound then you can predict the octanol-water coefficient for derivatives. If you know a coefficient for benzene, you could predict that it would be for xylenes and butylbenzene because they show additive properties. You don't actually have to measure a lot of compounds as long as you measure the parent compound.

A. J. CANTY (University of Tasmania): Your comment just then and the partition coefficient for dimethylmercury relate to some work that we have been doing investigating bodily distribution of phenyl compounds. You would expect a similar relationship of partition coefficients for diphenylmercury to dimethylmercury. We find that intraperitoneal injection of diphenylmercury, compared with monophenylmercury, shows a 10 to 20 times higher concentration in fatty tissue which would correspond fairly well with your partition coefficients.

M. L. GOOD (University of New Orleans): Are these head-space experiments the same techniques that you use for determining the volatile materials from a bacterial methylation procedure?

WASIK: Like Dr. Brinckman does?

GOOD: Yes.

WASIK: I believe his head-spaces are in equilibrium with the system. Do you circulate the volatile gases?

BRINCKMAN: We have done both static and circulation measurements. Our problem is detector sensitivity. With some microorganisms, particularly the plasmid-altered microorganisms, such as we have acquired from Professor Simon Silver's laboratory (Washington University, St. Louis), the *E. coli* which are good methylators of mercury, we might do direct measurements. You have to dedicate the apparatus because the experiments require some period of time and sterile conditions. Also, one experiences surface effects from the vessel.

WASIK: One of the most important things about this type of partition coefficient measurement is that these are at infinite dilution, and from that you can measure activity coefficients which give you more freedom to make new correlations.

RECEIVED November 3, 1978.

20

Aspects of Mercury(II) Thiolate Chemistry and the Biological Behavior of Mercury Compounds

ALLAN J. CANTY

Chemistry Department, University of Tasmania, Hobart, Tasmania, Australia

Complex formation between mercury compounds and thiols, e.g. cysteine, is believed to play a major role in the biological chemistry of mercury(1). The greater affinity of Hg(II) and MeHg(II) for thiols than other possible biological donor ligands has been well documented by stability constant studies in aqueous solution (2,3). Our interest in mercury(II) thiolates stems from studies of the chemistry of the antidote British anti-Lewisite which indicated that the structure and reactivity of simple thiolate complexes was little understood. In this review our recent work on the interaction of inorganic and organomercury compounds with British anti-Lewisite, simple thiols and sulphur containing amino acids is discussed, followed by an account of animal studies of the distribution and metabolism of phenylmercury compounds. In discussing the implications of chemical results, e.g. reactivity of thiolates, for the biological behaviour of mercury compounds it is assumed here that chemical studies provide only plausible pathways for biological behaviour.

In recent years other workers have reported studies of mercury thiolates that are related to the work described here, in particular nuclear magnetic resonance studies of the interaction of MeHg(II) with thiols (4-9) and the preparation (10-16) and X-ray structural analysis of key complexes of Hg(II), MeHg(II), and PhHg(II) with sulphur containing amino acids (10-15).

Complexes of British anti-Lewisite and other Thiols

British anti-Lewisite [dimercaprol, 2,3-dimercaptopropanol; abbreviated $BALH_3$ to indicate loss of thiol protons on complex formation, e.g. Hg(BALH)] has been used for the treatment of mercury poisoning in humans (17,18) and has been studied extensively in animal experiments (18-24). Although it may be eventually replaced by a more satisfactory treatment, e.g. hemodialysis (25,26), it is successful for poisoning by

0-8412-0461-6/78/47-082-327$05.00/0

inorganic mercury (17,18) and is the most satisfactory antidote for phenylmercury(II) poisoning [animal experiments only to date (18)], but has no therapeutic effect for methylmercury(II) poisoning in humans or animals (18). For PhHg(II) poisoning $BALH_3$ greatly increases the amount of mercury in the brain compared with the bodily distribution in the absence of $BALH_3$ treatment (18,19,20,21), and for MeHg(II) it merely hastens the distribution of mercury and may increase the amount of mercury in the brain (18). An increased mercury content in the brain is undesirable, as it attacks the central nervous system. $BALH_3$ also increases the amount of mercury in the brain following its administration for inorganic mercury poisoning (22,23,24), but this effect has been explained in terms of the timing and dosage of $BALH_3$ (24).

Isolation of Hg(BALH) (27,28) and evidence for the formation of $[Hg(BALH)_2]^{2-}$ (27), $(PhHg)_2BALH$ (28), and $(RHg)_nBALH_{3-n}$[n = 1 (29), 2 (28); R = $CH_2CH(OMe)CH_2R'$] were reported by several workers soon after the introduction of $BALH_3$ as an antidote for heavy metal poisoning. Mercuric chloride reacts immediately with $BALH_3$ in water to form a white solid identified as Hg(BALH) (27,28,30,31).

$$HgCl_2 + BALH_3 \rightarrow Hg(BALH) + 2HCl$$

Crystal structures of simple thiolates $Hg(SR)_2$ reveal either linear monomers [R=Me (32), Et (33)] (Figure 1) or a polymeric structure with tetrahedral mercury ($R=Bu^t$) (34) (Figure 2). Infrared and Raman spectra indicate that highly insoluble Hg(BALH) has a polymeric structure based on linear coordination for mercury (31) (Figure 3), rather than the cyclic structure usually presented (Figure 4). Thus, Hg(BALH) has ν_{as}(SHgS) 348 and ν_s(SHgS) 298 cm^{-1}, similar to that of $Hg(SMe)_2$ (377 and 297 cm^{-1}) and well removed from tetrahedral mercury in $Hg(SBu^t)_2$ (172 and 188 cm^{-1}) (31). Spectroscopic properties appropriate for identification of Hg(II) thiolates, e.g. infrared, Raman, and nuclear magnetic resonance, are presented elsewhere (31,35,36,37,38,39).

The simple thiolates $Hg(SR)_2$ are insoluble in water but soluble in organic solvents, e.g. $Hg(SR)_2$ ($R=Et,Bu^t,Ph$) are monomeric in chloroform. Hg(BALH) is insoluble in water, even at concentrations of ca. 10^{-4}M (40). An impure form of Hg(BALH) can be isolated by reaction of mercuric acetate with $BALH_3$ in pyridine (35). This solid is soluble in pyridine, and the related complex of 1,3-dimercaptopropanol, Hg(DMPH), can be isolated from water and forms a dimer in pyridine (35). The structure of Hg(DMPH) in pyridine is unknown but presumably involves pyridine coordination, $[Hg(DMPH)py_x]_2$, as it crystallizes as $Hg(DMPH)py_{1.5}$ containing coordinated pyridine. The solubility of impure Hg(BALH) in pyridine is of interest as Hg(BALH) is presumably formed in many "environments" *in vivo*,

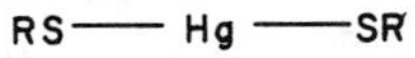

$RS—Hg—SR'$

Figure 1.

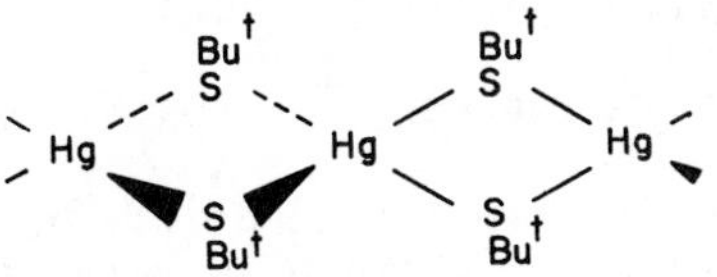

Bu^t Bu^t
S S
Hg Hg Hg
S S
Bu^t Bu^t

Figure 2.

— S S — Hg — S S — Hg —

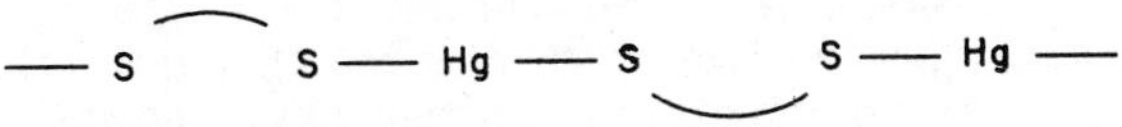

Figure 3.

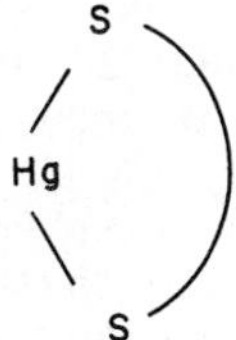

Figure 4.

and pyridine solubility suggests higher solubility in lipid tissue than more aqueous regions. The neutral complex may be present as a dimer $[Hg(BALH)L_x]_2$ related to Hg(DMPH) in pyridine, or possibly as the cyclic complex (Figure 4) with additional ligands coordinated to mercury.

In alkaline solution Hg(BALH) dissolves on addition of excess $BALH_3$ suggesting (27) formation of $[Hg(BALH)_2]^{2-}$, and addition of $BALH_3$ to a solution of impure Hg(BALH) in pyridine results in an increase in conductivity (35). Stability constants for formation of the neutral and ionic complexes in water have recently been determined by potentiometric titration (40), and the very high values contribute to the effectiveness of British anti-Lewisite as an antidote.

$$Hg^{2+} + BALH^{2-} \rightleftharpoons Hg(BALH) \qquad \text{Log } K = 25.74 \pm 0.45$$

$$Hg(BALH) + BALH^{2-} \rightleftharpoons [Hg(BALH)_2]^{2-} \qquad \text{Log } K = 8.61 \pm 0.10$$

Organomercury derivatives of $BALH_3$ may be obtained by reaction with phenylmercuric acetate in water and methylmercuric acetate in benzene (35).

$$2RHgO_2CMe + BALH_3 \rightarrow (RHg)_2BALH + 2MeCO_2H$$

Infrared and Raman spectra of these complexes and other organomercury thiolates indicate monomeric structures in the solid state as ν(Hg-S) values (326-388 cm^{-1}) are in the region expected for linear coordination for mercury, and coincidence of infrared and Raman values indicate absence of a centre of symmetry at mercury (Figure 5,6) (35), thus excluding dimeric structures similar to that formed by related PhHg(II) alkoxides in benzene (Figure 7) (41).

^1H NMR spectroscopy is particularly useful for characterization of organomercury compounds. Thus, $(MeHg)_2BALH$ has $J(^1H-^{199}Hg)$ 169 Hz for the MeHg(II) group, and PhHg(II) thiolates have $J(^{ortho}H-^{199}Hg)$ 144-158 Hz and $J(^{ortho}H-^{meta}H)$ 6-8 Hz (35).

The complexes $(RHg)_2BALH$ (R=Me,Ph) are insoluble in water but dissolve in pyridine and dimethylsulphoxide, and the related thiolate of lower molecular weight, $PhHgSCH_2CH_2OH$, is soluble and monomeric in chloroform. However, organomercury thiolates formed from naturally occurring thiols *in vivo* are likely to be water soluble, e.g. the L-cysteine complexes $MeHgSCH_2CH(NH_3)CO_2 \cdot H_2O$ and $PhHgSCH_2CH(NH_3)CO_2$ contain hydrophilic zwitterionic groups and crystallize from aqueous ethanol (12,36). Thus, displacement of biological thiol ligands with $BALH_3$ is expected to form more lipid soluble complexes, as suggested by Berlin *et al.* (20), and may account for higher concentrations of mercury in brain tissue of animals administered $BALH_3$ after injection of organomercury compounds when compared

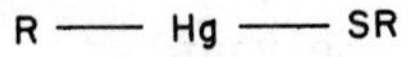

$$R - Hg - SR$$

Figure 5.

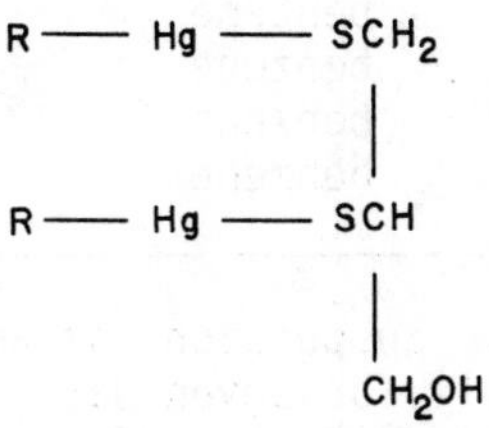

$$\begin{array}{l} R - Hg - SCH_2 \\ \qquad\qquad\quad | \\ R - Hg - SCH \\ \qquad\qquad\quad | \\ \qquad\qquad\quad CH_2OH \end{array}$$

Figure 6.

$$Ph - Hg \begin{array}{c} R \\ O \\ \diagup \quad \diagdown \\ \diagdown \quad \diagup \\ O \\ R \end{array} Hg - Ph$$

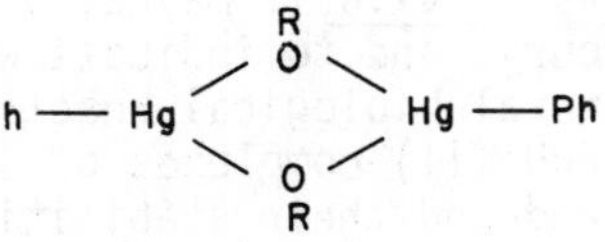

Figure 7.

with concentrations in the absence of $BALH_3$ treatment.

It was found that $(PhHg)_2BALH$ decomposes at ambient temperature in acetone, benzene, and methanol to form Ph_2Hg (30,35) (Table I).

$$(PhHg)_2BALH \rightarrow Ph_2Hg + Hg(BALH)$$

Table I
Decomposition of Some Phenylmercury (II) Thiolates[a]

Complex	Solvent	Yield of Ph_2Hg(%)
$(PhHg)_2BALH$	acetone	96
$(PhHg)_2BALH$	benzene	100
$PhHg(H_3cyst)$	benzene	55
$PhHg(H_3pen)$	benzene	81
$(PhHg)_2(H_2cyst)\cdot H_2O$	benzene	44
$(PhHg)_2H_2pen$	benzene	43

[a]From references 35,36. Suspensions at ambient temperature were stirred magnetically for seven days. Ph_2Hg was isolated as a pure solid from the filtrate.
[b]Yield of Ph_2Hg based on 'Ph'.
[c]$H_3cyst = SCH_2CH(NH_3)CO_2$; $H_2cyst = SCH_2CH(NH_2)CO_2$; similarly for $HSCMe_2CH(NH_3)CO_2$, DL-penicillamine.

If this reaction occurs *in vivo* it may also contribute to redistribution of mercury, and to indicate whether Ph_2Hg formation may be a general biological reaction in the absence of $BALH_3$ a series of PhHg(II) complexes of sulphur-containing amino acids was prepared and their stabilities studied (36). The complexes were synthesized by reaction of phenylmercuric acetate with the amino acids in aqueous ethanol, e.g.

$$2PhHgO_2CMe + H_4cyst \rightarrow (PhHg)_2(H_2cyst)\cdot H_2O + 2MeCO_2H$$

The DL-penicillamine complexes have been prepared by other workers, but the stability of the complexes toward decomposition had not been studied (16).

The amino acid complexes were found to decompose in benzene to form Ph_2Hg (Table I). The importance of these reactions, and decomposition of $(PhHg)_2BALH$, is difficult to assess as they are solvent dependent and rates of decomposition vary, e.g. $(PhHg)_2BALH$ and amino acid complexes may be readily prepared in aqueous solution, they decompose slowly in benzene, and when $PhHgO_2CMe$ and $BALH_3$ are reacted in ethanol immediate precipitation occurs and Ph_2Hg may be obtained from the filtrate on filtration. If Ph_2Hg is formed in vivo then the biological behaviour of Ph_2Hg is of interest as phenylmercury compounds, e.g. $PhHgO_2CMe$, are still widely used in agriculture and medicine. It has been reported that Ph_2Hg in "scarcely detectable" concentration formed by degradation of phenylmercuric acetate (formerly contained in derelict steel drums), was sufficiently toxic to kill fish within a few hours in the Boone Reservoir, Tennessee Valley (43).

Biological Behaviour of Diphenylmercury

Diphenylmercury has quite different physical and chemical properties than PhHg(II) compounds, e.g. it is a neutral non-polar molecule insoluble in water but soluble in organic solvents and is thus expected to be lipid soluble (44), and in contrast to PhHg(II) compounds (45,46,47) it interacts only weakly with donor molecules (48,49,50). Similarly, Me_2Hg does not form complexes (45) but MeHg(II) forms stable complexes, e.g. $[MeHgL]^+$ with pyridine (51,52), 2,2'-bipyridyl (51,52,53, 54), and 1,10-phenanthroline (52,53).

In distribution and metabolism studies we have injected ethanol solutions of mercuric chloride, phenylmercuric acetate, or Ph_2Hg intraperitoneally into rats (55,56). The rats were sacrificed at intervals ranging from 20 min. to 7 days and samples of blood, brain, liver, kidney, muscle, fat, and spleen were analysed for mercury. In another series of experiments faecal and urinary excretion was monitored for several days after injection.

During the first few days after injection, urinary excretion of mercury was much higher for the diphenylmercury-injected rats than for the phenylmercuric acetate or mercuric chloride-injected rats, with mercuric chloride having the lowest rate of excretion. Faecal excretion was similar for the three compounds, with phenylmercuric acetate being more rapidly excreted (Table II).

Analyses of blood and tissues for total mercury indicated that after initial marked differences in brain and fatty tissue concentrations, the distribution of mercury for Ph_2Hg resembled those of the other compounds after 1 day, but concentrations were generally lower than for the other compounds (55,56). The lower concentrations are explained by the more rapid excretion of mercury from Ph_2Hg.

During the first hour after injection mercury from Ph_2Hg accumulated at a higher concentration in the brain than from the other compounds, but after 6 hours these concentrations had decreased considerably (55,56). The concentration of mercury in fatty tissue was 10-20 times higher for diphenylmercury-

Table II
Urinary and Faecal Excretion of Mercury from Rats within Two Days of Injection[a]

	$HgCl_2$		$PhHgO_2CMe$		Ph_2Hg	
	Percentage of dose excreted					
Urinary Excretion:	2.5	2.2	4.8	8.0	30.5	38.3
Faecal Excretion:	5.2	4.5	12.2	8.4	5.6	3.6
Urinary + Faecal:	7.7	6.7	17.0	16.4	36.1	41.9

[a]From reference 56. Analyses for total mercury, as described elsewhere (55). Two rats were injected intraperitoneally with each compound, dose 248 mg mercury, all rats of weight 160 g.

injected rats at 20 min. after injection, but then rapidly dropped to values similar to the other mercury compounds (Table III). The much higher concentration of mercury in brain and fatty tissue immediately after Ph_2Hg injection is consistent with distribution of mercury as Ph_2Hg, and this was confirmed by thin-layer chromatography. A sample of fatty tissue taken from a diphenylmercury-injected rat 20 min. after injection was blended with benzene using a small Waring blender, and thin-layer chromatography showed the presence of diphenylmercury (ultraviolet irradiation); the silica gel of the plate at the R_f value of Ph_2Hg contained 5.19 mg. of Hg/g of silica gel compared with 0.15 mg/g for silica gel at lower R_f value on the same plate.

It has been established that phenylmercury is degraded to inorganic mercury in a few days in rats (57,58,59,60,61). Daniel *et al.* (60) represent this breakdown as

$$C_6H_5Hg^+ + H^+ \rightarrow C_6H_6 + Hg^{2+}$$

A similar breakdown may occur for Ph_2Hg, presumably via PhHg(II), as the initial high concentrations of mercury in brain and fatty tissue fall to values similar to that obtained with the other compounds after 6 hr. and 1 hr., respectively. Thus, if Ph_2Hg is formed *in vivo* its biological effects are difficult to evaluate as it is more rapidly excreted than PhHg(II) and apparently broken down by the body, but has a quite different initial distribution. However, it is of interest to note that although mercury vapour is oxidized to Hg(II) in ca. 30 sec. in blood this is sufficient time for mercury (from vapour) to

Table III
Concentration of Mercury in Brain and Fatty Tissues of Hooded Wistar Rats Injected Intraperitoneally with Mercury Compounds[a]

Time	Brain	Fat
A. Dose of 6 mg. of Hg/kg of rat.		
	mercuric chloride	
20 min. (2)	0.16 ± 0.02	10.5 ± 2.8
1 hr. (2)	0.33 ± 0.07	4.8 ± 2.1
6 hr. (2)	0.16 ± 0.01	3.4 ± 1.3
1 day (2)	0.24	15.7 ± 5
	phenylmercuric acetate	
20 min. (2)	0.14 ± 0.04	5 ± 2
1 hr. (2)	0.47 ± 0.03	4.4 ± 1.2
6 hr. (2)	0.9 ± 0.3	4.7 ± 0.2
1 day (2)	0.65 ± 0.02	3.5 ± 0.3
	diphenylmercury	
20 min. (2)	0.9 ± 0.2	147 ±13
1 hr. (2)	0.7 ± 0.2	10.4 ± 3.2
6 hr. (2)	0.26 ± 0.01	10.1 ± 3.4
1 day (2)	0.20 ± 0.03	3.6 ± 0.2
B. Dose of 1.5 mg. of Hg/kg of rat.		
	mercuric chloride	
20 min. (1)	0.04	2.34
	phenylmercuric acetate	
20 min. (1)	0.01	0.9
	diphenylmercury	
20 min. (1)	0.3	27.8

[a]From reference 56. Recorded as μg of Hg/g tissue, wet weight, and the range of values is indicated. The number of rats in each category is given in parentheses with the time.

achieve an ca. ten-fold higher accumulation in the brain than from inorganic mercury poisoning (27,62) leading to higher toxicity of mercury vapour.

Acknowledgements

I thank the National Health and Medical Research Council and the Commonwealth Development Bank for financial support, and my co-workers, R. Kishimoto and R. S. Parsons, for their contributions to work reviewed here.

Literature Cited.

1. See, e.g. Clarkson, T. W., Ann. Rev. Pharmacol. (1972), 12, 375.
2. Simpson, R. B., J. Am. Chem. Soc. (1961), 83, 4711.
3. Schwarzenbach, G., and Schellenberg, M., Helv. Chim. Acta. (1965), 48, 28.
4. Simpson, P. G., Hopkins, T. E., and Haque, R., J. Phys. Chem. (1973), 77, 2282.
5. Rabenstein, D. L., J. Am. Chem. Soc. (1973), 95, 2797.
6. Fuhr, B. J., and Rabenstein, D. L., J. Am. Chem. Soc. (1973), 95, 6944.
7. Rabenstein, D. L., and Fairhurst, M. T., J. Am. Chem. Soc. (1975), 97, 2086.
8. Fairhurst, M. T., and Rabenstein, D. L., Inorg. Chem. (1975), 14, 1413.
9. Sugiura, Y., Hojo, Y., Tamai, Y., and Tanaka, H., J. Am. Chem. Soc. (1976), 98, 2339.
10. Wong, Y. S., Chieh, P. C., and Carty, A. J., J.C.S. Chem. Comm. (1973), 741.
11. Wong, Y. S., Taylor, N. J., Chieh, P. C., and Carty, A. J., J.C.S. Chem. Comm. (1974), 625.
12. Taylor, N.J., Wong, Y. S., Chieh, P. C., and Carty, A. J., J.C.S. Dalton (1975), 438.
13. Wong, Y. S., Carty, A. J., and Chieh, P. C., J.C.S. Dalton (1977), 1157.
14. Wong, Y. S., Carty, A. J., and Chieh, C., J.C.S. Dalton (1977), 1801.
15. Carty, A. J., and Taylor, N. J., J.C.S. Chem. Comm. (1976), 214.
16. Hojo, Y., Sugiura, Y., and Tanaka, H., J. Inorg. Nucl. Chem. (1976), 38, 641.
17. Clarkson, T. W., and DiStefano, V., in "Drill's Pharmacology in Medicine", DiPalma, J. R., and Joseph, R., Eds., Ch. 53, McGraw-Hill, 4th Edn. (1971).
18. Swensson, A., and Ulfvarson, U., Int. Arch. Gewerbepath. Gewerbehyg. (1967), 24, 12.
19. Berlin, M., and Ullberg, S., Nature (1963), 197, 84.
20. Berlin, M., Jerksell, L.-G., and Nordberg, G., Acta. Pharmacol. Toxicol. (1965), 23, 312.
21. Berlin, M., and Rylander, R., J. Pharmacol. Exptl. Therap. (1964), 146, 236.
22. Fitzsimmons, J. R., and Kozelka, F. L., J. Pharmacol. Exptl. Therap. (1950), 98, 8.
23. Berlin, M., and Lewander, T., Acta. Pharmacol. Toxicol. (1965), 22, 1.
24. Magos, L., Brit. J. Indust. Med. (1968), 24, 152.
25. Kostyniak, P. J., Clarkson, T. W., Cestero, R. V., Freeman, R. B., and Abbasi, A. H., J. Pharmacol. Exptl. Therap. (1975), 192, 260.

26. Kostyniak, P. J., Clarkson, T. W., and Abbasi, A. H., J. Pharmacol. Exptl. Therap. (1977), 203, 253.
27. Gilman, A., Allen R. P., Philips, F. S., and St. John, E., J. Clin. Invest. (1946), 25, 549.
28. Benesch, R., and Benesch, R. E., Arch. Biochem. Biophys. (1952), 38, 425.
29. Lehman, J. F., Barrack, L. P., and Lehman, R. A., Science (1951), 113, 410.
30. Canty, A. J., and Kishimoto, R., Nature (1975), 253, 123.
31. Canty, A. J., Kishimoto, R., Deacon, G. B., and Farquharson, G. J., Inorg. Chim. Acta. (1976), 20, 161.
32. Bradley, D. C., and Kunchur, N. R., J. Chem. Phys. (1964), 40, 2258.
33. Bradley, D. C., and Kunchur, N. R., Can. J. Chem. (1965), 43, 2786.
34. Kunchur, N. R., Nature (1964), 204, 468.
35. Canty, A. J., and Kishimoto, R., Inorg. Chim. Acta. (1977), 24, 109.
36. Canty, A. J., and Kishimoto, R.,Aust. J. Chem. (1977), 30, 669.
37. Canty, A. J., Kishimoto, R., and Tyson, R. K., Aust. J. Chem. (1978), 31, 671.
38. Canty, A. J., and Tyson, R. K., Inorg. Chim. Acta. (1977), 24, L77.
39. Canty, A. J., and Tyson, R. K., Inorg. Chim. Acta. (1978), in press.
40. Coates, R. L., and Jones, M. M., J. Inorg. Nucl. Chem. (1977), 39, 677.
41. Bloodworth, A. J., J. Chem. Soc. (C) (1970), 2051.
42. Norseth, T., Acta. Pharmacol. Toxicol. (1973), 32, 1.
43. Derryberry, O.M., in "Environmental Mercury Contamination" (Hartung, R., and Dinman, B. D., eds.) p.76 Ann Arbor Science, Ann Arbor, Michigan.
44. Neville, G. A., and Berlin, M., Environ. Res. (1974), 7, 75.
45. Beletskaya, I. P., Butin, K. P., Ryabtsev, A. N., and Reutov, O. A., J. Organometal. Chem. (1973), 59, 1.
46. Canty, A. J., and Deacon, G. B., Aust. J. Chem. (1968), 21, 1757.
47. Canty, A. J., Fyfe, M., and Gatehouse, B. M., Inorg. Chem. (1978), 17, in press.
48. Canty, A. J., and Deacon, G. B., J. Organometal. Chem. (1973), 49, 125.
49. Canty, A. J., and Gatehouse, B. M., Acta. Cryst. (1972), B28, 1872.
50. Puhl, W. F., and Henneike, H. F., J. Phys. Chem. (1973), 77, 558.
51. Coates, G. E., and Lauder, A., J. Chem. Soc. (1965), 1857.
52. Canty, A. J., and Marker, A., Inorg. Chem. (1976), 15, 425.
53. Anderegg. G., Helv. Chim. Acta. (1974), 57, 1340.

54. Canty, A. J., and Gatehouse, B. M., J. C. S. Dalton (1976), 2018.
55. Canty, A. J., and Parsons, R. S., Toxicol. Appl. Pharmacol. (1977), 41, 441.
56. Canty, A. J., and Parsons, R. S., unpublished results.
57. Miller, V. L., Klavano, P. A., and Csonka, E., Toxicol. Appl. Pharmacol. (1960), 2, 344.
58. Berlin, M., and Ullberg, S., Arch. Environ. Health (1963), 6, 602.
59. Gage, J. C., Brit. J. Indust. Med. (1964), 21, 197.
60. Daniel, J. W., Gage, J. C., and Lefevre, P. A., Biochem. J. (1972), 129, 961.
61. Takeda, Y., Kunugi, T., Hoshino, O., and Ukita, T., Toxicol. Appl. Pharmacol. (1968), 13, 156.
62. Berlin, M., Jerksell, L.-G., and von Ubisch, H., Arch. Environ. Health. (1966), 12, 33.

RECEIVED August 22, 1978.

Mercury, Lead, and Cadmium Complexation by Sulfhydryl-Containing Aminoacids. Implications for Heavy-Metal Synthesis, Transport, and Toxicology

ARTHUR J. CARTY

Guelph–Waterloo Centre for Graduate Work in Chemistry, Waterloo Campus, University of Waterloo, Waterloo, Ontario, N2L 3G1, Canada

According to the A and B classification of Ahrland and Chatt (1) or the Hard and Soft Acid and Base nomenclature of Pearson (2) the polarisable heavy metal ions Hg^{2+}, Cd^{2+} and Pb^{2+} should form their strongest complexes with donor atoms from elements in the second or subsequent periods of the periodic table. It comes as no great surprise therefore that these elements appear in nature predominantly as their sulfides and that the interaction of sulfur containing molecules and ions with these metals plays a major role in their environmental and bio-chemistry.

The advent of man-made heavy metal pollution of natural waters has brought to light significant new aspects of heavy-metal sulfur chemistry and re-emphasised the need for a clear understanding of metal ion behaviour towards sulfur containing ligands. The purpose of this article is to summarise and contrast recent results on the complexation of methylmercury (CH_3Hg^+) inorganic mercury (Hg(II)), cadmium (Cd(II)) and lead (Pb(II)) as well as methyllead (IV) with small sulfhydryl containing aminoacids and to place these results in an environmental, biochemical and toxicological context.

For mercury there is much evidence pointing to the relevance of mercurial complexes with sulfur aminoacids in the microbiological transformation of Hg^{2+} to CH_3Hg^+ and in the biotransport, metabolism and toxicity of both inorganic and methyl mercury. Thus homocysteine ($HSCH_2CH_2CH(NH_3)COO$) and cysteine ($HSCH_2CH(NH_3)COO$) complexes may be implicated in one mechanism for the methylation of Hg(II) (as $HgCl_2$) by aerobic cultures of *Neurospora Crassa* (3) while the synthesis of methylmercury thiomethyl (CH_3HgSCH_3) in shellfish may involve an L-cysteine complex of methylmercury (Scheme 1) (4). There is little doubt that the simple alkylmercurials (e.g. CH_3Hg^+, $C_2H_5Hg^+$) exert their toxic effects principally by inactivating specific sulfhydryl sites of cysteine residues in proteins and enzymes. Indeed, the methylmercury cation CH_3Hg^+ has been used for decades by biochemists as a highly selective probe for sulfhydryl groups owing to the specific 1:1 stoichiometry of CH_3Hg^+/sulfhydryl complexation (5).

0-8412-0461-6/78/47-082-339$05.00/0

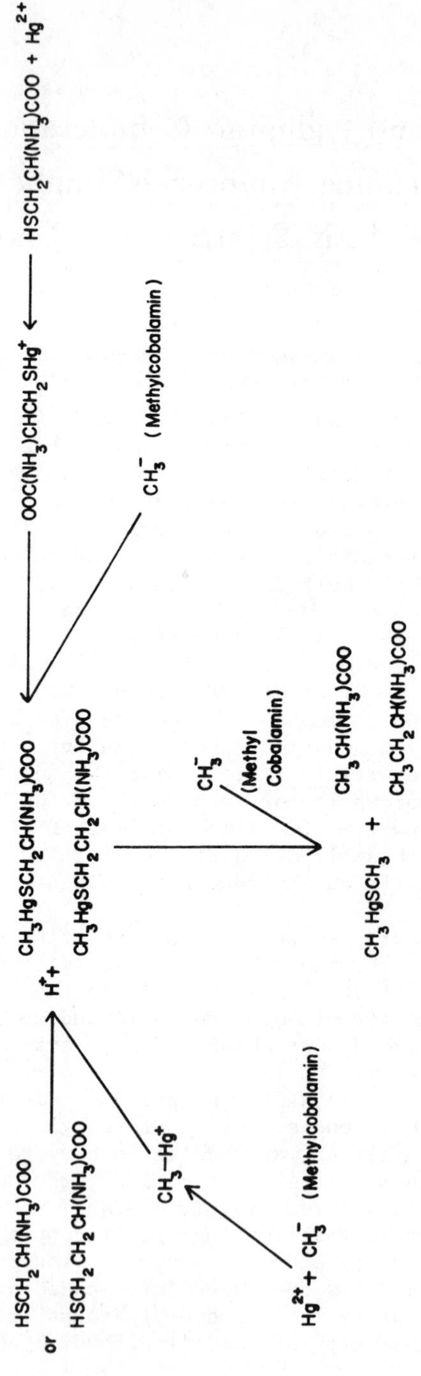

Scheme 1. Environmental synthesis of CH_3HgSCH_3

As an example, human albumin has been shown to contain 0.65 - 0.70 S-H groups per mole as determined by reactions with organic mercurials (6). The strength and stability of mercury-sulfur interactions in simple thiol complexes is also significant in the context of chelation therapy treatment for "inorganic" mercury poisoning and development of potential antidotes to CH_3Hg^+ poisoning, subjects dealt with elsewhere in this volume (7). Despite the extremely high values of the formation constants for CH_3Hg^+/thiol complexes

$$(\log \frac{[CH_3HgL]}{[CH_3Hg^+][L]} \approx 15.8 \text{ for } CH_3HgSCH_2CH(NH_3)COO \ (8)$$

and $\approx$ 22 for the CH_3Hg^+ - mercaptalbumin complex (9) there is evidence that exchange of the methylmercury cation between sulfhydryl sites may be rapid. Thus free glutathione anion (G^-) rapidly exchanges with bound ligand in the CH_3Hg G complex (10) presumably via nucleophilic attack at 2-coordinate mercury generating a pseudo 3-coordinate intermediate. Although the residual Lewis acidity in CH_3HgSR complexes is low (vide infra) it is clearly of major significance in the context of methylmercury mobility in biological systems. Small molecular weight sulfhydryl containing molecules such as glutathione facilitate mercurial transport in blood; albumin, which is present at the 5 gram per cent level in mammals may also play a role as a mercurial carrier (11).

Inorganic mercury also forms exceedingly strong bonds with sulfur anions. The stability constant log k for the 1:1, Hg^{2+}/L-cysteine complex is 45.4 a value which can be appreciated when compared to log k = 6.74 for chloride ion (12). Bis (mercaptides) of mercury are usually stable over the entire pH range and it is evident that the complexation of Hg^{2+} by L-cysteine residues in proteins and enzymes dominates the biochemistry and toxicology of inorganic mercury. "Bis(cysteinato) mercury" is a metabolic product of certain mercurial diuretics (13) and mercury may be bound to two (or possibly three) cysteinyl sulfur atoms in kidney metallothionein (14). Considerable evidence exists for biologically and environmentally significant Hg^{2+} complexes of the type RSHgX (X = anionic or neutral non-thiol ligand) (13,15), and it is likely that structural, physical and chemical differences between RHg^+ and Hg^{2+} complexes of L-cysteine and glutathione are responsible in large part for differences in organ distribution, toxicity and biological behaviour for these mercurials.

In contrast to the biochemists, inorganic chemists showed little interest in the simple aminoacid, peptide or thiol complexes of RHg^+ or Hg^{2+} until very recently. A 1972 review (16) of Hg^{2+}/aminoacid complexes is almost totally devoid of reliable structural information either in solution or the solid state. We thus set out to synthesize model compounds with the following goals in mind: (a) to provide quantitative information on the mercury-sulfur interaction, (b) to investigate the importance of secondary mercury-ligand binding, (c) to contrast the structures and properties of CH_3Hg^+ and Hg^{2+} complexes with sulfur amino-acids in a search for features pertinent to mercurial transport and toxicity, and (d) to fully characterise mercury species directly implicated in environmental synthesis of CH_3Hg^+ and metabolism of CH_3Hg^+ and Hg^{2+}.

L-Cysteine and DL-Homocysteine Complexes

Strange as it may seem the crystalline compound $CH_3Hg\ SCH_2$-$CH(NH_3)COO\cdot H_2O$ was first described only in 1974 (17) although the formation constant for 1:1 CH_3Hg^+/L-cysteine was reported in 1961 (8). NMR data have shown that CH_3Hg^+ binds to the sulfhydryl site over the entire pH range 1 - 13 (10, 18). The structure of $CH_3Hg\ SCH_2\ CH(NH_3)COO\cdot H_2O$ (Figure 1) solved by X-Ray analysis in our laboratory (19) illustrates a number of important points: (i) mercury is linearly 2-coordinate with strong C-Hg (2.09(4) Å) and Hg-S (2.35(1) Å) bonds, (ii) the molecule is essentially molecular with only weak intermolecular hydrogen bonds between cysteines and water molecules, and (iii) secondary intramolecular interactions are weak (Hg-O (carboxylate) = 2.84(2) Å). It is unlikely that the Hg...O interaction is maintained in solution (18). This structure illustrates nicely the weak residual Lewis acidity associated with a mercury atom coordinated by one RS^- group and an alkyl group, a key feature of CH_3Hg^+ coordination chemistry in biological systems. The molecule $CH_3Hg\ SCH_2\ CH(NH_3)COO$ is obviously a reasonable model for tissue bound CH_3Hg^+, and displacement of L-cysteine from this complex might well be used as an initial test of antidote effectiveness. Finally $CH_3\ Hg\ SCH_2\ CH(NH_3)COO$ is a key intermediate en route to $CH_3\ Hg\ SCH_3$ (Scheme 1) although, to our knowledge, the experiments to confirm the conversion to methylmercury thiomethyl via methylcobalamin generated CH_3^- have not yet been carried out.

Although the dominant feature of CH_3Hg^+ binding to proteins is the formation of strong Hg-S (cysteine) bonds, the nature of weak secondary interactions from neighbouring sites is pertinent to the impact of CH_3Hg^+ on protein structure and conformation. Our single crystal X-Ray diffraction results for the related molecule $CH_3Hg\ SCMe_2\ CH(NH_3)COO$ (Fig. 2) show that intermolecular Hg .....S (thio-ether type) contacts are barely significant (20). The shortest of these Hg...S distances (Hg...S of 3.35(1) Å) is the same as the sum of Van der Waals radii for Hg and S (3.35 Å). A recently solved structure of $CF_3\ Hg\ SCMe_2\ CH(NH_3)COO\cdot 0.5\ H_2O$ (21) (Fig. 3) illustrates rather dramatically the increase in residual Lewis acidity of mercury which accompanies replacement of an alkyl by a fluoroalkyl group. The arrangement of mercury and sulfur atoms in this structure resembles a distorted cubane

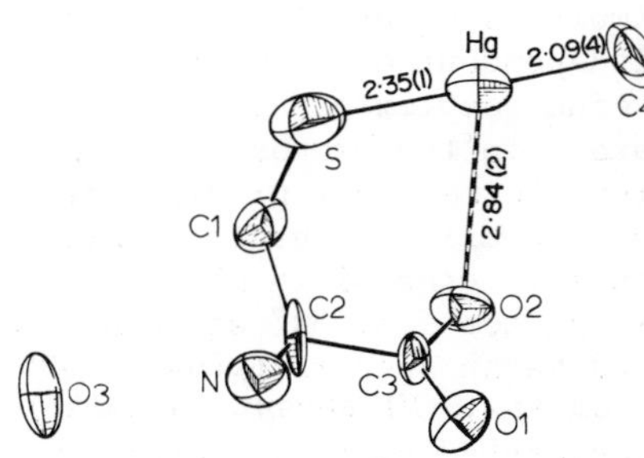

Figure 1. The structure of CH_3HgSCH_2-$CH(NH_3)COO \cdot H_2O$. The dashed line from Hg to O(2) represents a weak intramolecular interaction and O(3) is the oxygen atom of the water molecule of crystallization.

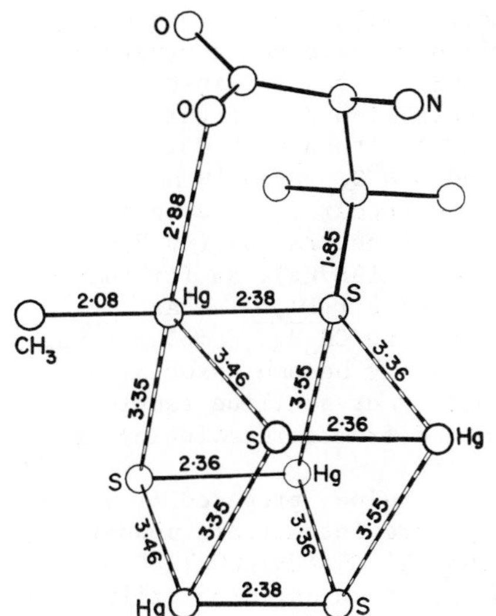

Figure 2. A portion of the crystal structure of $CH_3HgSCMe_2CH(NH_3)COO \cdot 0.5$ H_2O drawn to show the nature of weak inter- and intramolecular interactions. For clarity only on penicillamine molecule is illustrated.

although only the vertical mercury-sulfur edge bonds are strong. Secondary Hg...O and Hg....S interactions are both stronger here than in the DL-penicillamine derivative. Refinement of this structure (current R of 0.10 based on 1462 independent observed reflections measured on a GE-XRD-6 Datex automated diffractometer) is continuing.

Our efforts to synthesise CH_3Hg^+ and $HgCl_2$ derivatives of DL-homocysteine $HSCH_2CH_2CH(NH_3)COO$ were motivated by the observation that this aminoacid stimulates methylation of Hg^{2+} in Neurospora Crassa, a cobalamin independent organism (3). Methyl group transfer to a homocysteine complex of Hg^{2+} would generate methylmercury homocysteinate in the proposed mechanism (3). The 1:1 $MeHg^+$ complex, formed as a precipitate on adding MeHgOH (0.86 g) in ethanol (80 ml) to a basic solution of DL-homocysteine (0.5 g in a mixture of ethanol (5 ml) and water 50 ml), recrystallised as thin plates from 30% aqueous ethanol. Microanalyses, infrared and Raman spectra (ν(Hg-C) 535 ir, 538 R cm^{-1}; ν(Hg-S) 334 ir, 336 R cm^{-1}; ν(S-H) absent in Raman) of $CH_3HgSCH_2CH_2CH(NH_3)COO$ and the deuterated complex, together with the ^1H nmr in D_2O which showed a large downfield shift of the H_γ protons ($\Delta\delta$ = +0.51) and a $^2J_{CH_3-^{199}Hg}$ of 186 Hz, confirmed a structure essentially the same as that of the L-cysteine derivative (10, 19). Under similar conditions but employing 2:1 molar ratios of CH_3Hg^+: homocysteine, a crystalline 2:1 complex $(CH_3Hg)_2S\ CH_2CH_2CH(NH_2)COO$ was obtained. Preliminary X-Ray measurements have established that this compound is monoclinic, space group C2/c with a = 28.33, b = 11.24, c = 7.31 Å; β = 87° 11' and Z = 8. From ^1H nmr measurements (22) the two methyl groups appear to be associated with one sulfhydryl and one amino site at neutral pH ($\Delta\delta\ H_\gamma$= + 0.53 ppm; $\Delta\delta H_\alpha$=+ 0.37 ppm; $J_{CH_3-^{199}Hg}$ = 198 Hz), as for the corresponding DL - penicillamine derivative (10,20). As observed by Rabenstein (10) for glutathione/CH_3Hg^+ complexation, at acid pH $<$ 4 both CH_3Hg^+ groups probably become associated with the sulfhydryl group. Unfortunately, crystalline samples of $(CH_3Hg)_2\ SCH_2\ CH_2\ CH(NH_2)COO$ obtained from acid solution are unsuitable for intensity data collection.

The "inorganic" complexes of homocysteine, prepared from dilute aqueous solutions at neutral pH are intractable powders. Compounds of apparent formulae $Hg[SCH_2\ CH_2\ CH(NH_3)COOH]_2\ Cl_2$ and $HgCl[SCH_2\ CH_2\ CH(NH_3)COO]\cdot \frac{1}{2}\ H_2O$ have been satisfactorily analysed from 2:1 and 1:1 mixtures of aminoacid and $HgCl_2$. Using $HgBr_2$ and DL-homocysteine under strongly acid conditions a complex of unusual stoichiometry, $HgBr_2\ [SCH_2\ CH_2CH(NH_2)CO]$ 2HBr was obtained (22). A full, 3-dimensional X-Ray study revealed a structure consisting of polymeric $HgBr_4^{2-}$ ions sharing two bridging bromides packed together with cations of homocysteine lactone $\overline{SCH_2\ CH_2CH(NH_2)C}O^+$ (Figure 4). Clearly on acid treatment, cleavage of strong Hg-S (homocysteine) bonds occurs to generate the lactone, in marked contrast to the

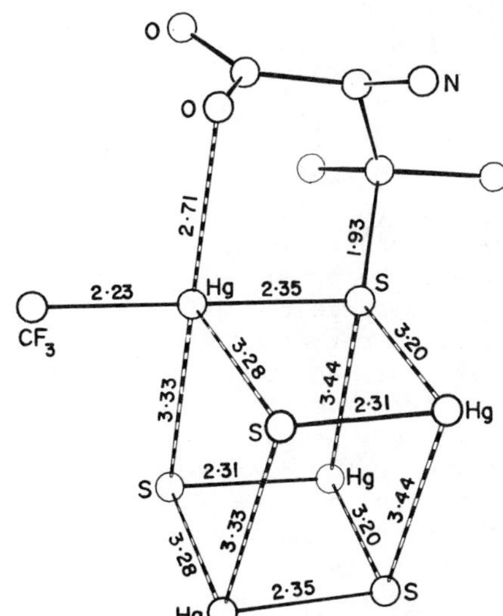

Figure 3. A portion of the crystal structure of $CF_3HgSCMe_2CH(NH_3)COO \cdot 0.5\ H_2O$ showing inter- and intramolecular interactions. Only one amino acid is illustrated.

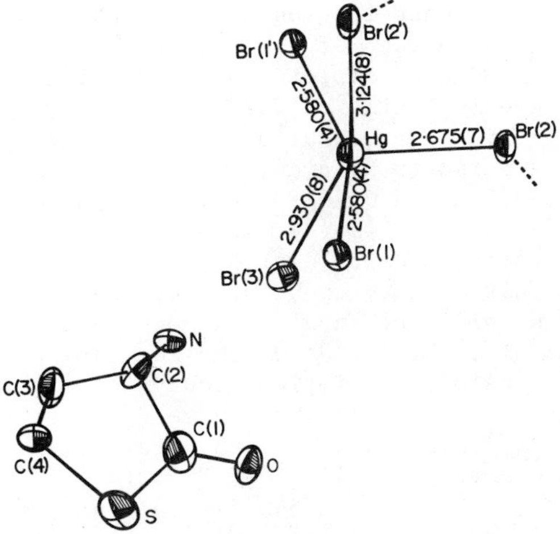

Figure 4. An ORTEP II plot of the molecular structure of $[\overline{SCH_2CH_2CH(NH_3)CO}]_2 \cdot HgBr_4$ showing the atomic numbering. Bridging bromine atoms are indicated by an additional dashed bond.

L-cysteine and DL-penicillamine complexes of Hg^{2+}.

In view of difficulties in structurally characterising the "inorganic" complexes of homocysteine purportedly involved in CH_3Hg^+ synthesis we began work on the L-cysteine and DL-penicillamine derivatives of $HgCl_2$. Compounds of formulae $Hg[SCH_2CH(NH_3)COO][SCH_2CH(NH_3)COOH]\cdot Cl.\ 0.5\ H_2O$ (1), $HgCl_2[SCH_2CH(NH_3)COOH]$(2), $Hg[SCH_2CH(NH_2)COO]$(3) $[HgCl_2]_2[SCMe_2\ CH(NH_3)COOH]\ 2H_2O$ (4), $[Hg\{SCMe_2\ CH(NH_3)COOH\}_2\ Cl]Cl.H_2O$ (5), and $HgCl_2[SCMe_2CH(NH_3)COOH]$(6) have been synthesized and characterized by spectroscopic methods. Synthetic and spectroscopic details will be published elsewhere (23, 24) but it is pertinent here to summarise the major structural features of these compounds as revealed by single crystal X-Ray diffraction. Pictorial representations of 1, 2, 4 and 5 are shown in Figure 5. There are several important structural features to emphasize: (i) In 1 and 5, neglecting the relatively weak interactions between chloride ions and mercury, the basic structural units present are the classical two coordinate ~S-Hg-S~ stereochemistry long thought to represent the bound state of Hg^{2+} in proteins and dimercaptides. Significantly, loss of H^+ and Cl^- from 1 would generate "mercury cysteinate", $Hg[SCH_2CH(NH_3)COO]_2$ (13, 16). The environment of mercuric ion in human kidney metallothionein, which binds Hg^{2+} more strongly than either Zn^{2+} or Cd^{2+}, may resemble that in 1 and 5 with two very strong Hg-S(cysteine) interactions providing the main binding. If bond-lengths are used as a rough criterion of bond strength, comparison of Hg-S distances in the "2-co-ordinate" Hg^{2+} and CH_3Hg^+ complexes of L-cysteine and DL-penicillamine, reveals little difference between the affinities of CH_3Hg^+ and Hg^{2+} for deprotonated sulfhydryl groups. (ii) Two types of mercapto bridges, one highly unsymmetrical in 4 (Hg(1)S of 2.822(5)Å and Hg(2)-S of 2.356(5)Å) and one almost symmetrical in 2 (Hg-S = 2.490(4), Hg-S' = 2.453(4)Å) are evident. Clearly thiols are quite capable of providing sites for two mercury atoms. This ability to bridge mercury atoms may be mechanistically significant in site exchange and transport of inorganic mercurials. It is noteworthy that sulphur bridging is also a recurring feature of simple mercury-thiol chemistry, leading to structures in which the degree of polymerization varies considerably (25,26). (iii) Chloride ion interactions of varying strengths are present (cf Hg-Cl(1) of 2.582(4) in 2, Hg-Cl(1) of 2.850(5)Å in 5, Hg(1)-Cl(3) of 2.348(5)Å in 4, and Hg(1) - Cl(2') of 3.335(6)Å in 4) suggesting that chloride ion can compete favourably for mercurial coordination sites and hence modify the polarity and solubility of the bio-complexes. The different distribution ratios of Hg^{2+} and CH_3Hg^+ between plasma and red blood cells (27,28) may well reflect the greater Cl^- content in plasma. Transportation and membrane permeability of different mercurials depend on mercurial polarity and solubility, two parameters sensitive to halide ion interactions.

With the complete characterization of the Hg^{2+}/L-cysteine derivatives above, we feel confident that the homocysteine complexes $Hg[SCH_2CH_2CH(NH_3)COOH]Cl_2$ and $HgCl[SCH_2CH_2CH(NH_3)COO]\cdot 0.5\ H_2O$ have closely related structures based essentially on digonal S-Hg-S or Cl-Hg-S bonds. In line with our aim to assess the relevance of these cysteine and homocysteine compounds to environmental mercury methylation, we are carrying out methylation studies using active sediments spiked with the mercury complexes and also with cultures of Neurospora Crassa growing in the presence of these compounds.

Comparison of Binding Preferences for Inorganic Mercury, Cadmium and Lead (11).

A comparison of solubility products for M^{2+}/S^{2-} or M^{2+}/OH^- and stability constants for ethylenediamine complexes (Fig. 6) gives an indication of the relative affinities one might expect to find for mercury, cadmium and lead in their aminoacid complexes. Mercury clearly has a very high affinity for sulfur, with cadmium and lead having approximately equal but much lower attractions. As a converse to this one might expect that in complexes with sulfur aminoacids where Hg-S bonding would predominate, cadmium and lead might exhibit considerably more diverse behavior, binding simultaneously to several different sites. This is exactly the picture which has emerged from X-Ray studies of mercury, cadmium and lead complexation with sulfur aminoacids. In Figure 7 we illustrate schematically the modes of interaction recently found in a variety of complexes. Some related bond distances are shown in Table I. Although direct comparisons between metals are difficult owing to different complex stoichiometries and states of aminoacid ionization, the structure of $Cd[SCMe_2\ CH(NH_2)COO]\ H_2O$ (31) illustrates the ability of cadmium to interact with S, N and O sites of the aminoacid. Comparison of the Cd-S distances (av 2.565(7)Å) with Hg-S bond lengths in Table I show that the weaker cadmium-sulfur interactions in the cadmium species are compensated by bonding capacity directed to oxygen and nitrogen donors. This tendency of cadmium to prefer a combination of binding sites implies a much lesser specificity than mercury for sulfhydryl sites in proteins and is consistent with the failure of reagents such as D- penicillamine to act as good antidotes to cadmium poisoning. These observations also suggest that the stereochemistry of Cd^{2+} in metallothionein is likely to differ significantly from that of Hg^{2+} in the same metalloprotein.

The lead (II) compound $Pb[SCMe_2\ CH(NH_2)COO]$ (29) closely resembles $Cd[SCMe_2CH(NH_2)COO]\ H_2O$ in stoichiometry, and each metal ion is pseudooctahedrally coordinated by ligand atoms, with a 3S, 2O, N donor set for lead and 3O, 2S, N for cadmium. Like cadmium, lead seems to favour binding to a combination of S, N and O sites in aminoacids and thus it also differs from mercury in this respect.

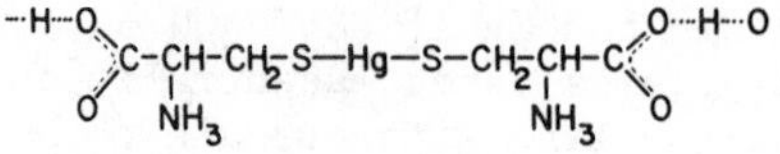

Hg(L-cystH)(L-cystH$_2$).Cl. 0·5H$_2$O (1)

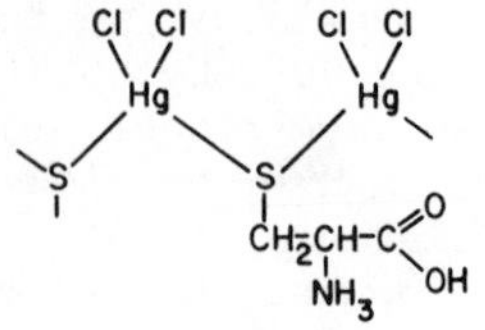

HgCl$_2$(μ-S-L-cystH$_2$) (2)

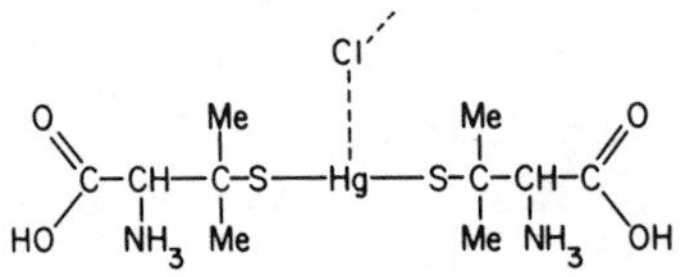

HgCl(penH$_2$)$_2$.Cl.H$_2$O (5)

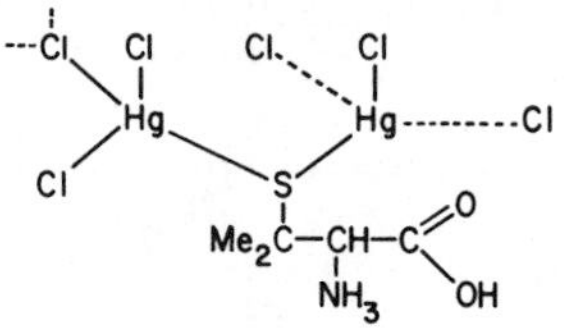

(HgCl$_2$)$_2$(μ-S-penH$_2$).2H$_2$O (4)

Figure 5. Schematic representations of the crystal structures of Hg[SCH$_2$-CH(NH$_3$)COO][SCH$_2$CH(NH$_3$)COOH]Cl · 0.5 H$_2$O (1), HgCl$_2$[SCH$_2$CH-(NH$_3$)COOH] (2), [HgCl$_2$]$_2$[SCMe$_2$CH(NH$_3$)COOH] · 2H$_2$O (4), and [Hg-{SCMe$_2$CH(NH$_3$)COOH}$_2$Cl] Cl · H$_2$O (5).

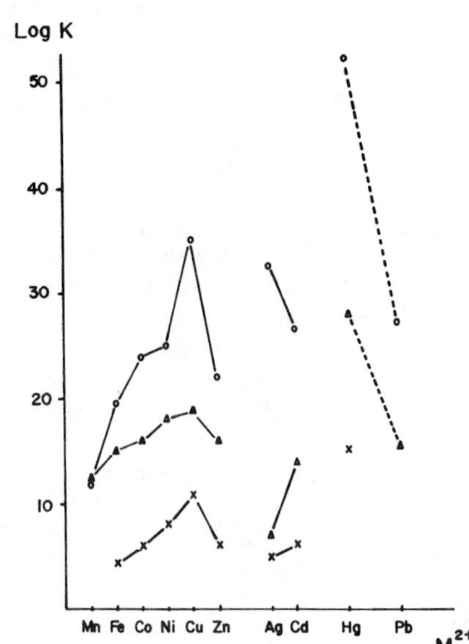

Figure 6. Stabilities of divalent metal–ligand complexes as indicated by stability constants or solubility products. For sulfides and hydroxides log k refers to solubility products; and for ethylenediamine complexes log k refers to stability constants. Key: (○) sulfide; (△) nitrogen; (×) oxygen.

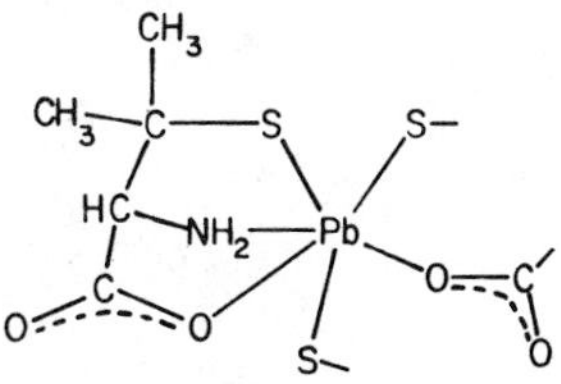

Pb(D-pen) (Ref. 29)

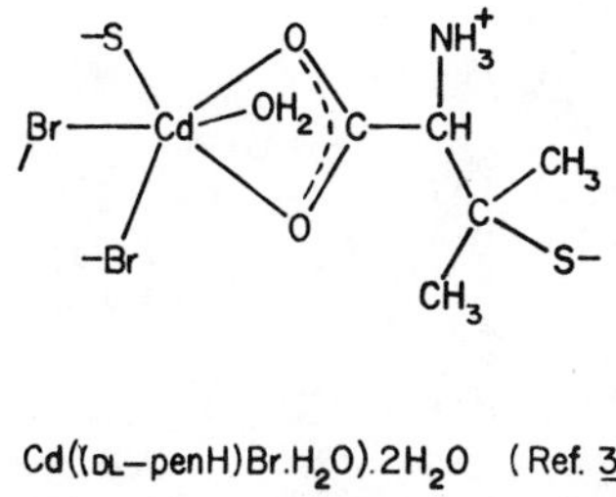

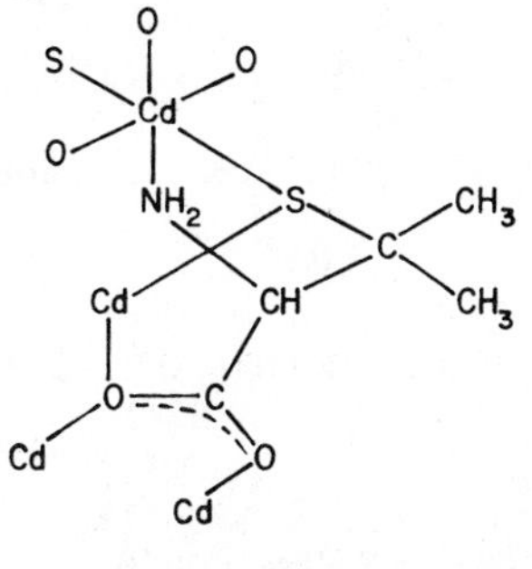

Cd((DL-penH)Br.H_2O).$2H_2O$ (Ref. 30)

Cd(D-pen). H_2O (Ref. 31)

Figure 7. Pictorial representations of metal–ligand interactions in lead and cadmium–penicillamine complexes. For comparison with mercury complexes, see *Figure 5.*

Table I

M-X Bond Lengths in Amino Acid Complexes

Compound	M-S	M-N	M-O	Ref.
$CH_3Hg(SCH_2CH(NH_3)COO).H_2O$	2.352(12)		2.84(2)	19
$HgCl_2[SCH_2CH(NH_3)COOH]$	2.490(4)			24
	2.453(4)			
$Hg[SCH_2CH(NH_3)COO]_2HCl\cdot 0.5\ H_2O$	2.355(3)			24
	2.329(5)			
$(HgCl_2)_2[SC(CH_3)_2CH(NH_3)COOH]2H_2O$	2.822(5)			23
	2.356(5)			
$CdBr[SC(CH_3)_2CH(NH_3)COO]3H_2O$	2.444(2)		2.262(5) 2.715(5) 2.490(6)	30
$Cd[SC(CH_3)_2CH(NH_2)COO]H_2O$	2.563(7)	2.38(2)	2.57(2) 2.51(2)	31
	2.567(7)		2.40(2)	
$Pb[SC(CH_3)_2CH(NH_2)COO]$	2.716	2.444	2.444 2.768	29
$Me_2Pb[SCH_2CH(NH_2)COO]$	2.50(1)	2.46(3)	2.55(4)	39

METHYLATION AND DEMETHYLATION OF LEAD

Recently there has been a major upsurge in interest in environmental lead chemistry with reports that trimethyllead(IV) (32) and lead(II) nitrate (33) can be biologically methylated to tetramethyl lead contrary to earlier predictions. The methylation of environmental lead to toxic Me_4Pb would pose a potential health threat if proven, since lead, particularly in the form of lead(II), continues to accumulate in the environment as a product of internal combustion engines. Although an initial controversy (34) arose concerning the chemical (i.e. disproportionation) or biological nature of processes effecting the $Me_3Pb(IV)$ → Me_4Pb conversion, recent experiments appear to have conclusively proven the existence of a biological methylation sequence (35). For Pb(II) salts there are, on the surface, cogent reasons for rejecting the possibility of biomethylation to tetramethylead. With the exception of a few air and moisture sensitive compounds with bulky groups (36), lead(II) alkyls are unknown. Even monomethyllead(IV) species are unstable in the absence of stabilising ligands. Furthermore, the oxidation reduction potential for Pb(II)/Pb(IV) disfavours electrophilic mechanisms for Pb(II), pertinent in biological methylation of other metals (e.g., Hg(II), Pd(II)) (37). Thus transfer of a methyl group as CH_3^- (say from methylcobalamin) would yield a highly unstable $Me\text{-}Pb(II)^+$ species from which $MePb(IV)^{3+}$ could only be generated by an unfavourable 2-electron oxidation or via disproportionation. In our view a much more likely mechanism is what can essentially be described as oxidative addition of CH_3^+ to an uncharged (or anionic) lead(II) complex:

$$CH_3^+ + [:Pb^{II} L_n]^{(2-n)+} \rightarrow [CH_3 - Pb^{IV} L_n]^{(3-n)+}$$

(L = uninegative ligand)

A variety of biological methylating agents (compare for example the recent paper of McBride and Cullen on arsenic methylation (38)) capable of transferring carbonium ions are potentially capable of synthesising $CH_3Pb(IV)$ in this way. The Pb(II) - penicillamine complex (Figure 7) is a potential model for the lead(II) chelate. However, it would clearly be advantageous if the lead(II) lone pair was stereochemically available for facile methylation, as in 7 where X simply represents the backbone of a tripod-like tridentate ligand.

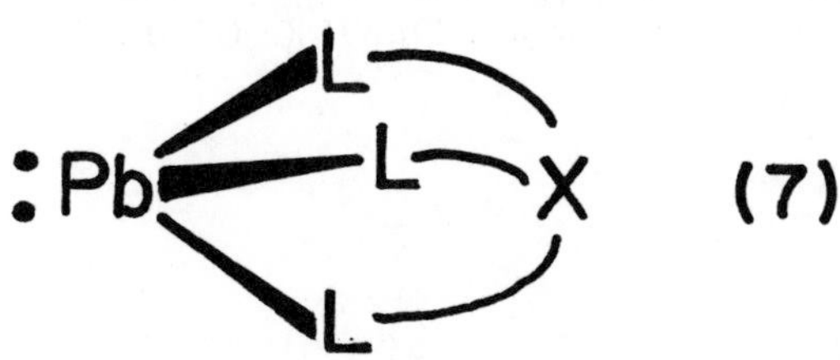

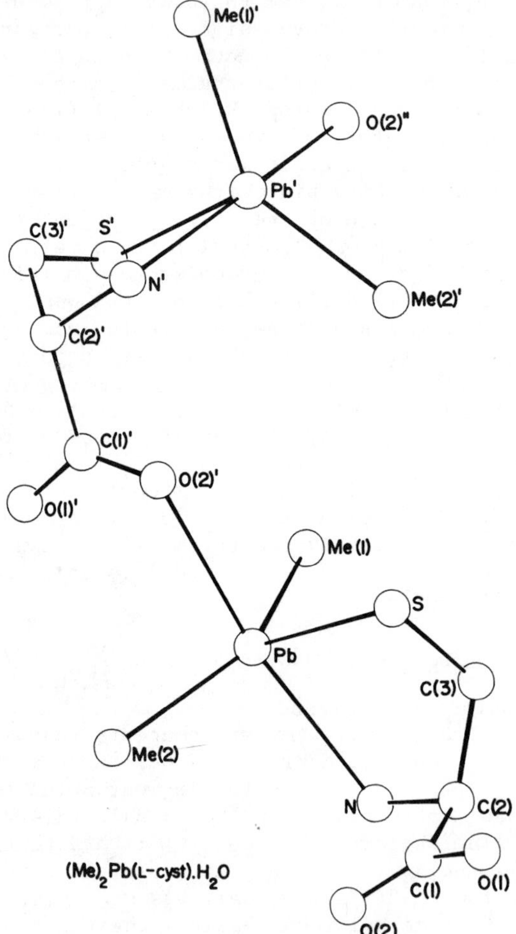

Figure 8. A perspective view of the molecular structure of $(CH_3)_2Pb[SCH_2CH(NH_2)COO] \cdot H_2O$

Experiments designed to establish the best choice of ligand and methylation conditions for the Pb(II) → Me_4Pb interconversion are currently in progress in our laboratory in collaboration with Drs. Y.K. Chau and P.T.S. Wong at CCIW Burlington.

Although lead(II) salts are the major lead pollutants in the environment, occasionally high concentrations of tetramethyllead or trimethyllead species accumulate from gasoline spills, improperly burnt fuel, etc. In such cases, the influence of natural ligands on the rate of environmental degradation to less toxic lead(II) salts is pertinent. There is evidence that sulfur ligands may facilitate the cleavage of lead(IV)-alkyl bonds. Thus $(CH_3)_3$ PbOAc reacts slowly with L-cysteine or DL-penicillamine in aqueous solution to give dimethyllead(IV) cysteinate (Fig. 8) or penicillaminate (39). By contrast, trimethyllead in acidic, neutral and basic aqueous solution shows very little decomposition to inorganic lead over a period of several months (40). The dimethyllead(IV) cysteinate is polymeric with each cysteinate coordinated to one lead atom via S and N sites and to a second lead atom via a carboxylate oxygen. This conversion of $(CH_3)_3$PbIV to $(CH_3)_2$ Pb(cyst) under mild conditions may have biological implications; alkylleads are potent neurotoxins, and may exert their effects via binding to active sulfhydryl sites in tissues (*cf* Fig. 8).

Acknowledgements I would very much like to acknowledge the major contributions which Dr. Nicholas J. Taylor a research associate at Waterloo, has made to this work. Most of the synthetic and X-Ray structural work quoted in this article has been carried out by Dr. Taylor. Initial studies of methylmercury complexation were undertaken by a graduate student (now Dr) Y.S. Wong and the homocysteine work in progress is due to L.F. Book. I would also thank my colleague Dr. P.C. Chieh who has collaborated with me in some of this work. Financial support has been provided by Environment Canada, principally through the Water Resources Support Program.

Literature Cited

1. Ahrland S., Chatt J., and Davies N.R., Quarterly Rev. Chem. Soc., (1958), 12, 265.

2. Pearson R.G., J. Amer. Chem. Soc., (1963), 85, 3533.

3. Landner L., Nature, (1971), 230, 452.

4. Wood J.M., Adv. Environmental Sci. Technol. (1971), 2, 39.

5. Hoch F.L., and Vallee B.L., Arch. Biochem. Biophys., (1960), 91, 1.

6. Hughes W.L. Jr., J. Amer. Chem. Soc., (1947), 69, 1836.

7. Canty, A.J., "Chemical Problems in the Environment: Occurrence and Fate of Organoelements" ACS Advances in Chemistry Series, F.E. Brinckman and J.M. Bellama, Eds., (1978)

8. Simpson R.B., J. Amer. Chem. Soc., (1961), 83, 4711.

9. Hughes W.L. Jr., Cold Spring Harbour Symposium on Quantitative Biology, (1950), 14, 79.

10. Rabenstein D.L. and Fairhurst M.T., J. Amer. Chem. Soc., (1975), 97, 2086.

11. Webb J.L., "Enzyme and Metabolic Inhibition", Vol. 2, Academic Press, New York (1969).

12. Schwarzenbach G. and Schellenberg M., Helv. Chim. Acta., (1965), 48, 28.

13. Weiner I.M., Levy R.I. and Mudge G.H., J. Pharmacol. Exp. Ther. (1962) 138, 96.

14. Kagi J.H.R. and Vallee B.L., J. Biol. Chem., (1961), 236, 2435.

15. Hilton B.D., Man H., Hsi E., and Bryant R.G., J. Inorg. Nucl. Chem., (1975), 37, 1073.

16. McAuliffe C.A., and Murray S.G., Inorg. Chim. Acta Rev., (1972) 6, 103.

17. Wong Y.S., Taylor N.J., Chieh P.C. and Carty A.J. J. Chem. Soc. Chem. Commun., (1974), 625.

18. Rabenstein D.L., Accounts Chem. Res., (1978), 11, 100.

19. Taylor N.J., Wong Y.S., Chieh P.C. and Carty A.J., J. Chem. Soc. Dalton Trans., (1975), 438.

20. Wong Y.S., Carty A.J. and Chieh C., J. Chem. Soc. Dalton Trans., (1977), 1801.

21. Carty A.J. and Taylor N.J., unpublished results.

22. Book L.F., Chieh C., and Carty A.J., unpublished results.

23. Taylor N.J., Carty A.J. and Wong Y.S., J. Chem. Soc. Dalton Trans., to be published. See also Taylor N.J. and Carty A.J., J. Chem. Soc. Chem. Commun., (1976) 214.

24. Carty A.J. and Taylor N.J., unpublished results. For a preliminary account see: Carty A.J. and Taylor N.J., J. Amer. Chem. Soc., (1977), 99, 6143.

25. Canty A.J. and Kishimoto R., Inorg. Chim. Acta., (1977) 24, 109.

26. Canty A.J. and Tyson R.K., Inorg. Chim. Acta., (1977) 24, L77.

27. Aberg B., Ekman L., Falk R., Grietz U., Pearson G., and Snihs J.O., Arch. Envir. Health., (1969) 19, 478.

28. Lundgren K.D., Swensson A. and Ulfvarson, U. Scand J. Chim. Lab. Invest., (1967), 20, 164.

29. Freeman H.C., Stevens G.N., and Taylor I.F. Jr., J. Chem. Soc., Chem. Commun., (1974), 366.

30. Taylor N.J. and Carty A.J., Inorg. Chem., (1977), 16, 177.

31. Freeman H.C., Huq F., and Stevens G.N., J. Chem. Soc. Chem. Commun., (1976), 90.

32. Wong P.T.S., Chau, Y.K. and Luxon, P.L., Nature (1975), 253, 263.

33. Schmidt U., and Huber H., Nature, (1976) 259, 177.

34. Jarvie A.W.P., Markall R.N. and Potter H.R., Nature (1975), 225, 217.

35. Chau Y.K. and Wong P.T.S., "Chemical Problems in the Environment: Occurrence and Fate of Organoelements" ACS Advances in Chemistry Series, F.E. Brinckman and J.M. Bellama, Eds., (1978)

36. Davidson P.J. and Lappert M.F., J. Chem. Soc., Chem. Commun., (1973), 317.

37. Ridley W.P., Dizikes L.J. and Wood J.M., Science (1977), 183, 1049.

38. Cullen W.R., Froese C.L., Lui A., McBride B.C., Patmore D.J., and Reimer M., J. Organometal. Chem., (1977), 139, 61.

39. Carty A.J. and Taylor, N.J., unpublished results.

40. Sayer T.L., Backs S., Evans C.A., Millar E.K. and Rabenstein D.L., Can. J. Chem., (1977) 55, 3255.

Discussion

F. E. BRINCKMAN (National Bureau of Standards): To emphasize your last point concerning the appropriate coordination of lead, the Jarvie mechanism [Nature, (1975) 255, 217], involves an abiotic redistribution, presumably to form tetramethyllead. We reported that in Chesapeake Bay anoxic sediments with high biogenic sulfur content, not only did gaseous elemental mercury form from biological activity, but mercury also partitioned into the lipid or oil phases in the sediment body. The mercury was principally in soluble form; about 2% or so as the element and/or methylmercurials but principally in much larger molecules not yet characterized by us or anybody else. But the point here is bioavailability. I think we will find a number of coordination sites which will hold the metal available for other activity besides simple mineralization as the sulfide. Sulfur doesn't seem to be the only key element in understanding this coordination chemistry and the bioavailability in these systems.

R. H. FISH (University of California, Berkeley): Have you ever looked at thioethers?

CARTY: We looked at both the methylmercury and the inorganic complexes in methionine, and the crystalline methylmercury complex contains the methionine bound to mercury by the nitrogen. In solution it is linear. It's with this zwitterionic complex with a positive charge on mercury and a negative charge on the carboxylate.

FISH: What about $HgCl_2$?

CARTY: In Hg^{2+}, the perchlorate contains the methionine bound through two thioethers and two carboxylates.

FISH: Yes, because we looked at a pyridyl ethyl-cysteine molecule. We found a nine-numbered ring with the pyridyl-nitrogen and the amino function. The sulfur never interacted.

CARTY: I think the thioether interactions are much less strong. In fact, in solutions of methionine with CH_3Hg^+, the thioether coordination only occurred at pH of less than 2, otherwise it is NH_2 or carboxylate that interacts.

F. HUBER (University of Dortmund): We examined compounds of organolead and organotin mercaptocarboxylic acids. We observed that the lead species can coordinate to the sulfur, or to oxygen, or to even both. We did experiments with mercaptopropionic acid as a model compound for cysteine, without an NH_2 group, and observed that in many cases there is coordination between lead and sulfur. The carboxylic acid didn't react. With trimethyllead(IV) and dimethyllead(IV) compounds, especially with the trimethyllead compounds, there was an acidolysis reaction giving methane and dimethyllead(IV).

CARTY: In solutions of trimethyllead acetate and L-cysteine, if you look by nmr techniques at the reaction with a pH of 6 or 5, you find only weak coordination of the sulfhydryl group. At alkaline pH, dimethylation occurs.

HUBER: We try to work in neutral solutions.

W. P. RIDLEY (University of Minnesota): Your suggestion that we should not ignore the possibility that lead(II) could accept a carbonium methyl group from some biological methyl donor should certainly be investigated. Thallium(I) could be methylated in the same fashion, assuming a proper complex could be formed, i.e., the transfer of CH_3^+ to Tl(I) to give monomethylthallium(III). The standard reduction potentials of the lead(II) to lead(IV) and thallium(I) to thallium(III) couples are very similar, suggesting that they may proceed in a similar fashion.

HUBER: On the possibility of methylation of Pb(II), in many complexes we do not see an available lone pair; the Pb(II) adopts octahedral coordination. Nonetheless, I think the possibility exists for this reaction.

W. R. CULLEN (University of British Columbia): If the sulfur was already coordinated to the lead, then methylation of the sulfur could offer means for transfer of methyl onto the lead or onto the thallium. This might be a viable route by which oxidative addition might take place.

HUBER: When sulfur is coordinated to lead, there might be a possibility that the lone pair could be stereochemically directed. Halide complexes usually have an octahedral geometry, but there are some complexes, especially with thiophosphates, where the lone pair is active.

CARTY: Certainly, with the ability that inorganic chemists have these days to tailor-make ligands, one would not be surprised to find that you can make a ligand which would actually tie the bonds back and let the lone pair be free for methylation.

G. E. PARRIS (Food and Drug Administration): Potassium antimony tartrate has a structure in which one side of the antimony is open, and it is of the few soluble compounds of inorganic antimony. The possibility of transferring CH_3^+ to this potential nucleophile is interesting. However, since antimony Mössbauer has shown that the lone pair is probably more in a S-type orbital, as it is in trimethylstibine, there is some question whether or not these are good nucleophiles for accepting CH_3^+.

M. L. GOOD (University of New Orleans): That's a good point; the Mössbauer is definitive on the character of those electrons.

CARTY: There are some alkyl compounds of lead(II), containing bulky alkyl groups, which have been prepared by Professor M.F. Lappert (University of Sussex). In those compounds the Pb(II) behaves as a donor to transition metals, so the lone pair is available.

RECEIVED August 22, 1978.

22

Release Mechanisms of Organotin Toxicants from Coating Surfaces: A Leaching Model for Antifouling Coatings

CHARLES P. MONAGHAN, VASANT H. KULKARNI, and MARY L. GOOD

Department of Chemistry, University of New Orleans, Lakefront, New Orleans, LA 70122

Fouling of ship hulls and marine installations by vegetable and/or animal organisms is a well-known, worldwide problem which results in significant economic loss due to structural damage or fuel requirements (1,2,3). Thus any program for fuel economy and preservation of marine structures by the Navy, the shipping industry, or the fishing fleet must contain the development of effective antifoulant procedures as a high priority. The only practical solution to the fouling problem through the years has been the utilization of paints and coatings having antifoulant activity. Most of the successful coatings have been paint formulations containing a toxic component. The most widely used toxicant has been cuprous oxide although various organometallic compounds of arsenic, mercury, lead and tin have been used. Studies have indicated that the organotin antifoulants show particular promise, since they exhibit control over a variety of fouling organisms for relatively long periods of time and they do not promote corrosion when applied over conductive substrates (4,5). However, if these organotin containing coatings find widespread application in the marine industry and in naval operations, their long term environmental impact on harbors and shipyards will become most important. In addition, the need for laboratory testing techniques for coating evaluation and comparison will be critical. Thus the development of laboratory procedures for the chemical speciation of the toxicant released to the environment and methods for evaluating the leaching parameters becomes an area of fertile research. This paper will report our initial efforts to develop in detail a leaching model for the loss of organotin toxicants from conventional antifouling coatings. Work on the speciation of the released toxicant moieties will be presented in a subsequent paper.

For an antifouling paint to be effective, the toxicant must leach from the coating at a rate high enough to repel or kill organisms at the coating surface but not so high as to cause rapid depletion and early failure of the coating. De la Court and de Vries (6,7) and Marson (8,9) have outlined those parameters

0-8412-0461-6/78/47-082-359$05.00/0

which are necessary to describe the leaching of toxicant from an insoluble matrix containing cuprous oxide (Cu_2O). The models developed by these authors show considerable insight into the leaching process.

Coatings in which the toxicant particles and soluble pigment particles are dispersed throughout the coating matrix in such a manner that the particles are always in contact are called "continuous contact coatings". In these coatings, it is assumed that when the particles dissolve, pores are developed through which the toxicant diffuses to the surface of the coating and the porous exhausted matrix remains intact after the particles have dissolved. Several steps are required to describe the leaching process. The solvent diffuses through the exhausted matrix to the leaching zone where it dissolves the toxicant or soluble pigment, leaving a pore. The solvated toxicant molecule diffuses through the exhausted matrix to the surface and then diffuses through the quiescent layer into the bulk of the solution (sea or harbor water). Cuprous oxide dissolves in sea water to establish a reaction equilibrium as follows:

$$1/2\ Cu_2O + H^+ + 2Cl^- \rightleftharpoons CuCl_2^- + 1/2\ H_2O$$

The rate determining steps for toxicant leaching are considered to be the diffusion of $CuCl_2^-$ through the exhausted matrix and the subsequent diffusion of $CuCl_2^-$ through the quiescent layer. If these assumptions are correct, a concentration approaching saturation will develop in the leaching zone. A stationary state will be established where the concentration of the toxicant is constant throughout any cross section of the exhausted matrix and throughout the quiescent layer. Under these conditions the flux of toxicant through a unit cross section is the same in both the exhausted matrix and the quiescent layer. It is evident from the formulation of this model that a detailed study of the matrix is necessary to characterize the leaching process.

There are several points where coatings containing organotin compounds differ from those containing cuprous oxide. These differences complicate the study of the leach rates and makes examination of the surface mandatory. The most effective organotin toxicants, of general formula R_3SnX, exist either as solids or as liquids. For conventional coatings where the toxicant is physically mixed with the paint matrix in a common solvent, the liquid organotins would be expected to disperse homogeneously throughout the coating; whereas, the solid materials may disperse as fine particles forming a heterogenous mixture. The usual pigments, carbon, ferric oxide and titanium dioxide, are insoluble in most coating formulations and are also dispersed as fine particles. In successful organotin containing coatings, the percentage of toxicant is about ten percent or less as contrasted to the forty to fifty percent toxicant composition commonly used in cuprous oxide coatings. Thus the "continuous contact" leaching models are not appropriate for these materials. It has been proposed (10, 11, 12, 13) that some of the matrix must dissolve for leaching to occur in a coating containing organometallic toxicants. To evaluate this proposition the coating surface must be examined in detail to characterize the aging mechanisma (dissolution of coating matrix, cracking, sloughing off of exhausted matrix, etc.) and/or the leach solution must be analyzed for dissolved matrix. It has also been proposed (10) that the leachate species for all of the coatings containing R_3SnX is the same molecule, namely R_3SnOH, and that the leaching rate depends not only on the matrix, but also on the physical state of the toxicant in the coating.

The model developed below has been designed with the view that a practical goal of any leaching model is to predict the useful lifetime of an antifouling coating with a minimum amount of laboratory and field testing. Thus the model presented here is a general empirical approach to the leaching problem with little emphasis on microscopic detail. Laboratory tests can be performed and the model can be used to determine the amount of organotin remaining in the matrix for a given set of experimental conditions. A reasonable extension of the model is the prediction of minimum coating lifetimes by defining the extreme conditions that a coating is likely to encounter and then designing laboratory testing procedures which simulate those conditions.

Details Of The Leaching Model

At any time during the lifetime of the antifouling coating, toxicant molecules diffuse out of the coating in a direction normal to the coating plane. The molecules diffuse through a thin quiescent layer of solution adhering to the surface and into a well-mixed turbulant layer of solution (the environment). This diffusion process is described by Fick's First Law of Diffusion. According to this Law (14)

$$J = -D\left(\frac{\delta c}{\delta x}\right)_{y,z} \tag{1}$$

where D is the diffusion coefficient and J, the flux, is the quantity of substance diffusing per unit time through a unit area. Thus,

$$J = \frac{1}{A}\frac{dm}{dt} \tag{2}$$

Therefore,

$$\frac{dm}{dt} = -DA\left(\frac{\delta c}{\delta x}\right)_{y,z} \tag{3}$$

Let

$$\frac{(c-c_s)}{x} \approx \left(\frac{\delta c}{\delta x}\right)_{y,z} \tag{4}$$

where c is the concentration in the bulk solution, c_s is the organotin concentration at the coating surface, and x is the thickness of the quiescent layer which is a function of the water velocity moving pass the surface. See Figure 1 for a pictorial representation of the leaching process. Thus,

$$\frac{dm}{dt} = -\frac{DA}{x}(c-c_s) \tag{5}$$

For an experimental container of volume v, one obtains

$$\frac{dc}{dt} = -\frac{DA}{xv}(c-c_s) \tag{6}$$

If an organotin compound is a liquid, it will mix thoroughly with its matrix forming a homogeneous system. If the organotin compound is a solid, it may dissolve somewhat in the matrix and solvent forming a low polymers of associated toxicant molecules, or it may remain as solid particles dispersed throughout the matrix (15,16). The regions of surface that are active in the leaching process will depend on the development of pores and channels through the matrix as the organotin compound and/or matrix dissolve and leach into the environment. If the area of the surface plane is A_o and if f is the fraction of the plane that is active, then

$$A = f\,A_o \tag{7}$$

and

$$\frac{dc}{dt} = -\frac{DfA_o}{xv}(c-c_s) \tag{8}$$

Integrating, one obtains

$$c = c_s - c_s \exp\left(-\frac{DfA_o}{xv}t\right) \tag{9}$$

The characterization of a coating involves determining c_s and Df/x by measuring the bulk concentrations at various times and using the least squares technique to fit the theoretical equation to the experimental data. The general form of the equation is

$$c = c_s - c_s \exp\left(-\alpha\frac{A_o}{v}t\right) \tag{10}$$

where α is Df/x and is characteristic of the coating for a given temperature and flow condition.

Now consider obtaining an expression for the lifetime of the coating knowing c_s and α:

$$\frac{dm}{dt} = -\frac{DfA_o}{x}(c-c_s) \tag{11}$$

In a large body of water (ocean or harbor), c is always neglible and A_o is the surface area of the ship continuously exposed to the water. Then

$$\frac{dm}{dt} = \alpha A_o c_s \tag{12}$$

Integrating, one obtains

$$m = \alpha A_o c_s \tau \tag{13}$$

where m is the mass of organotin leached out of the coating. Define a critical density, ρ_τ, which is achieved at the time τ where fouling becomes appreciable (say 50%). The mass of organotin lost from the coating is given by

$$m_\tau = \alpha A_o c_s \tau, \tag{14}$$

and

$$\rho_\tau = \rho_o - \alpha c_s \tau \tag{15}$$

where ρ_o is the original mass of organotin per unit area. The minimum useful lifetime of the coating is then given by

$$\tau = \frac{1}{\alpha c_s}(\rho_o - \rho_\tau) \tag{16}$$

Experimental Section

Static leach tests were performed on aluminum panels coated with Alum-A-Tox (a commerical coating obtained from Standard Paint and Varnish Company of New Orleans) containing an organotin compound as toxicant. Alum-A-Tox is a vinyl-type coating containing titanium dioxide as pigment. Either bis(tributyltin) oxide, tributyltin chloride, tributyltin acetate, triphenyltin acetate, or triphenyltin chloride was mixed into the coating so that a coating composition of 3.6% Sn by weight was obtained. The aluminum panels (3" x 4" x 1/16") were sanded and cleaned with carbon tetrachloride. They were sprayed with the paint on one side and were allowed to air-dry for at least three weeks. Each panel was then immersed in 550 mL artificial sea water (17) in a beaker and maintained at 21±2°C for the duration of the experiment. Solutions were replaced at the time of sampling (about every two weeks). The leachates were analyzed for tin using colorimetric methods.

The organotin concentration in the leachate was determined

after extracting the organotin compounds from the aqueous phase. The 550 ml of leachate from a given sample were diluted to 1000 mL with distilled water, and 100 mL was then pipetted into a separatory funnel. After adding 10 mL of chloroform to the separatory funnel and shaking vigorously for one minute, the contents were allowed to separate into two layers. The chloroform layer was removed, the extraction procedure was repeated three to four times to achieve complete extraction of the organotin compounds, and all of the chloroform portions were combined. Tin present in the chloroform was determined by the dithizone colorimetric method (18).

Any tin species remaining in the aqueous phase were subjected to an oxidation procedure before measuring the tin concentration. The aqueous phase was acidified with 16M HNO_3 and evaporated cautiously to a small volume. After adding 5-6 mL of 18M H_2SO_4, the solution was evaporated to the incipient fumes of SO_3. The residue was dissolved in distilled water and the resulting solution was diluted to 100 mL. The tin present in this solution was determined using the phenylfluorone colorimetric method (19). No tin above background was found in the aqueous phase for any sample. This suggests that the tin in the leachate is solely in the organotin form and not in the inorganic form.

It was assumed that the leaching would be slow enough that

$$\frac{\Delta c}{\Delta t} \approx \frac{dc}{dt}$$

where Δc is the measured organotin concentration and Δt is the time of accumulation. The leach rate - time data were fitted with a quadratic equation, and this empirical equation was integrated to obtain concentration-time data. Initial guesses for c_s and α were obtained after comparing the concentration-time data with simulated data generated for trial values of c_s and α by a program called LEACH2. The best values of c_s and α resulting from the previous procedure were further refined by a least squares program called LEACH1. As the concentration-time data were obtained from smoothed data, uncertainties in c_s and α have not been calculated.

To determine the solubilities of the organotin compounds used in this study, 50 mg of each compound were mixed with 100 mL of artificial sea water. The solutions were maintained at 21±2°C for two days with occasional stirring. The amount of organotin dissolved in the sea water was determined by the dithizone colorimetric method (18).

Results and Discussion

The leaching results for the tributyltin compounds and the triphenyltin compounds are shown in Figure 2 and Figure 3

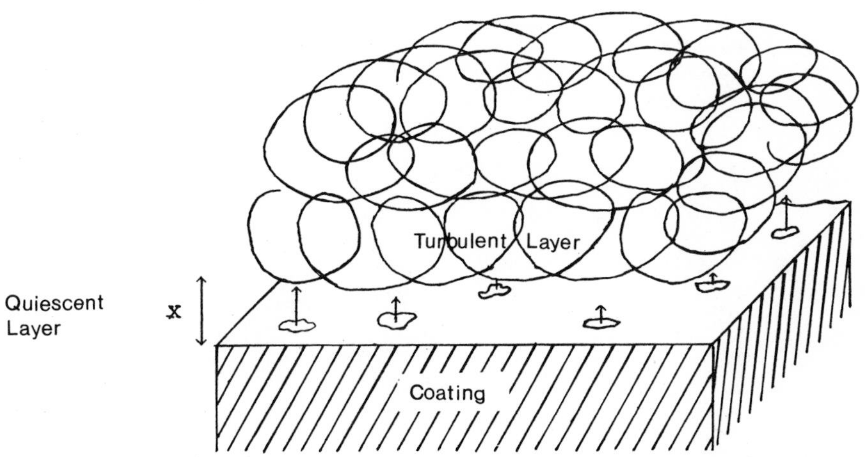

Figure 1. Pictorial representation of the leaching of toxicant molecules from an antifouling coating

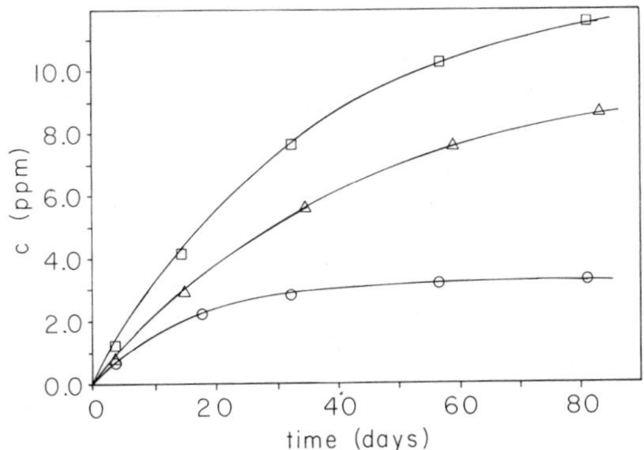

Proceedings of the 4th Annual Controlled Release Pesticide Symposium: Anti-Fouling Section

Figure 2. Leaching of tributyltin compounds from Alum-A-Tox coatings: (□) = $(C_4H_9)_3SnOAc$, (△) = $[(C_4H_9)_3Sn]_2O$, and (○) = $(C_4H_9)_3SnCl$. Each solid line represents the least squares fit to Equation 10.

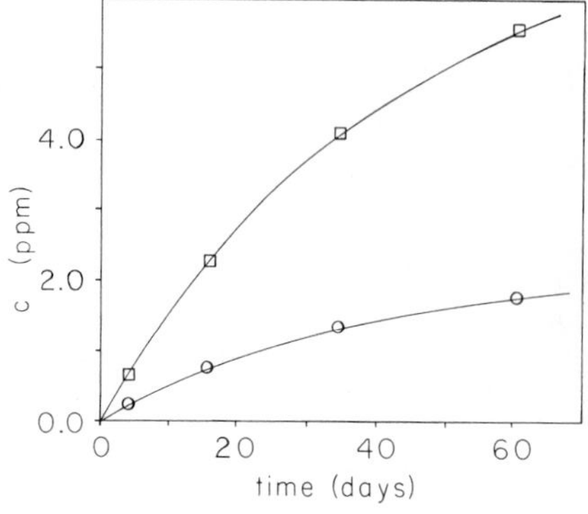

Proceedings of the 4th Annual Controlled Release Pesticide Symposium: Anti-Fouling Section

Figure 3. Leaching of triphenyltin compounds from Alum-A-Tox coatings: (□) = ϕ_3SnOAc and (○) = ϕ_3SnCl. Each solid line represents the least squares fit to Equation 10.

respectively. The parameters giving the best fit to the concentration-time data and the solubilities are given in the Table.

TABLE I

Parameters for Alum-A-Tox Coatings Containing Organotin Compounds

Toxicant	c_s (ppm)	α(cm day^{-1})	Solubility (ppm)
ϕ_3SnOAc	7.2	0.16	3.34
ϕ_3SnCl	2.2	0.18	4.22
$(C_4H_9)_3$SnOAc	12.8	0.19	1.88
$(C_4H_9)_3$SnCl	3.3	0.45	1.57
$[(C_4H_9)_3Sn]_2O$	9.9	0.17	5.92

As t goes to large values, c approaches c_s (see equation 10). The maximum concentration of organotin in sea water is its solubility. From this argument c_s, the concentration of organotin at the surface, would have the solubility as its maximum value. From the table it can be seen that c_s is of the same order of magnitude as the solubility but the values of c_s are larger than the solubility in many cases.

The parameter α must be considered to be empirical. There are no diffusion constants available for organotin compounds, and it is impossible to measure the thickness of the quiesent layer. However, since the solvated organotin species are expected to be the same (for R_3SnX where R is constant) in all cases (10), the same diffusion constant would be applicable. Under constant flow conditions, the thickness of the quiescent layer would be the same for all the coatings. Values of α would therefore provide a relative measure to rank or order the fraction of the active surface between the various coatings. This fraction of the surface that is active in the diffusion process is expected to be dependent on the physical state of the organotin compound in the coating as well as the porosity of the matrix. However, since only one matrix and a few selected toxicant compounds were studied, any further discussion based on the magnitude of c_s and α would be entirely speculative.

These preliminary results are promising in that an empirical diffusion model has been developed which appears to be generally applicable for describing the leaching process associated with a series of organotin toxicants contained in a commercial coating vehicle. The results do indicate that it should be feasible to design laboratory testing procedures for making relative comparisons of the properties of a series of coatings of different formulations. However, further studies must be carried out before the optimum range of c_s and α for satisfactory antifouling performance can be ascertained. In addition, further testing is necessary before additional refinements to the model can be made.

The experimental design for this initial study was not optimum in that the leaching rates were faster than anticipated. This makes suspect the assumption that the leach rate can be approximated by the measured organotin concentration divided by the accumulation time. The second iteration of this experiment will include other types of coating matrices and a new sampling procedure which will allow concentration-time data to be acquired directly. After these static tests have been completed, the effect of the depth of the quiescent layer and the possibility of accelerated leaching will be evaluated by carrying out all of the measurements on coatings mounted on rotating drums designed for a variety of speeds.

Acknowledgements

This work is a result of research sponsored by the Office of Naval Research and the NOAA Office of Sea Grant, Department of Commerce, under Grant No. R/MTR-1. The Alum-A-Tox coating base was supplied by Standard Paint and Varnish Company of New Orleans.

Literature Cited

1. Starbird, E. A. and Sisson, R. F., National Geographic (1973) November, 623.
2. Starbird, E. A., The Readers Digest (1974) March , 123.
3. Murray, C., Chemical and Engineering News (1975) January 6, 18.
4. Evans, C. J., J. Paint Tech. (1970) 34, 17.
5. Vizgirda, R. J., Paint and Varnish Prod. (1972) 62, 25.
6. de la Court, F. H. and De Vries, H. J., Prog. Org. Coat. (1973) 1, 375.
7. de la Court, F. H. and De Vries, H. J., J. Oil Colour Chem. Assoc. (1973) 56, 388.
8. Marson, F., J. Appl. Chem. (1969) 19, 93.
9. Marson, F., J. Appl. Chem. Biotechnol. (1974) 24, 515.
10. Monaghan, C. P. Hoffman, J. F., O'Brien, E. J., Jr., Frenzel, L. M. and Good, Mary L., "Proceedings of the 4th Annual Controlled Release Pesticide Symposium: Antifouling Section", R. L. Goulding, Ed., Corvallis, Oregon, August, 1977.
11. Kronstein, M., Mod. Paint and Coat. (1977) 67 (8) 57.
12. Carr, D. S. and Kronstein, M., Mod. Paint and Coat. (1977) 67 (11) 41.
13. Kronstein, M., Am. Chem. Soc., Div. Org. Coat. Plast. Chem. Pap. (1978) 38, 705.
14. Crank, J., "The Mathematics of Diffusion," Oxford at the Clarendon Press, 1956.
15. Hoffman, J. F. Kappel, K. C., Frenzel, L. M. and Good, M. L., "Organometallic Polymers, " C. E. Carraher, Jr., J. E. Sheats, and C. U. Pittman, Jr., Eds., p. 195, Academic Press, New

York, 1978.
16. O'Brien, E. J., Monaghan, C. P. and Good, M. L., "Organometallic Polymers," C. E. Carraher, Jr., J. E. Sheats, and C. U. Pittman, Jr., Eds., p. 207, Academic Press, New York, 1978.
17. LaQue, F. L., "Marine Corrosion: Causes and Prevention," p.98, John Wiley and Sons, New York, 1975.
18. Kulkarni, V. H. and Good,M. L., under preparation.
19. Kulkarni, V. H. and Good, M. L., Anal. Chem., in press.

Discussion

C. MATHEWS (Naval Civil Engineering Laboratory, Port Hueneme, California): You said that you didn't find any inorganic tin after a stated length of exposure. I believe Dr. Zuckerman said that it was rather nice that this thing broke down to the inorganic compound. Did I get that right?

GOOD: Yes, you did. It's not a conflict. What he is saying is that it breaks down in the atmosphere. These are not atmospheric tests; these are tests done in a laboratory where I don't have a lot of ultraviolet light. The question of pesticides is one thing, but boat coatings is another. The pesticides actually sit out on the ground, where they get quite a lot of sunlight; the organotin leachates are not going to have such UV exposure when they come off. I'm saying nothing here about the ultimate species that may or may not be in the environment. I'm simply telling you what comes off the surface.

M. KRONSTEIN (Manhattan College, New York): If you use real coating materials the reaction mechanism is complete different. The organotin interacts with vehicle of the paint, in particular with the low polymer fraction of the vehicle, even in the case of vinyl polymers. This can easily be determined by extracting leachate water with ethyl ether, then dissolving the residue in carbon tetrachloride. With an infrared spectrum and with an atomic absorption spectrum you will see the amount of organic moiety and tin that are in the water. Now they leach out bound and not separate. If the water is exposed to sunshine or to radiation in the laboratory, the released reaction product continues to polymerize, turns insoluble, then precipitates out. If you analyze the precipitate you observe organic material and tin (or the copper or the lead if they were used in the paint). The important point is that the organotin may come out as tin hydroxide or tin oxide and may not be as toxic. If it is still part of the polymer, it may be as toxic as it was in the paint. I will give a paper on the lead case tomorrow [cf. Abstr. Papers, 175th Nat'l Meeting ACS, Anaheim, CA., ORPL-135].

GOOD: I specifically chose a commercial paint which is sold

and used in South Louisiana. The material which comes off in this particular case is not part of the polymer; it is just the toxicant. It is true that in some materials the organotin coating can react with the polymer. I can tell you which ones do; it is very clear from the Mössbauer and the infrared spectra. In this particular case they do not. The model may actually work for those kinds of materials involved in either case, provided we determine what the actual bond-breaking mechanism is. Frankly, we don't quite know the mechanism for all those polymeric ones yet.

F. E. BRINCKMAN (National Bureau of Standards): I'm intrigued by your extension over the many models we have seen in the past involving diffusion or controlled release with near zeroth order kinetics. It is your conception that the time-to-failure will depend on some concentration of some species of known bioactivity. Is that what you finally see as being that measure? If so, we are talking about a potential standard.

GOOD: That is what one assumes. There is a difference between assumption in this case and knowing that for certain because the problem is that the kill mechanism or the control mechanism is not totally understood.

BRINCKMAN: That's right, because the most critical part of the analytical expression is the size of x. The issue is that diffusion gradient, or that concentration gradient, immediately above the surface at instantaneous time when that microorganism comes in.

GOOD: This particular experiment has another possible aspect, a biological one. You can determine x graphically and get a reasonable value for it. You ought to be able to do that if you can determine the diffusion coefficients by some other method. Then, if you could determine x for a static experiment or a rotating experiment, then we might be able to answer the question how close the organism has to get to the toxicant before it's effective. You remember, we don't know what the depth of the toxicant layer has to be before you get control.

G. E. PARRIS (Food and Drug Administration): With regard to the leach rate and toxicity to organisms, I think this is an appropriate approach in terms of diffusion through a quiescent layer. However, in terms of time-to-failure, this reminds me of the system which recently received attention, the leaching of tris(2,3-dibromopropyl)phosphate from polyester fibers. There, isn't the limiting characteristic the diffusion within the solid phase rather than the liquid phase? In the "tris" case, you can wash the fiber several times to get rid of a surface loading and establish some sort of a steady-state leaching condition. If you take the fiber out of solution and let it sit, the toxicant migrates very slowly to the surface by completely different kinetics.

GOOD: This model at this moment does not include that. It looks only at what is coming off at the surface involving toxicant islands. As I discussed, this model may be all right because you are not talking about leaching through cracks and crevices in big polymers.

PARRIS: You used a correction factor to account for the fact that you do have islands and not a continuous surface. What concentration do you use, the concentration in the island or some average concentration?

GOOD: It's an average concentration.

PARRIS: Isn't that contrary? You are correcting for size, but then you are assuming that the concentration is continuous.

GOOD: You mean that the concentration is continuous at the surface? You can't do much about that unless you make a much more completed model. We can check that very easily experimentally by going to less and less concentrated systems. That is, make the islands further apart.

PARRIS: It is then a question of how you interpret the diffusion coefficient?

GOOD: That's true. What we would see on this model is an effective diffusion coefficient which is not precisely the same as a diffusion coefficient involving the total solid and an interface.

RECEIVED August 22, 1978.

23

Metal-Ion Transport Mediated by Humic and Fulvic Acids

GORDON K. PAGENKOPF

Department of Chemistry, Montana State University, Bozeman, MT 59717

The movement of trace metals through the water environment involves a large number of chemical reactions. In some natural waters one particular type of reaction will be dominant whereas in others a balance will be established. This report presents a brief look at the interactions between trace metals and two commonly encounted natural chelating agents.

Humic and Fulvic Acid

In many cases most of the organic material found in natural waters is derived from natural sources. The leaching of soil and decaying plants accounts for most of this material making the number of possible organic compounds virtually limitless. One group, the humates, appears to play a role in metal transport and has been widely studied (1). These materials are dark colored and in many cases comprise a majority of the organic material found in natural waters. The substances have been divided into two general classes on the basis of solubility. The first class, which is designated as humic acid, is that material soluble in base but insoluble in acid and/or alcohol. Fulvic acid, the second class, is soluble in both base and acid. The isolation of these materials from natural waters can be tedious since the normal concentration range is 1-5 mg/l. There are reports of higher values however (2,3). Both of these acids may be precipitated from solution with excess barium.

A selected group of elemental analyses for humic (HA) and fulvic acid (FA) is shown in TABLE I. It should be noticed that the oxygen content of fulvic acid is significantly greater than the values observed for humic acid and the carbon content of humic acid is greater than that of fulvic acid.

The oxygen functionality includes carboxyl, phenolic, alcoholic, carbonyl, and methoxyl groups. The total acidity is considered equal to the sum of the carboxyl and phenolic acidity. These groups also appear to play a major role in the binding of metal ions. TABLE II summarizes some representative acidity

0-8412-0461-6/78/47-082-372$05.00/0

TABLE I

Elemental Analysis of Humic and Fulvic Acids (1).

	%C	%H	%O	%S	%N
Soil HA					
	56.4	5.5	32.9	1.1	4.1
	53.8	5.8	36.8	0.4	3.2
	60.4	3.7	33.6	0.4	1.9
Coal HA					
	64.8	4.1	28.7	1.2	1.2
Soil FA					
	42.5	5.9	47.1	1.7	2.8
	47.6	4.1	47.3	0.1	0.9
	50.9	3.3	44.8	0.3	0.7
Water FA					
	46.2	5.9	45.3	--[a]	2.6

a) not determined

TABLE II

Acidity Values for Humic and Fulvic Acids

	Total Acidity meq/g	Total Acidity g/eq	Carboxyl meq/g	Phenolic meq/g	Ref.
Soil HA					
	6.6	151	4.5	2.1	1
	8.7	114	3.0	5.7	1
	5.7	175	1.5	4.2	1
	10.2	98	4.7	5.5	1
	8.2	122	4.7	3.6	1
Coal HA					
	7.3	137	4.4	2.9	1
	7.4	135	3.4	4.0	4
Soil FA					
	14.2	70.4	8.5	5.7	1
	12.4	80.7	9.1	3.3	1
	11.8	84.7	9.1	2.7	1
	10.2	97.9	7.0	3.2	5

values. The data is presented as meq/g and g/eq.

These natural complexing agents exhibit considerable variation in nitrogen content. Studies with humic acid and fulvic acid, which contain 4% nitrogen, indicate that approximately one-fourth of the total nitrogen is present as amino acid nitrogen and a comparable amount is present as ammonia nitrogen (6). This amount of nitrogen could provide additional coordination sites.

A variety of techniques have been utilized to obtain the average gram formula weight of these acids. The values range from a few hundred to several million. In general, the molecular weights of fulvic acid are less than humic acid, with values lying in the range of a few hundred to a few thousand. Complexometric studies utilizing ion selective electrodes have provided FA acid values of 600, 1200, 1750, 2000, and 2775 daltons (5,7). Similar studies with coal humic acid have provided a value of 6761 daltons (4). In addition, it has been observed that 90% of the HA in Florida river waters lies within the 1000-5000 range (8). Many of the problems associated with obtaining values by more conventional methods lie with the obtainment and utilization of appropriate standards. The ion selective electrode technique, using cadmium, may prove to be a reliable method for obtaining these data.

Potentiometric titrations of HA and FA with strong base suggests that a large fraction of the ionizable protons are titrated by pH 7 (4,5,9,10,11). A combination of the g/eq data that is summarized in TABLE II and the observed average gram formula weights provides a large number of ionizable protons per molecule. For example, the FA characterized by Brady (5) which has an average gram formula weights of 2775 and a total acidity of 10.2 meq/g will have approximately 28 ionizable protons, and if more than half of these are titrated by pH 7, this molecule will have a high net negative charge which will facilitate complexation.

The exact description of these ligands has not as yet been assigned but a reasonable qualitative picture of the molecules would be a polymeric polyelectrolyte. There are many acidic groups capable of complexing metals and probably a few nitrogen donor sites. The spatial orientation of the complexation sites is not known; however, it is presumed that several donors are available for simultaneous coordination of metal ions. Charge neutralization is important in determining the solubility of these molecules as evidenced by the fact that HA is insoluble in strong acid and that FA can be precipitated by excess barium.

Complexation

A variety of techniques have been utilized to evaluate the conditional stability constants for the complexation of trace metals by humic and fulvic acids. These include potentiometric titrations, ion exchange, spectrophotometry, electrochemical measurements, metal-ion specific electrodes, nuclear magnetic resonance, and others. A selected sample of the conditional stability constants is listed in TABLE III. This listing is by no means complete, but it does provide a summary of the stability constant variation with pH and metal ion.

One of the first proposed models for the complexation of metal ions by HA and FA is similar to bidentate coordination by salicylic acid (9,20). The ligands certainly contain sufficient functionality to accommodate this mode of complexation. The pH profile data indicate that protons are displaced by metal ions

$$\text{Ar(OH)}-\text{C(=O)}-\text{O}^{(-)} + \text{M} = \text{Ar}(-\text{O}-\text{M}-\text{O}-\text{C(=O)}-) + \text{H}^{+}$$

(21). Since the pKa values of phenolic oxygens are usually much greater than carboxylic groups, there is some question as to the extent of phenolic oxygen participation in binding reaction (4,5).

Another mode of complexation (22) would involve bidentate carboxylic acid groupings. This grouping would also exhibit hydrogen ion dependence.

$$>\text{C}\begin{matrix}\text{CO}_2^- \\ \text{CO}_2\text{H}\end{matrix} + \text{M} = >\text{C}\begin{matrix}\text{C(=O)}-\text{O} \\ \text{C(=O)}-\text{O}\end{matrix}>\text{M} + \text{H}^{+}$$

A comparison to model compounds such as oxalic and malonic acid predicts sizable complex stability. This type of coordination has been utilized to rationalize the increase in average gram formula weight of the ligands that occurs coincident with metal ion facilitated precipitation (22). In this case, the metal is thought to bridge two or more polymeric chains together.

The complexation of more than one metal ion per ligand, each with comparable stability, requires the presence of repeating units within the large polymeric molecule. The magnitude of the stability constants is similar to that expected for dicarboxylate coordination, and the influence of changing pH in the near neutral region may be assigned to a net negative charge increase in the ligand as the pH increases. The formation of these multimetal complexes suggests many interesting phenomenon regarding trace metal carrying capacity and diverse metal ion competition for donor sites.

The stability constants obtained at low pH are not large, most lying between 100 and 1000, and it is probable that ion-pairing or monodentate coordination is operative. Recent nmr studies (23) with Mn(II) and fulvic acid may be interpretated by the formation of electrostatic outer sphere complex. The stability of the complex is not too much larger than a comparable one with K^+. Similar studies with Fe(III) substantiate an inner

TABLE III

Selected Conditional Stability Constants

Complex	pH	Log K	ref	Complex	pH	Log K	ref
CuFA	3.0	3.3	10	CdHA	5.5	4.9	4
CuFA	5.0	4.2	10	CdHA	6.0	5.3	4
$CuHA_2$	4.0	5.9	12	CdHA	7.0	5.9	4
$CuHA_2$	4.0	5.5	12	CdHA	7.5	6.3	4
$CuFA_2$	4.0	5.4	12	Cd_2HA	6.0	9.2	4
CuFA	3.0	3.2[a]	13	Cd_2HA	7.0	10.6	4
CuFA	4.0	3.6[a]	13	Cd_2HA	7.5	11.7	4
CuFA	5.0	4.0[a]	13	Cd_3HA	6.0	13.7	4
CuFA	3.0	5.8	14	Cd_3HA	7.0	15.7	4
CuFA	6.0	5.0	7	Cd_3HA	7.5	16.5	4
$CuFA_2$	6.0	9.5	7	Co(II)FA	3.0	2.8	10
CdFA	5.0	3.0[a]	13	Co(II)FA	5.0	4.1	10
CdFA	6.0	3.4[a]	13	Co(II)FA	4.5	4.0	15
CdFA	5.7	5.3	5	Co(II)FA	6.0	6.6	15
CdFA	6.7	5.6	5	Co(II)HA	4.5	6.0	15
CdFA	7.7	6.0	5	Co(II)HA	6.0	8.3	16
Cd_2FA	5.7	9.8	5	Co(II)HA	4.7	4.6	16
Cd_2FA	6.7	10.6	5	Co(II)HA	5.8	5.5	10
Cd_2FA	7.7	10.7	5	NiFA	3.0	3.2	10
Cd_3FA	5.7	14.0	5	NiFA	5.0	4.2	10
Cd_3FA	6.7	15.5	5	ZnFA	3.0	2.2	10
Cd_3FA	7.7	15.4	5	ZnFA	5.0	3.6	10
ZnFA	3.5	1.7	14	Fe(III)FA	1.0	4.45	19
ZnHA	5.0	6.0	17	Fe(III)FA	1.5	4.18	19
ZnHA	6.0	6.1	17	Fe(III)FA	2.5	4.18	19
ZnHA	3.6	4.4	18	Fe(III)FA	3.5	5.1	14
ZnHA	5.6	6.2	18	Hg(II)FA	3.0	4.9[a]	13
ZnHA	7.0	6.8	18	Hg(II)FA	4.0	5.1[a]	13
ZnHA	4.7	4.2	16	PbFA	3.0	2.7	10
ZnHA	5.8	5.3	16	PbFA	5.0	4.0	10
MnFA	3.0	2.2	10	PbFA	6.8	5.5	7
MnFA	5.0	3.7	10	$PbFA_2$	6.8	10.4	7
Fe(III)FA	1.7	6.1	10	CaFA	3.0	2.7	10
Fe(II)FA	4.5	5.4	15	CaFA	5.0	3.3	10
Fe(II)FA	6.0	5.6	15	MgFA	1.9	2.1	10
Fe(III)HA	4.5	6.8	13	AlFA	2.35	3.7	10
Fe(III)HA	6.0	7.2	13				

a) Values estimated from Figure 1 in reference 13.

sphere mode of complexation. As the pH increases from near 2 to 8, the conditional constants for inner sphere complexation will become more favorable and it is expected that inner sphere complexation will predominate at the higher pH values. Electron paramagnetic resonance studies of a Cu-HA complexes suggest that copper is bound in a porphyrin type of donor network (23). These complexes should be more stable than those previously presented and in addition the metal ions may not be labile.

The binding of copper by a coal humic acid exhibits two major types of interaction (25) in the pH 1-3 region. One involves hydrogen exchange comparable to those previous presented and the other is nonexchange. The latter is most important at low pH and does not fit basic adsorption theory. At these pH values the ligand may be aggregated by a metal or hydrogen ion which ultimately provides flocculation. The formation of intermolecular bridges is believed to facilitate the aggregation. The coagulation of peat humic acids through the addition of metal ions is presumed due to the formation of a coiled hydrophobic conformation and concurrent expulsion of water from the organic matrix (26).

At low pH the interactions are dominated by adsorption and ion exchange that may or may not involve inner sphere coordination. As the pH is increased, the more conventional mode of complexation involving the inner sphere of the metals and oxygen and nitrogen donors takes over. The observed stabilities are comparable to multidentate coordination.

Rate of Complex Formation and Dissociation

It is sometimes reported that complexation of trace metals by humic and fulvic acid results in inert complexes which render the trace metals unavailable for transport. There is limited experimental evidence to support this concept and in fact many of the interactions should be labile. The only detailed study involves the complexation of Fe(III) with fulvic acid (19).

This reaction was studied over the pH range of 1 to 2.5. The rate of formation of the complex is first order in each reactant and the observed rate constant is $196 M^{-1} sec^{-1}$. This rate constant is a factor of thirty less than the rate constant for the reaction

$$Fe^{3+} + SO_4^{2-} \rightarrow FeSO_4^{+}$$

which is 6370 sec^{-1} (27). This later value is in agreement with the predicted value utilizing the rate of water loss from Fe(III) and the outer sphere association constant (28). It is difficult to predict the outer sphere association constant for FA, but from these results it appears that the value is approximately 0.1 at low pH. It is believed (19) that coordination of the first FA binding site is associated with the rate determining step. If this is so, and if it holds for other similar complexes, one can make a reasonable estimate for the rate of formation of all metal-FA complexes.

The rate constant for the dissociation of Fe(III)-FA is calculated to be 0.007 sec^{-1}. The half-life for dissociation is in excess of 10 seconds, which is a fairly slow rate by some standards; however, it is fast when compared to geochemical life-time.

It should be emphasized that these are homogeneous phase reactions. If solid particles are present where other processes are required to liberate the metal from the complexation site, the reactions may be much slower.

The rate constants for the formation of a variety of metal complexes may be estimated from the relationship (29)

$$k_f = K_{os} k_{m-H_2O}$$

where k_f is the rate constant for the formation of the complex, K_{os} is the outer sphere association constant (28) and k_{m-H_2O} is the rate constant for loss of water from the particular metal ion. Water loss values for selected metals are listed in TABLE IV (30).

With exception of Al(III) all metal ions should react more rapidly than Fe(III). As the pH increases the stability constants increase which is associated with an increased forward rate constant due to a favored electrostatic attraction between the metal ion and the progressively deprotonated ligand. The deprotonation of the ligand should also decrease the rate constant for ligand dissociation. Since $K_{eq} = k_f/k_r$ both of these

TABLE IV

Characteristic Water Loss Values

Metal	Log k, sec^{-1}	Metal	Log k, sec^{-1}
Cu(II)	8.5	Mn(II)	6.6
Cd(II)	8.3	Fe(III)	2.0-3.3
Co(II)	5.5	Fe(II)	6.1
Ni(II)	4.2	Hg(II)	9.3
Zn(II)	7.5	Pb(II)	7.7
Mg(II)	5.1	Ca(II)	8.5
Al(III)	0.5		

changes tend to increase the values of the equilibrium constants. As was mentioned previously for iron(III), the considerations only apply for homogeneous phase reactions. These reactions should be of major importance in waters of near neutral pH with

HA and FA acting to buffer the trace metal concentrations in solution. If the mode of complexation is not dominated by carboxylate donors, as in the porphyrin complexes, the rate of metal exchange may be significantly less. This could lead to the identification of "labile" and "inert" types of complexation within a given ligand.

Heterogeneous Phase Interactions

The mode and extent of trace metal exchange between the solution and solid phases is critical to the interpretation of trace metal transport. There are a variety of controlling processes that have been identified. These include ion exchange, precipitation, and absorption onto the surfaces of clays, hydrous oxides (31), and solid organic materials. As these processes remove trace metals from a natural water, dissolution and complexation tend to solubilize the metals and return them to the solution phase. These exchange interactions are of course complicated by the diversity of the systems; however, several laboratory studies have provided insight into what may be expected in a natural system.

A laboratory model (32) that simulates many of the controlling processes included three solid phases; potassium bentonite, hydrous MnO_2, and solid humic acid. Carbonate, bicarbonate, tannic acid, and soluble humic acid were the solubilizing ligands. The observed distribution of Cu(II), Cd(II), and Zn(II) is pH dependent with fifty percent of the total copper being absorbed at pH 6.5. The copper in solution was completely complexed as determined by anodic stripping voltammetry. The total concentrations of the metals were higher by a factor of 10-100 than those commonly observed in natural waters.

A variety of studies have focused on the interaction of FA with clays (33, 34). It is observed that FA can readily solubilize Cu(II) absorbed onto montmorillonite (35). Over the pH range of 2.5 to 6.5 the amount copper solubilized by 50mg of FA doubled, increasing from .12 to .26 mg. At the higher pH value, virtually all of the soluble copper was complexed by FA. The amount solubilized at pH 6.5 represented 49% of the original Cu(II) absorbed onto the clay.

The absorption of Pb-FA by three soils has been studied (36) and the results are presented in Table V. The soils include a montmorillonite clay, a loam, and a fine silty sand. The concentration of the lead complex varied from 1×10^{-7} M to 1×10^{-5} M, and the amount of the complex absorbed was determined by measuring both lead and fulvic acid concentrations in the solution phase. Correction for absorption to the reaction flasks has been made. The loam and the fine-silt do not exhibit the absorption capacity demonstrated by montmorillonite for Pb; however, they do absorb a sizable fraction of the lead complex. The montmorillonite absorbs more and is capable of reducing the concentration of the complex to less than 10^{-7}M which corresponds to less than 20μg/l Pb. The fulvic acid data indicate that there is partial dissociation of the complex with the free Pb being preferentially absorbed. Similar Pb absorption studies without any fulvic acid resulted in much lower concentration of lead in solution.

From this study it appears that fulvic acid is capable of elevating the concentration of lead in solution; however, the amount in solution is very dependent upon the physical characteristics of the heterogeneous phase. For clays that demonstrate a sizable absorption capacity for organic compounds, the amount of fulvic acid in solution will be small, less than 1 mg/l. For other soils the interaction is less. As a consequence, more of the complexing ligand will remain in solution and be available for the mobilization and transport of trace metals.

TABLE V

Absorption of Lead-Fulvic Acid Complex (36)

$Pb\text{-}FA_T$ M	Marias Montmorillonite	Holloway Loam	Amsterdam Fine-Silt
	Percent Absorbed (As Per Pb Analysis)		
1.0×10^{-7}	62	—[a]	—[a]
3.0×10^{-7}	82	84	88
1.0×10^{-6}	77	84	85
3.0×10^{-6}	97	53	86
1.0×10^{-5}	98	45	81
	Percent Absorbed (As Per FA Analysis)		
1.0×10^{-5}	75	21	44
pH of solutions	7.6	6.5	8.3

a) Not determined.

Chemical Speciation

Many of the trace metals that are found in natural waters are extensively complexed. With pH values near or above neutral and bicarbonate concentrations of 1.5×10^{-3}M, the hydroxide and carbonate complexes may dominate the speciation. In waters with high sulfate and chloride ion concentrations, these complexes may also represent a sizable portion of the total soluble metal. One method of predicting the concentrations of trace metals in solution is to assume that the concentration of the free metal ion is regulated by the solubility of an insoluble carbonate or hydroxide salt. By specifying the pH and alkalinity the total metal in solution is readily predicted utilizing the appropriate equilibrium constants. The results of such calculations for copper, lead, cadmium and zinc is shown in TABLE VI.

TABLE VI

Predicted Total Trace Metal Concentrations[a]

Metal	Free M^{2+} µg/l	Total Metal µg/l	Observed Conc. Range (37) µg/l
Copper	.06	11	9.5
Cadmium	6.00	9	15.0
Lead	0.7	4	23.0
Zinc	65.00	190	64.0

a) Solid phases are CuO, $CdCO_3$, $Pb(OH)_2$, pH = 8.3, $[HCO_3^-] = 1.5 \times 10^{-3}$

For copper the concentration of the free metal ion represents a small fraction of the total. The other metals are not as extensively complexed.

The concentration of organic material in natural water often ranges from 1-5 mg/l. If an average gram formula weight per complexing site is assumed to be 1,000, one would predict an effective complexation concentration of $1\text{-}5 \times 10^{-6}$M. A survey of the effective stability constants listed in TABLE III indicates that the values are too small at low pH to make a significant contribution to the total complexation. As the pH is increased the conditional stability constants increase, and by pH 7 the values have reached 10^6. Constants of this magnitude are probably a minimum for any significant contribution to the speciation since 10^{-6} of free complexation sites would be required to achieve a 1 to 1 distribution of M^{2+} to the complexed form.

At present, the available data for the metals that form the most stable complexes (copper, nickel, iron(III), and lead) suggest conditional stability constants greater than 10^6 at neutral pH. If these values are as high as 10^8, the ratio of free metal to organic complexed metal could be as high as 100. This is comparable to the ratio observed for some of the hydroxo and carbonate complexes (see TABLE VI). With stability constants as large as 10^8, the complexing capacity of the HA and FA would be equal to the effective complexation site concentration.

Transport

Some of the factors that influence the transport of trace

metals by humic and fulvic acid have been discussed. Past and current studies indicate that HA and FA form complexes with sufficient stability to complex sizable fractions of the trace metals at concentrations commonly found in natural waters. The rates of formation and dissociation are predicted to be rapid on the geochemical time scale and thus the systems may, in many cases, approach dynamic equilibrium. The interactions between the solution and solid phases are primarily responsible for the regulation of the solution concentrations. These interactions not only provide a source of material to be desorbed but also control the maximum concentrations that will occur in solution. The system can be thought of as a large dynamic metal-ion buffer. Since there is a continual input of organic material, its concentration remains fairly constant and available to mobilize or release trace metals. Both of which occur to some degree depending upon the particular system.

The mean life time of a metal complex in the system is regulated by the rate of complex formation, dissociation, and exchange with the solid phase. At this time there is minimal information pertaining to the latter. In one field study approximately 28 percent of the total available manganese was transported through a watershed by humic material (38). Studies of Fe, Ni, Cu, Cr, Co, and Mn transport (39) suggest a major role being played by crystalline material and surface coatings. The same distribution is not observed in other systems (40). In any event the controls on the amount of organic material and the stability of formed metal complexes are such that metal-fulvic acid complexes may play an important role in the transport of trace metals. It is likely, given system diversity, that the maximum enhanced carrying capacity due to organic complexation will be approximately a factor of 10 greater than that predicted by a totally inorganic model. This would place the concentration of the trace metals in the 10^{-6} to 10^{-5} molar concentration range.

The trace metals will compete with each other for the available complexation sites. Iron(III) forms some of the more stable complexes and thus this metal may "saturate" the HA and FA coordination sites rendering them unavailable to metals such as lead and cadmium. This competition would as a consequence greatly reduce the effectiveness of HA and FA in the transport of trace metals through natural water systems.

Literature Cited

1. Schnitzer, M., and Khan, S.U., "Humic Substances in the Environment," Marcel Dekker, Inc., New York, 1972.

2. Steelink, C., J. Chem. Ed., 54, 599 (1977).

3. Weber, J. H., and Wilson, S.A., Water Res., 9, 1079 (1975).

4. Whitworth, C., and Pagenkopf, G.K., J. Inorg. Nucl. Chem., 1978.

5. Brady, W., and Pagenkopf, G.K., Can. J. Chem., 1978.

6. Khan, S.U., and Sowden, F.J., Can. J. Soil Sci., 51, 185 (1971); ibid. 52, 116 (1972).

7. Buffle, J., Greter, F., and Haerdi, W., Anal. Chem., 49. 216 (1977).

8. Blount, W., Dissertation Abstracts, B. 32 (12), 7200 (1972).

9. Khan, S.U., Soil Sci. Soc. Am. Proc., 33, 851 (1969).

10. Schnitzer, M., and Hansen, E.H., Soil Sci. 109, 333 (1970).

11. Gamble, D.S., Can. J. Chem., 48, 2662 (1970).

12. Stevenson, F.I., Krastanov, S.A., and Ardakani, M.S., Geoderma, 9, 129 (1973).

13. Cheam, V., and Gamble, D.S., Can. J. Soil Sci., 54, 413 (1974).

14. Schnitzer, M., and Skinner, S.I.M., Soil Sci., 102, 361 (1966).

15. Malcolm, R.L., The Geological Soc. Am., Inc. Memoir 133, 79 (1972), U.S.G.S.

16. Chmielcwska, B., Polish J., Soil Sci., 2, 107 (1969).

17. Deb, D.L., Kohli, C.B.S., and Joshi, O.P., Fert. Technol., 13, 25 (1976).

18. Randhawa, N.S., and Broadbent, F.E., Soil Sci., 99, 362 (1965).

19. Langford, C.H., and Khan, T.R., Can. J. Chem., 53, 2979 (1975).

20. Gamble, D.S., Schnitzer, M., and Hoffman, I., Can. J. Chem., 48, 3197 (1970).

21. Schnitzer, M., and Skinner, S.I.M., Soil Sci., 96, 86 (1963).

22. Stevenson, F.J., "Environ. Biogeochem., Proc. Int. Symp., 2nd ed., 1975 Vol. 2 Edited by J.O. Nriagu. Ann Arbor Sci., Ann Arbor, MI., 1976 pp 519-540.

23. Gamble, D.S., Langford, C.H., and Tong, J.P.K., Can. J. Chem., 54, 1239 (1976).

24. Goodman, B.A., and Cheshire, M.V., Nature, New Biol., 244, 158 (1973); ibid, J. Soil Sci., 27, 337 (1976).

25. Green, J.B., and Monahan, S.E., Can. J. Chem., 55, 3248 (1977).

26. MacCarthy, P., and O'Cinneidi, S., J. Soil Sci., 25, 429 (1974).

27. Davis, G.G., and Mac F. Smith, W., Can. J. Chem., 40, 1836 (1962).

28. Beck, M.T., Coordn. Chem. Revs., 3, 91 (1968).

29. Eigen, M., Bunsenges. physik. Chem., 67, 753 (1963).

30. Geier, G., Bunsenges. physik. Chem., 69, 617 (1965).

31. Jenne, E.A., In "Trace Inorganics in Water" Edited by R.F. Gould, Adv. Chem. Ser. No. 73, A.C.S., Washington, D.C. 1968, p. 389.

32. Guy, R.D., Chakrabarti, C.L., and Schramm, L.L., Can. J. Chem., 53, 661 (1975).

33. Kodama, H., and Schnitzer, M., Can. J. Soil Sci., 51, 509 (1971).

34. Schnitzer, M., and Kodoma, H., Clays and Clay Minerals, 20, 359 (1972).

35. Schnitzer, M., and Kodama, H., Clays and Clay Minerals, 22, 107 (1974).

36. Pagenkopf, G.K., and Brady, W., Unpublished experimental observations, 1978.

37. Kopp, I.F., and Kroner, R.C., NTIS Report PB 2156080. 1968.

38. Zajicek, O.T., and Pojasek, R.B., Water Resources Res., 12, 305 (1976).

39. Gibbs, R.J., Science, 180, 71 (1973).

40. Perhoc, R.M., U.S. Nat. Tech. Inform. Serv., PB Rep. 1974 No. 232427/5GA 45 pp.

Discussion

J. H. WEBEk (University of New Hampshire): I have a question about molecular weights for fulvic acids; you mentioned a 1000 to 10,000 range. The samples we have done both for solid and for aquatic fulvic acids, and also recalculation of Schnitzer's data, showed a value of about 650. Thus, we get a value much lower than you mentioned.

PAGENKOPF: It demonstrates that these natural chelating agents are repetitive polymeric units that are degraded to smaller units. We have characterized a fulvic acid that we isolated from an Oregon soil that has an average formula weight of 2700. There's another value from France of around 1800, and your values are down in the 600 range. I don't disagree with that at all. If you take humic acids and you break them up, correspondingly smaller chelates are obtained.

WEBER: The second question is how did you measure stability constants for cobalt and nickel?

PAGENKOPF: Those cobalt and nickel values are actually Malcolm's data, not mine. Our cadmium values were obtained by ion selective electrodes.

WEBER: How did you treat your data? That's a very difficult thing in this sytesm.

PAGENKOPF: When utilizing the cadmium ion selective electrodes, we have a balance between complexation, stability, hydrolytic properties of metals, and a workable range or a reasonable concentration range of reactants. For cadmium we can work in the 10^{-7} or 10^{-5} M L^{-1} range fairly well. Through mass balance, we can graphically evaluate our average gram formula weight with a plot of the amount of material that complexed as a function of inverse of free metal concentration. We assume a stoichiometry of 1 to 1 which is justified on the basis of mass balance. We extrapolate that to infinite free metal concentration and that gives us a measure of the average gram formula weight. Then what we do is utilize that value and we go on to subsequently evaluate complexes of higher stoichiometry.

A. AVDEEF (University of California, Berkeley): Is there any evidence that any of these systems contain dihydroxy units; catechols, for example?

PAGENKOPF: Yes, I am sure there are some. From interpretation of conditional stability constant data and corresponding pH change you can infer something about the deprotonation of the groups. One of the first models that was proposed, and still a viable one, is that mode of coordination comparable to salicylic acid, where you have phenolic oxygen and a carboxylic group. In the pH profiles two things can happen. The acidity of these phenolic groups is higher than we would expect, or they don't participate to the degree that we think they do. In other words, if you look at the change in stability constants over a pH range of 5 to 7.5, and if a hydrogen ion is replaced for every metal, it may be even two metals, then you would expect to see sizable changes in the conditional stability constants. Over this range, we see a change of 1.5 or so pH units. Not quite as much as expected if the pK's of these phenolic groups are as high as 9 or 10. If the pK's are low then it is a situation different from anything we really know much about. But I'm sure that in some systems of the donors involve phenolic oxygen.

AVDEEF: What color are these complexes at pH 7 to 10?

PAGENKOPF: The ligands themselves are dark brown. We work at a range of a few milligrams per liter, so we are working with dark brown, reddish-brown solution.

AVDEEF: Are these things toxic?

PAGENKOPF: Let me describe the experiments we have done. We tried to establish the toxicity of trace metal complexes to fish. We find that they really don't bother the fish very much. In fact, to establish toxicity you have to get the concentration of the toxicant so high you can't see the fish anymore.

M. P. EASTMAN (University of Texas, El Paso): Do you clean up these humic acids before you start working on them? That is, how do you make sure you haven't got metal contamination?

PAGENKOPF: That's a very good point. It's difficult to address because we start out with a fairly sizable ash content. We go through a whole sequence of steps involving strong acid treatment, dialysis against good chelating agents, and ion chromatography, until we get the ash content down as low as possible. We also check for iron content to make sure we got out everything we can, otherwise we would have all our sites filled up.

EASTMAN: What are some typical values for ash content iron?

PAGENKOPF: Well, we start out ash content of 11%. You can get it down to less than 1% of you keep working on it.

WEBER: I have a comment about purifying these with respect to metal ions. We did get our materials' ash content down to about 1%. Emission spectrocopy on our samples showed predominately silicon, aluminum, and iron. What we found is probably clay. By more careful centrifuging, and occasionally a treatment with HF, we find out we can get rid of the silicon, aluminum, and iron. There are very few trace metals, it seems.

PAGENKOPF: This alludes to the point concerning transport. Aluminum and iron can compete successfully with these other trace metals.

BRINCKMAN: Do you have any speculation or any knowledge of work relating to coordination of organometal ions to humates or fulvates? I'm thinking of course of these ions described [cf. Tobias this volume] as hard acids. They possibly could compete rather selectively with some of the donor types that you have indicated. Would they compete even with iron?

PAGENKOPF: I can't recall any work. It is conceivable that they could show the same types of stability, particularly with multiple donor sites or anything not coordinately saturated.

RECEIVED August 22, 1978.

24

Organotins in Biology and the Environment

J. J. ZUCKERMAN
Department of Chemistry, University of Oklahoma, Norman, OK 73019

ROBERT P. REISDORF, HARRY V. ELLIS III, and RALPH R. WILKINSON
Midwest Research Institute, 425 Volker Boulevard, Kansas City, MO 64110

Tin in the form of the metal and its alloys has greatly affected the course of human history (1) from the Bronze Age of antiquity to the Tin Drum (2) of today. Implements of tin alloy dating from about 3000 BC have been found at Ur, and tin is mentioned in the Bible (3). Tinplate, tinware, pewter, bronze and brass, solder, bearing, bell and type metal, and toothpaste tubes are all examples of current use, and we are in contact with tin all through our lives, from our tinplated babyfood cans through stannous fluoride-containing toothpastes and the dental amalgam in our teeth to the tinplate on our casket linings.

Introduction

Tin chemistry has been the subject of much research effort in the last few years. Elemental tin possesses two allotropic modifications, forms a wide variety of intermetallic phases and inorganic compounds, and has an extensive organometallic chemistry. Basic studies in organotin chemistry are stimulated by the obvious close analogies and interesting differences with the even more extensive chemistry of the cognate organosilicon compounds, and with the vast body of organic chemistry, and also by the success with which a large number of modern physical techniques can be applied to organotin compounds. Tin possesses, for example, two spin of one half nuclides, tin-117 and tin-119, which have become important in nuclear magnetic resonance studies of tin-proton and tin-carbon coupling constants as well as more recently in the tin chemical shifts themselves (4, 5); ten stable isotopes (the largest number for any element), which allows easy identification of tin-bearing fragments in the mass spectrometer (6); the isomeric nuclide tin-119m, which with iron-57 is one of the two easiest to observe nuclear gamma ray (Mössbauer) resonances (7); and easily assignable tin-carbon stretching frequencies in the infrared and Raman (8). Useful chemical information is routinely

0-8412-0461-6/78/47-082-388$09.25/0

obtained from this battery of powerful techniques in combination with others such as X-ray and UV photoelectron spectroscopy (9). X-ray, neutron and electron diffraction techniques have been applied to solve the structures of nearly 600 inorganic and organotin compositions (10).

Tin and its compounds are found in two stable oxidation states, tin(II) and tin(IV), and assume a wide variety of structural types from the allotropic modifications of the element and its alloys to compounds in which the tin atom is two- to eight- coordinated in neutral, cationic and anionic species with intra- and intermolecular association to give dimers and higher oligomers and one-, two- and three-dimensional polymeric arrays. Tin alloys and intermetallic phases exhibit superconductivity (Nb_3Sn) (11), ferromagnetism (Cu_2MnSn) (12), and semiconductivity ($ZnSnAs_2$) (13). Polyatomic anions, the Zintl phases, have been characterized to Sn_9^{4-} (14). Organotin complex ions can be cationic or anionic, and both can exist in one crystal as in the terpyridyl complex of dimethyltin dichloride which contains both a tin-bearing anion and cation (15). Associated polymers can be one- $[(CH_3)_3SnCN]$ (16), two- $[(CH_3)_2SnF_2]$ (17, 18) and helical, three-dimensional $[(CH_3)_2Sn]_2NCN$ (19). Organotin compounds can take two- $[(h^5\text{-}C_5H_5)_2Sn]$ (20), three- $[\{[(CH_3)_3Si]_2CH\}_2Sn]_2$ (21), four- $[R_4Sn]$ (10), five- $[(CH_3)_3SnCl\cdot C_5H_5N$(trigonal bipyramidal) (22) or $[(C_6H_5)_3Sn]_3SnNO_3$(square pyramidal) (23)], six- $[[(CH_3)_2SnCl_4]^{2-}$ (24)] and seven-$[CH_3Sn(NO_3)_3$ (25)], coordinated forms. The structure of tin(II) derivatives show the effect of the stereochemically active lone pair of electrons present in this oxidation state as a space in the coordination sphere (26).

The Organotin Literature

The literature of organotin compounds is now so vast that some guidance for the interested reader is in order. Two important single-author works are available (27, 28) along with a three-volume edited work (29), but these all date from 1970-1971. Extensive reviews of the formation and cleavage of the bonds of tin to carbon (30), the halogens and the halogenoids (31) were published in 1968 and 1972, respectively. A more recent work brings together the papers delivered at a 1976 Symposium on chemistry and applications (32). The structural data for over 250 organotin compounds have been reviewed in 1978 (10). However, the definitive compilation of information on organotin compounds is to be found in the Gmelin Handbuch der Anorganischen Chemie. The first four volumes covering R_4Sn (1975) (33), R_3SnR'(1975) (34), $R_2SnR'_2$, $R_2SnR'R''$, $RR'SnR''R'''$, heterocycles and spiranes (1976) (35) and organotin hydrides (1976) (36) have been published thus far. The publishing program under the authorship of H. Schumann and I. Schumann will eventually include all compounds in which tin is bound to carbon. In the meantime, readers can keep abreast of the literature using the Annual

Surveys for each year since 1964 published in various forms associated with the Journal of Organometallic Chemistry (37-48), and using the yearly volumes on Organometallic Chemistry in the Specialist Periodical Reports of the Chemical Society of London (49-55). The manufacture and use of selected alkyltin compounds have been reviewed in 1976 in a U.S. Government report (56).

This review paper is based upon two U.S. Government commissioned reports: Criteria for a recommended standard... Occupational Exposure to Organotin Compounds, from the National Institute for Occupational Safety and Health (NIOSH), November, 1976, under the editorship of E.S. Flowers (57). The Stanford Research Institute staff developed the basic information for this report. The second report, Assessment of the Need for, the Character of, and the Impact Resulting from Limitations on Selected Organotins. Phase I. Assessment of the Need for Limitations on Organotins, from the Environmental Protection Agency (EPA), Office of Toxic Substances (OTS), July, 1977, under the editorship of R.R. Wilkinson and I.C. Smith of the Midwest Research Institute whose staff developed the basic information and wrote the report (58). The EPA-OTS project officer, P. Hilgard, has kindly given permission to quote from this preliminary draft report before its release by the agency. One author (J.J.Z.) was a consultant in the preparation of these documents for the government agencies. In addition, much useful information was shared by the participants at a Workshop on Organotins organized by M.L. Good in February, 1978 under the sponsorship of the U.S. Office of Naval Research (59).

Historical

Until recently it was generally believed that organometallic compounds, which are usually air- and moisture sensitive, were purely artificial and synthetic, and a text published in 1964 stated, "The situation regarding applications may be summed up by saying that there are no organometallic compounds in nature, there seemingly being no mechanism for their formation..." (60). However, it had been recognized early in the 19th century that cases of arsenic poisoning could be traced to the use of domestic wallpapers containing Scheele's green (copper hydrogen arsenite) and Paris or Schweinfürter's green (cupric aceto-arsenite) dyes. Gmelin in 1839 described a garlic odor in rooms in which the symptoms developed (61), and it was demonstrated in 1872 that moldy green wallpaper liberated an arsenic-containing gas (62), but it was the definitive experiments of Gosio in 1897 that established this "Gosio-gas" as an organometallic compound, an alkylarsine. It remained for Challenger to identify the garlic odor component as $(CH_3)_3As$ in 1933 (64), and to coin the term "biological methylation" for its production.

Despite continued publication by Challenger (65, 66) on the

subject of biomethylation, the crystal structure of the vitamin B-12 (cobalamin) coenzyme 5,6-dimethylbenzimidazolyl-cobamide in 1961, which showed the presence of a cobalt-carbon bond (67), was viewed as a curiosity at the time. Meanwhile, the tragic deaths of residents at Minimata Bay, Japan became recognized as owing to ingestion of methylmercury derivatives formed from the inorganic mercury effluent of a nearby factory (68), and in 1969 it was demonstrated that living organisms have the ability to methylate mercury (69).

Now the methylation of metals and metalloids in the environment is recognized as an important route in their mobilization. The role played by microorganisms in the biochemical transformations whereby elements are methylated is being elucidated, and some of the molecular mechanisms are beginning to be grasped (68, 70-78).

Organotins in the Environment

Production. The annual world consumption of tin in all forms was *ca.* 200,000 tons (*ca.* 400 million lbs. or *ca.* 180 million kg) in 1976, but of this total only *ca.* 55 million lbs. (*ca.* 25 million kg) was in the form of organotin compounds. These compounds were composed of a weighted average of *ca.* 31% tin, and so only *ca.* 17 million lbs. (*ca.* 7.7 million kg) of tin or 4.26% of the annual world production of tin is produced in this form. The U.S. consumption of organotin compounds was *ca.* 24 million lbs. (*ca.* 11 million kg) in 1976. Midwest Research Institute forcasts call for an 11-13% annual growth for the next ten years (58).

Table I lists the estimated U.S. production of selected aklyltin compounds for the period 1965 to 1976. The total weight of these organotin compounds produced in this country during this 12-year period is *ca.* 185 million lbs. (*ca.* 84 million kg).

Use. Over two-thirds of the total world annual production of organotin compounds is devoted to the thermal stabilization of polyvinyl chloride (PVC) plastics (79). The mechanism of PVC breakdown is not entirely clear at present, but the most commonly accepted view is that the decomposition is related to a dehydrochlorination reaction at an allylic or tertiary chlorine site with the formation of a double bond. As the degradation continues, conjugated unsaturated systems are formed which diminish optical clarity and lend undesirable color to the plastic. As little as 0.1% decomposition can lead to blackening. The dehydrochlorination process is autocatalytic, and in the processing of unplasticized PVC resin, temperatures well in excess of those required for the initiation of the degradative process are attained (>200°C). The mode of action of the effective stabilizers is not completely known, but probably

TABLE I. Estimated Annual U.S. Production of Selected Alkyltin Compounds (Million Pounds Per Year)

Year	Bu IOMA	Me IOMA	Bu LM	Bu Maleate	Oct. IOMA	Oct. Maleate	Mixed metals	DBTDL	DBTH	TBTO®	TBTF	Total
1965	2.3	-	1.0	0.22	-	-	-	0.6	-	0.86	-	5.0
1966	4.6	-	1.0	0.22	-	-	-	0.9	-	0.93	-	7.6
1967	4.9	-	0.9	0.48	-	-	-	1.2	-	1.20	-	8.7
1968	6.2	-	0.9	0.52	0.16	0.02	-	1.6	-	1.32	-	10.7
1969	7.1	-	0.9	0.57	0.20	0.03	-	2.0	-	1.50	-	12.3
1970	8.3	0.7	1.1	0.59	0.32	0.05	-	2.4	-	1.67	0.01	15.2
1971	8.2	1.4	1.1	0.62	0.37	0.06	-	2.8	-	1.97	0.02	16.6
1972	10.5	2.9	1.3	0.26	0.56	0.08	-	2.5	0.8	1.61	0.05	20.6
1973	10.1	4.0	1.3	0.22	0.62	0.08	0.8	3.0	0.9	1.09	0.08	23.0
1974	9.3	4.5	1.3	0.21	0.37	0.07	0.5	3.5	1.1	2.07	0.12	23.0
1975	7.0	4.0	1.0	0.20	0.50	0.07	0.4	3.0	0.9	1.40	0.30	18.8
1976	9.0	4.5	1.1	0.22	0.60	0.07	0.5	4.3	1.0	2.30	0.50	23.5
1981	13.0 to 14.5	6.5 to 8.0	2.0	0.33	0.90	0.20	0.0	6.0	1.9	4.6	1.0	36.5 to 39.5
1986	15.0 to 18.5	8.0 to 10.0	3.0	0.45	1.20	0.30	0.0	8.0	2.5	9.0	2.0	49.5 to 55.0
Total (1965 - 1976)	87.5	22.0	12.9	4.33	3.70	0.53	2.2	27.8	4.7	18.81	1.08	185.5

Note: Bu IOMA = Butyltin isoctylmercaptoacetate plus blends
Me IOMA = Methyltin isooctylmercaptoacetate plus blends
Bu LM = Dibutyltin-bis(laurylmercaptide)
Bu Maleate = Dibutyltin alkylmaleate esters
Oct. IOMA = Di(*n*-octyl)tin-S,S'-bis(isooctylmercaptoacetate)
Oct. Maleate = Di(*n*-octyl)tin maleate polymers
DBTDL = Dibutyltin dilaurate
DBTH = Dibutyltin bis-(2 ethyl hexoate)
TBTO = Bis(tributyltin)oxide
TBTF = Tributyltin fluoride

taken from refs. 56 and 58.

involves exchange of allylic chloride atoms with the anionic portion of the organotin compound, and the absorption of liberated hydrogen chloride to release compounds which can add across the unsaturated centers.

Typical stabilizers are dialkyltin compounds containing thio or ester groups. The original patents issued to V. Yngve of Carbide and Carbon Chemicals Corporation in 1936 and based upon the discoveries of W.M. Quattlebaum, were for the application of dibutyl compounds, but dioctyltins are now in use as well. The addition of small amounts of monoalkyltins has a synergistic effect. The groups attached to the alkyltin moieties include the laurate and maleate, and several mercapto derivatives derived from octyl and isooctyl esters of thioglycolic acid. In the last decade there has been a shift of emphasis from flexible to rigid PVC products, the latter requiring much higher processing temperatures. The need for better stabilization has thus become more acute if PVC with the desired colorlessness and transparency for packaging films, piping, bottles and siding is to be produced. Dimethyltins, because of their extremely great thermal stability allow the use of high working temperatures and high working speeds, and their application may have economic advantages (80).

Organotin compounds are also used as catalysts in the production of rigid polyurethane foam and for the room temperature vulcanization of silicone elastomers. Biocidal and anthelmintic applications will be discussed separately, below.

Table II lists the estimated annual U.S. consumption of selected alkyltin compounds in the four use areas discussed above, plus miscellaneous uses and exports. The capital value of the organotin chemicals used in PVC heat stabilization was $50 million in 1976, with the total capital value for all uses estimated at $67.7 million.

Final Disposal. Consumption of commercial organotin chemicals is given by sector for 1976 in Figure 1. The route for each use area from introduction through consumption and release to the environment is given on a weight basis.

Industrial disposal options include evaporation, bacterial degradation settling ponds, sludge burial within property lines, use of contract disposal services, deep-well injection, hydrolysis and precipitation methods, discharge to municipal treatment plants, NPDES (National Pollution Discharge Elimination System) permits, and potential reclamation or recovery and recycling techniques. Midwest Research Institute estimates that only 0.5% of production, or *ca.* 0.12 million lbs. (*ca.* 0.054 million kg) are lost to the environment by these routes at the source of manufacture.

Consumer disposal options include municipal sewers, municipal waste collection which reaches landfill or incinerator, normal weathering of the consumer product, and

TABLE II. Estimated Annual U.S. Consumption of Selected Alkyltin Compounds By Use Area (Quantities in Million Pounds)

Year	PVC heat stabilizer	Catalysts	Biocidal	Anthelmintic	Miscellaneous excluding exports	Total
1965	3.5	0.5	0.50	0.15	0.36	5.0
1966	5.8	0.7	0.60	0.16	0.33	7.6
1967	6.3	1.0	0.70	0.17	0.50	8.7
1968	7.8	1.4	0.80	0.18	0.52	10.7
1969	8.8	1.8	0.90	0.19	0.60	12.3
1970	11.1	2.2	1.1	0.21	0.57	15.2
1971	11.8	2.6	1.3	0.22	0.67	16.6
1972	15.6	3.1	1.5	0.23	0.16	20.6
1973	17.1	3.7	1.7	0.24	0.26	23.0
1974	16.2	4.4	2.0	0.24	0.19	23.0
1975	13.2	3.7	1.5	0.20	0.20	18.8
1976	16.0	5.0	2.0	0.27	0.20	23.5
1981	23 to 26	7.5	5.0	0.35	0.60	36.5 to 39.5
1986	28 to 33.5	10.0	10.0	0.50	1.0	50 to 55
Total (1965-1976)	133.2	30.1	14.6	2.5	4.6	Grand total (1965-1976) 185.0

taken from refs. 56 and 58.

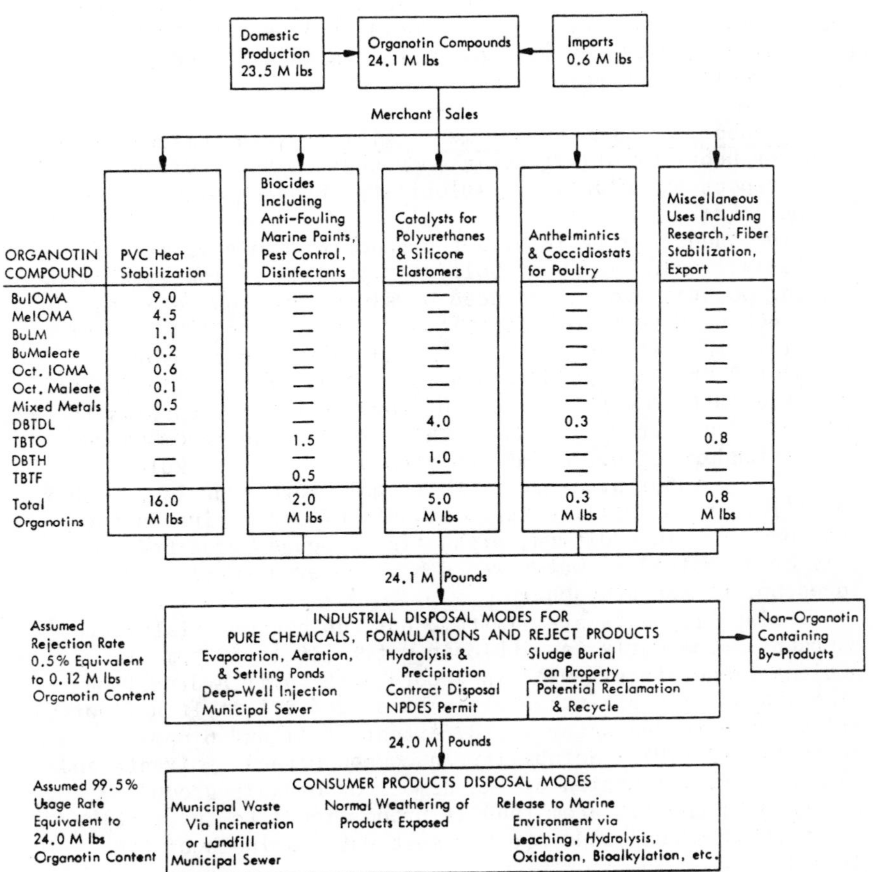

Figure 1. Organotin chemicals consumption in 1976 (58)

release to the marine environment through leaching. The 24 million lbs. (ca. 11 million kg) of organotin compounds annually manufactured in the U.S. will eventually reach the environment in some form. The 24 million lbs. of organotins are formulated or compounded into perhaps 2,400 million lbs. (ca. 1,100 million kg) of industrial and commercial products which are redistributed to the environment. Table III shows the estimated quantities of organotins released to the environment in 1976 by various routes which are shown schematically in Figure 2.

Transport in the Environment. This section discusses the transfer behavior of organotin compounds in the environment with respect to volatility, solubility, bioaccumulation and bioconcentration.

The industrially important organotin chemicals are generally liquids and waxy solids or powders of low volatility. Boiling points, even at reduced pressures, are quite high, e.g., 400°C/10 torr for dibutyltin dilaurate and 180°C/2 torr for bis(tributyltin)oxide (TBTO)®, and the vapor pressure calculated at 25°C based upon a molar heat of vaporization of 20.2 Kcal/m for TBTO® is 1.6 x 10^{-3}torr, which corresponds to 510 mgm/m^3 of air (58). While this exceeds the recommended NIOSH standard of 0.1 mg/m^3 analyzed as tin for workplace air (57), a vapor pressure of this small magnitude is probably insufficient to mobilize large amounts of organotins in the environment. In addition, organotin compounds adsorbed in soil or dissolved in water would have an even smaller tendency to escape to the atmosphere.

Solubility data are limited, but the higher trialkyltin compounds are nearly insoluble in water [10 to 50 ppm at ambient temperatures (98)] in accord with their hydrocarbon-like character. Bis(tributyltin)oxide and tributyltin fluoride are soluble in sea water to the extent of 51 and 6 ppm, respectively (99). Solubility in common organic solvents and in lipids is a function of the number of organic groups attached to the tin atom, and is proportional to their bulk. The methyltin compounds are the most water soluble of the organotins.

Leaching from soils has been investigated using carbon-14 labeled compounds. Even hot methanol treatment fails to remove triphenyltin acetate completely, and neither this fungicide nor any of its abiotic transformation products are leached over a six-week period. Over 70% of the activity was found in the upper 4 cm of the packed soil column used (93).

Neutron activation analyses have been carried out on unspecified dried water plant material, sediments and fish samples taken from a Bavarian river both up- and downstream from a factory producing tin compounds and using a chlorine alkali-electrolysis process involving mercury. The results

TABLE III
Estimated Quantities of Organotins
Released to the Environment for 1976
Various Routes

Organotin bearing product	Organotin content (lb)	In use (lb)	Discarded (lb)	Environmental route and quantity (lb)	
Plastics, PVC, PU, silicones	21×10^6	16×10^6	5×10^6	0.5×10^6 4.5×10^6	incinerated landfill
Antifouling paints	0.8×10^6	0.4×10^6	0.4×10^6	0.4×10^6	ocean and estuarine water sediments
Industrial and consumer slimicides and other biocides	0.7×10^6	0.5×10^6	0.2×10^6	0.2×10^6	rivers and estuaries, sediments
Latex paints	0.5×10^6	0.475×10^6	0.025×10^6	2.5×10^3	dumped or leaked
All other applications including export	1.1×10^6	0.77×10^6	0.34×10^6	0.34×10^6	landfill

taken from ref. 58.

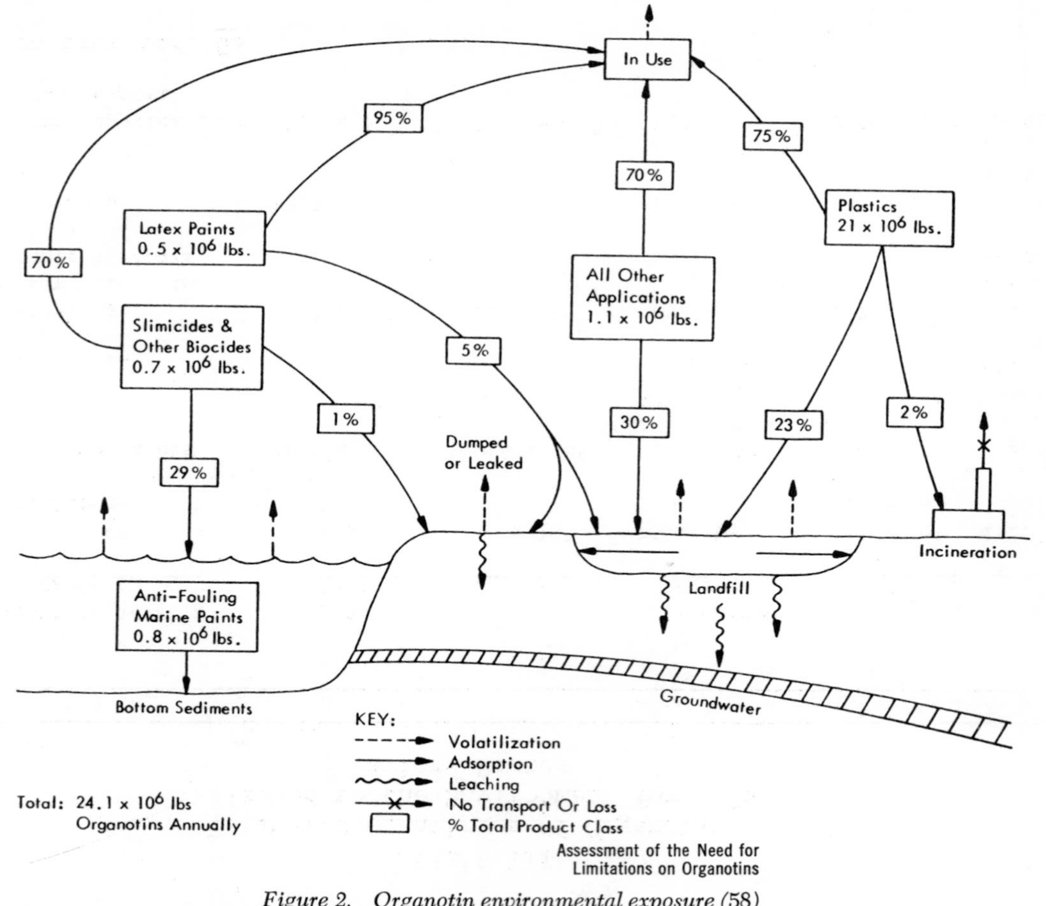

Figure 2. Organotin environmental exposure (58)

for the sediments were 0.9 and 112 ppm Hg and 16.6 and 69 ppm Sn up- and downstream, respectively, on a dry weight basis. The bioconcentration in the water plants is shown in Table IV.

TABLE IV

Element	Origin	Water plant samples (dry weight basis, ppm) Upstream	Downstream	Apparent magnification factor
Sn	Alz	13	3,100	240
Sn	Alzkanal	14	2,370	170
Hg	Alz	1.3	230	180
Hg	Alzkanal	0.6	295	500

Both tin and mercury were found in fish meat and livers at low (*ca.* 1 ppm) concentration. The species present were not identified (100).

Environmental Fate. Once in the environment the organotin compounds are leached from consumer products by environmental forces such as heat, light, and the action of water, oxygen, ozone, carbon dioxide and microorganisms. It has been estimated that *ca.* 5.0 million lbs. (*ca.* 2.3 million kg) of organotins in PVC are disposed of annually by landfill and incineration and that *ca.* 1.0 million lbs. (*ca.* 0.45 million kg) of biocides are released annually to the environment via municipal sewers, surface waterways, harbors and the oceans (58). The human and ecological significance of these inputs to the environment depend upon the persistence of the organotins, their mobility between media, and the available pathways to susceptible populations of the organotins themselves and their degradation products.

Organotin compounds are subject to both chemical and biological degradation.

Chemical Degradation. Carbon-tin bonds are thermally stable below 200°C, but are capable of polarization by attacking species in either direction. Organotin compounds are thus susceptible to attack at the carbon-tin bond by both nucleophilic and electrophilic reagents, leading to hydrolysis, solvolysis, acidic and basic reactions, halogenation, etc. (27-32). The results of scattered kinetic studies in the literature on the cleavage of alkyl-, unsaturated and aromatic groups from tin by hydrogen chloride and metal halides, CrO_3 in glacial acetic acid, alkali metal hydroxides in water

TABLE V.

Reaction	pH	Temperature (°C)	Reaction media
A. Alkyltin, saturated			
$(C_2H_5)_4Sn + HgCl_2 \xrightarrow[H_2O]{CH_3OH} (C_2H_5)_3Sn + C_2H_5HgCl$	-	25	H_2O/CH_3OH (30%/70%) $X_{CH_3OH} = 0.51$ [O.T.][a] = 1.25×10^{-3} M $[HgCl_2] = 1.00 \times 10^{-3}$ M
$(C_2H_5)_4Sn + HCl \xrightarrow{C_6H_6} (C_2H_5)_3SnCl + C_2H_6$	2	20	HCl in C_6H_6 [O.T.] = 2.18×10^{-1} M
$(C_4H_9)_4Sn + CrO_3 \xrightarrow{HOAc} (C_4H_9)_3SnOAc + C_4H_9$	< 0	20	CrO_3 in glacial HOAc $[CrO_3] = 3 \times 10^{-4}$ M [O.T.] = 1×10^{-3} M
B. Alkyltin, unsaturated			
$(CH_3)_3SnCH_2CH{=}CH_2 + HCl \xrightarrow[96\%]{CH_3OH} (CH_3)_3SnCl + CH_2{=}CH{-}CH_3$	2-4	25	H_2O/CH_3OH (4%/96%) $[HCl] = 10^{-2} - 10^{-4}$ M [O.T.] = 5×10^{-4} M
$(CH_3)_3SnCH_2{-}C(CH_3){=}CH_2 + HCl \longrightarrow (CH_3)_3SnCl + CH_3{-}C(CH_3){=}CH_2$	2-4	25	H_2O/CH_3OH (4%/96%) $[HCl] = 10^{-2} - 10^{-4}$ M [O.T.] = 5×10^{-4} M
$(C_2H_5)_3SnCH_2CH{=}CH\phi + NaOH \longrightarrow (C_2H_5)_3SnOH + \phi CH_2CH{=}CH_2 \xrightarrow{OH^-} \phi CH{=}CH{-}CH_3$	12.7	25-40	H_2O/CH_3OH (40%/60%) [O.T.] = 1×10^{-4} M
C. Alkyl-aryltins			
$(CH_3)_3Sn{-}\phi + HClO_4 \longrightarrow (CH_3)_3Sn{-}OH + C_6H_6$	0.77 and 0.86	50	H_2O/CH_3OH (40%/60%) [O.T.] 4.8×10^{-4} M
$(CH_3)_3Sn{-}\phi + NaOH \longrightarrow (CH_3)_3SnOH + C_6H_6$	> 14	50	H_2O/CH_3OH (40%/60%) $[OH^-] = 3$ M [O.T.] = 4.8×10^{-4} M
$(CH_3)_3SnCH_2\phi + KOH \longrightarrow (CH_3)_3SnOH + CH_3\phi$	12	75	H_2O/DMSO (14%/86%) $[OH^-] = 10^{-2}$ M [O.T.] = 1×10^{-4} M

a/ [O.T.] = Organotin, moles/ℓ.

Solvolysis of Organotins

Reaction rate	Half-life	References and other parameters
k_2 - 5.59 ℓ $mole^{-1}$ min^{-1}	143 min	81
k_2 = 7.5 x 10^{-4} ℓ $mole^{-1}$ sec^{-1}	350 sec	82
k_2 = 2.84 ℓ $mole^{-1}$ sec^{-1}	350 sec	82, 83
k_2 = 0.475 ℓ $mole^{-1}$ sec^{-1}	1.17 hr	84
k_2 = 24.8 ℓ $mole^{-1}$ sec^{-1}	80.7 sec	84
k_2 varies from 9.2 x 10^{-4} (25°C) to 46.0 x 10^{-4} ℓ $mole^{-1}$ sec^{-1} (40°C)	23 to 115 days	85
k_1 = 300 x 10^{-3} min^{-1} k_1 = 228 x 10^{-3} min^{-1}	2.3 min 3.0 min	86
k_1 = 12.1 x 10^{-3} min^{-1}	57 min	87
k_1 = 566 x 10^{-3} min^{-1}	1.2 min	88

and aqueous perchloric acid are displayed in Table V. Homolytic reactions involving organotin compounds with free radicals have been recently reviewed (81). All these studies were of course carried out in homogeneous media where it is found that the cleavage of the organic groups from tin is always first-order in each reactant. In polar solvents there is probably initial solvation of the tin compound, followed by electrophilic attack (S_E2) on a carbon atom adjacent to tin. With alkali there can be nucleophilic attack (S_N2) on the tin atom with expulsion of a carbanion. Although some organotin compounds will undergo unimolecular photolysis or thermolysis under mild conditions, free organotin radicals are usually formed by bimolecular reaction with some other radical. The attack can be at the tin-carbon bond, or elsewhere in the molecule (81). From these and other studies (90, 91) it can be generalized that the progressive cleavage of organic groups from tin is dependent upon the type of organotin compound, the number of organic substituents, and the solvolytic conditions. The relative ease of removal of aliphatic groups decreases with increasing size of the group, but unsaturated and aromatic groups are cleaved more rapidly. For the series:

$$R_4Sn \xrightarrow{k_4} R_3SnX \xrightarrow{k_3} R_2SnX_2 \xrightarrow{k_2} RSnX_3 \xrightarrow{k_1} SnX_4 \quad (1)$$

the reaction rates are $k_4 >> k_3 >> k_2 \approx k_1$. Laboratory solvolytic reactions generally represent extreme pH conditions (pH < 1 or > 14). Half-lives range from one minute to 115 days, depending upon these conditions and specific organotin compounds studied. The solvolysis of tetraalkyltins carried out under less severe conditions (pH = 4 to 10), may be several orders of magnitude slower (10^{-4} to 10^{-6}), and these tetraalkyltins will react 10 to 100 times faster than trialkyltins. The solvolysis rates of dialkyltins again approach those of the tetraalkyltins.

The inorganic anionic groups in the organotin compounds react with moisture and air to cleave from tin in an hydrolysis-oxidation to give stannols and oxides. In this way successive reaction of both parts of the molecule leads eventually to completely inorganic hydrated tin oxides.

Biological Degradation. Microbial action is also significant. Rate constants and half-lives have been estimated from laboratory studies in water and are listed in Table VI (92). Other studies have utilized carbon-14 labeled triphenyltin acetate in which the release of $^{14}CO_2$ from soil samples was monitored in the dark, and a half-life of 140 days determined for concentrations of 5 and 10 ppm (93). Using UV irradiation for 10 minutes to one hour, degradation into di- and monophenyltin compounds was observed for the acetate (93, 94) and hydroxide (94). The degradation of the chloride on sugar beet leaves in a green house environment has been investigated using a tin-113

TABLE VI

Rate Constants and Half-Lives at 20°C for Selected Organotin Compounds of Potential Commercial Interest in Distilled Water (92)

Compound	Concentration range (mg/ℓ)	Rate constants (day^{-1})	Half-life (days)
TBTO®	2-4	0.038	18.2
Tributyltin methacrylate	0.2-0.4	0.035	19.8
Bu IOMA	1.5-3	0.607	1.14
Diethyltin dicaprylate	1-2	0.311	2.22
Dioctyltin bis-(isobutylmaleate)	3-4	0.213	3.25

label. The pattern of conversion to inorganic tin can be seen in Table VII. No measurable amounts of acitivty were found in

TABLE VII

Degradation of Triphenyltin Chloride on Sugar Beet Leaves in a Greenhouse Environment (95)

		Percentage of total radioactivity applied				
Experiment No.	Days after application	ϕ_3Sn^+	ϕ_2Sn^{+2}	ϕSn^{+3}	Sn^{+4}	Not extracted[a/]
1	0	100	-	-	-	-
2	3	86	10	<1	3	1
3	7	67	13	1	16	3
4	14	47	9	1	38	5
5	21	33	7	1	55	4
6	28	26	4	1	60	9
7	35	22	2	1	65	10
8	42	19	<2	1	67	12

a/ Hydrolyzed and aged ϕSn^{+3} and Sn^{+4}.

the stems of the plants or in the sugar beets during the 42 days of the experiments. All the activity could be accounted for. The apparent half-life was initially two weeks, but the degradation rate slowed after this (95). Tributyltin fluoride, which is used as an antifouling agent, hydrolyzes very rapidly in low concentration in sea water to give the chloride and oxide. Carbon dioxide can react with the oxide to form the carbonate. Ultimately, through the action of sunlight and oxygen, hydrated inorganic tin oxide is formed in a stepwise degradation sequence (96). While the various results from separate studies disagree in their details, it is clear that the conversion to less alkylated and arylated tins proceeds sequentially to produce hydrated inorganic tin oxides with half-lives ranging from seconds to days, depending upon the conditions.

Tests on clay-based soils indicates that organotin compounds disposed of in landfills will be immobilized. Under simulated soil leaching conditions, using carbon-14 labeled compounds, high Freundlich isotherms were determined for different soil types suggesting retention of 94.9 to 99.2% of the organotin compounds at their original site of placement. No more than 0.2% of the test triorganotin compounds were found in the leachates (97).

Biomethylation of Inorganic Tin. Microbial aerobes can solubilize HgS (solubility product 10^{-53}M) in the form of Hg^{2+} by oxidizing the sulfide through sulfite to sulfate (101), whence Hg^{2+} can be reduced to mercury metal by bacterial enzymatic action and liberated to the atmosphere from the hydrosphere because of the considerable vapor pressure of Hg°. An alternative detoxification mechanism converts Hg^{2+} to methyl- and dimethylmercury (102, 103), the latter of which is volatile and lost to the atmosphere. The synthesis of dimethylmercury is *ca.* 6000 times slower than the synthesis of methylmercury (104).

The chemistry, biochemistry and toxicology of tin is somewhat different, however, beginning with the non-volatility of the metal itself. Microbial methylation of inorganic tin has been claimed for a *Pseudomonas* species isolated from Chesapeake Bay in a widely quoted conference report in 1974 (105). In the laboratory, tin(II) under nitrogen at pH 1 in 1.0 *M* aqueous salt solution in the presence of methylcobalamin at *ca.* 5 x 10^{-4}M with a single electron oxidizing agent such as iron(III) chloride or aquocobalamin is methylated during one day at 20°C. Carbon-14 labeled methylcobalamin showed no organic products resulting from the cleavage of the cobalt-carbon bond. Tin(IV) did not participate. The rate of the methylation reaction was measured in 10- to 100-fold excess of tin(II) as $1.4M^{-1}s^{-1}$ which is dependent upon the pH. In aqueous tartrate buffer at pH as high as 5, carbon-cobalt bond cleavage was found to occur (106). The mechanism for the methylation has been discussed in terms of the reduction potentials for the various species involved (76).

No evidence has been previously published either from the laboratory or Nature, however, for the methylation of inorganic tin by the species discussed above past the monomethyl stage (105, 106). Unlike the monomethylmercury derivatives which are deadly poisonous neurotoxins, the lipid solubility of the monomethyltins is too low, and they travel in the urinary tract where they can cause renal problems. The monomethyltins are expected to be more toxic generally than the higher aliphatic derivatives, but no data are yet available. Single-dose LD_{50} values for mice are in the range 1400-1520 mg/kg for n-butyl- and n-octyl tris(2-ethylhexylmercaptoacetate) and n-butyltin trichloride, and is >6,000 mg/kg for n-butyltin acid (107, 108). This compares with LD_{50} values in mice of 35 for the di-n-butyl-(109), and 117 for the tri-n-butyltin (110, 111) chlorides and 6,000 mg/kg for tetra-n-butyltin (117).

A complex biogeological cycle has been proposed for tin involving successive alkylations which are as yet undemonstrated to give di-, tri- and tetramethyltin species (76). The *Pseudomonas b.* strain of bacteria found in Chesapeake Bay gave evidence, however, of some conversion of hydrated tin(IV) chloride to dimethyltin dichloride after 10 days incubation at ambient temperature in >2% oxygen as determined by comparison with an authentic mass spectrum, but the parameters for reproducability of these experiments have not been worked out (105), and no refereed publication has yet appeared.

The Mercury-Tin Crossover System. The biologically produced methyltin species described above was shown to methylate Hg(II) ion after eight days of incubation as determined by mass spectrometry (105). This result is corroborated by an earlier observation that aqueous trimethyltin methylated mercury(II) ion (113), and the recent results of computer studies in which changes of pH, pCl, temperature and ionic strength were imposed on aqueous laboratory systems of pairs of tin(IV) and mercury(II) cations between which methyl groups were being transferred (114). A representation of the crossover scheme is shown in Figure 3.

Biological Effects of Exposure

Metabolism of Organotin Compounds. Tetraethyltin was shown to undergo destannylation *in vivo* by a rat liver microsomal enzyme system successively to the tri-(115), di-(116) and monoethyl (117) derivatives. Other trialkyltin compounds behave similarly (116, 118). The total metabolic yield was <10%, however. More recently, it was demonstrated that a

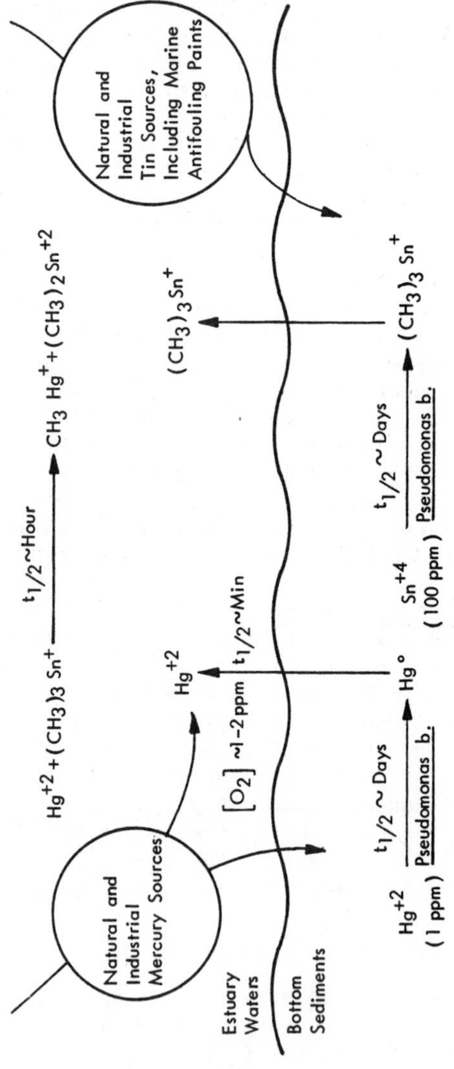

Assessment of the Need for Limitations on Organotins

Figure 3. Bimetallic mercury-tin "crossover" scheme (58)

cytochrome P-450 monooxygenase enzyme and not a lipid peroxidase system was responsible for metabolism, and that carbon hydroxylation occurred on the butyl groups of carbon-14 labeled tributyltin acetate at the α- to γ-positions (119, 120). The products were compared with authentic samples synthesized for the purpose, and similar *in vivo* reactions were shown to occur in mice (120). The primary hydroxylated metabolic products of tetrabutyltin-1-^{14}C were rapidly destannylated to the tributyltin derivatives (120). Successive *in vivo* destannylation of tricyclohexyltin hydroxide ultimately yields inorganic tin(IV) (118), but more recent findings reveal that a cytochrome P-450 dependent monooxygenase enzyme-induced hydroxylation is a major metabolic reaction yielding 2-, 3- and 4-hydroxycyclohexyltin derivatives as metabolites (121, 122). Tin-113 labeled triphenyltin acetate is apparently not hydroxylated *in vitro*, but rats can metabolize this organotin to give substantial amounts of di- and monophenyl-derivatives (120).

Exposure to Organotin Compounds. The toxic response of the common guppy, *Lebistes reticulatus*, to tributyltin compounds has been determined. Guppies appear to be fatally sensitive to less than 1 ppm of bis(tributyltin)oxide (123). Tests on *Lebistes* using triphenyltin hydroxide show that 0.1 ppm killed 43% after 19 hour exposure and 100% after 48 hours (124). Triphenyltin acetate used to control the operculate snail *Lanistes ovum* also had deleterious effects upon germinating swamp rice, and several other tributyltins have been tested for their phytotoxicities as well as their molluscicidal activity (125). Tributyl- and triphenyltins are also effective against algae and barnacles (126, 127, 128). Tripropyl-, butyl-, and triphenyltin derivatives are toxic to mollusks. Concentrations of trialkyltins of *ca.* 1.0 ppm are lethal to *Australorbis* (or *Biompharlaria*) *glabrata* (129). Values for bis(tributyltin) oxide range between 0.05 and 0.10 ppm (125). Maximum toxicity to all types of life occurs with the triorganotin derivatives, but there are important variations within this class of compounds. Mammaliam toxicity reaches a maximum with the ethyl group and falls off rapidly with increased chain length as shown in Table VIII. Insects, on the other hand, are most affected by the trimethyltins, and the tri-n-propyl and butyltins are most effective against fungi and bacteria (137, 137a, 138, 139). The tributyl and triphenyltin compounds, which are not particularly hazardous to mammals (LD_{50} 100-200 mg/kg), are very effective biocides against marine fouling organisms, for example, algae, barnacles, shrimp and tubeworms at levels of 0.1 to 1.0 ppm (125). The comparative bactericidal effects of the Group IV di- and tri-organo derivatives as shown in Table IX (140). Dibutyltin dilaurate controls the chicken parasite *Raillietina cesticillus* at dose

Table VIII. ORAL LD_{50} VALUES FOR ORGANOTIN COMPOUNDS

Compound	Species	LD_{50} (mg/kg)	References
Methyltin isooctylthioglycolate	Rat	1,261	Weisfeld (129a)
Butyltin trichloride	Mouse	1,400	Pelikan and Cerny (115, 116)
Butyltin isooctylmercaptoacetate	Mouse	1,520	Pelikan and Cerny (115, 116)
Dibutyltin bis(isooctylthioglycolate)	Rat	1,037	Weisfeld (129a)
Dibutyltin dichloride	Rat	100	Klimmer (130a)
Dibutyltin dichloride	Rat, male	182	Mazaev et al. (117)
Dibutyltin dichloride	Rat, female	112	Mazaev et al. (117)
Dibutyltin dichloride	Mouse	35	Mazaev et al. (117)
Dibutyltin dichloride	Guinea pig	190	Mazaev et al. (117)
Dibutyltin oxide	Rat	520	Klimmer (130)
Dibutyltin bis(isooctyl)	Rat	200	Calley et al. (120)
Dibutyltin dilaurate	Rat	175	Klimmer (130)
Dibutyltin bis(nonylmaleate)	Rat	120	Klimmer (130)
Dibutyltin bis(nonylmaleate)	Rat	170	Klimmer (130)
Dioctyltin oxide	Rat	2,500	Klimmer (130)
Dioctyltin maleate	Rat	4,500	Klimmer (130)
Dioctyltin maleate	Mouse	2,250	Pelikan et al. (119)
Dioctyltin bis(butylmaleate)	Rat	2,030	Klimmer (130)
Dioctyltin bis(butylmaleate)	Mouse	3,750	Pelikan et al. (110)
Dioctyltin bis(isooctylmaleate)	Rat	2,760	Klimmer (130)
Dioctyltin bis(isooctylmaleate)	Mouse	2,700	Pelikan et al. (110)
Dimethyltin bis(isooctylthioglycolate)	Rat	1,380	Weisfeld (129a)
Trimethyltin acetate	Rat	9	Barnes and Stoner (1958)
Triethyltin acetate	Rat	4	Barnes and Stoner (1958)
Triethyltin sulfate	Rat	6	Stoner et al. (1955)
Triethyltin sulfate	Rabbit	10	Stoner et al. (1955)
Tripropyltin acetate	Rat	118	Barnes and Stoner (1958)
Triisopropyltin acetate	Rat	44	Barnes and Stoner (1958)
Tributyltin acetate	Rat	380	Barnes and Stoner (1958)
Bis(tributyltin)oxide	Rat	132	Klimmer (1969)
Bis(tributyltin)oxide	Rat	180	Truhaut et al. (1976)
Bis(tributyltin)oxide	Rat	234	Sheldon (1975)
Trihexyltin acetate	Rat	1,000	Barnes and Stoner (1958)
Triphenyltin acetate	Rat	136	Klimmer (1963)
Triphenyltin acetate	Guinea pig	24	Kimbrough (1976)
Triphenyltin acetate	Guinea pig	21	Klimmer (1963)
Triphenyltin hydroxide	Rat, male	171	Marks et al. (1969)
Triphenyltin hydroxide	Rat, female	268	Marks et al. (1969)
Tetraethyltin	Rat, male	9	Mazaev et al. (1971)
Tetraethyltin	Rat, female	16	Mazaev et al. (1971)
Tetraethyltin	Mouse	40	Mazaev et al. (1971)
Tetraethyltin	Guinea pig	40	Mazaev et al. (1971)
Tetraethyltin	Rabbit	7	Mazaev et al. (1971)

levels of 75 mg/kg without harm to the chickens (141, 142), and the effect on egg production, fertility and hatchability was temporary (143). LD_{50} values for tricyclohexyltin hydroxide

TABLE IX
Comparative Bactericidal Effects of Group IVA Organometals
Figures represent the minimum concentration in g/ml needed to inhibit all growth on Streptococcus lortis (140).

	R_3MX			R_2MX_2		
	Ge	Sn	Pb	Ge	Sn	Pb
CH_3	>500	>500	200		200	1
C_2H_5	50	100	50	>500	50	5
$n-C_3H_7$	5	5	2		20	0.5
$n-C_4H_9$	1	5	1	>500	20	0.2
$n-C_5H_{11}$	2	10	5		50	0.2
$n-C_6H_{13}$	20	50	10		>500	1
C_6H_5	>500	5	1	>500	50	1

toward quail are in the range 255-390 mg/kg with no effect on egg production, fertility or hatching noted at the 20 ppm dietary level (144), and toward dogs, cats and monkeys at >800 mg/kg (145). Sheep receiving intrarumenal injections of tricyclohexyltin hydroxide in doses of 15 mg/kg experienced no ill effects; with 25 mg/kg a transitory anorexia set in; at 50 mg/kg reversible CNS depression and diarrhea, while four sheep dosed with 150 mg/kg died. Yearling cattle and goats can tolerate multiple treatments of up to 0.1% w/v of tricyclohexyltin hydroxide in a water spray on the hide with no behavioral effects, but one ewe aborted twin mid-term fetuses eight days after a single application of a 28% solution and died five days later (146).

Against the bollworm Heliothis zea and tobacco budworm H. virescens larvae, foliar spray of triorganotin compounds revealed LD_{50} values of 0.20 and 0.50 mg/kg, respectively (147). Bis(tributyltin)oxide at 1% w/v in wool killed 100% of clothes moths Tineola bisselliell larvae without injestion (148). Other triorganotin derivatives showed values of 0.25 to 0.30 ppm against the mosquito Culex pipiens larvae compared with 0.04 ppm for DDT (149, 150). Various triphenyltin derivatives act as

95% reproduction inhibitors on the housefly Musca domestica L. in as low as 62 ppm dietary dose levels, with LD_{50} values of 1,000 ppm (151), and act as chemosterilants toward the German cockroach Blattella germanica L. and the confused flour beetle Tribolium confusum at 0.1% by weight in the diet. Jacquelin du Val Antifeeding effects were also noted on various lepidoterous larvae including the cotton leafworm Prodenia litura F. and the Colorado beetle Leptinotarsa decemleneata Say and Agrotis ypsilon Rott (152). Sugar beet leaves treated with a 0.1% w/v solution of triphenyltin acetate and hydroxide stopped larvae feeding.

In 1973 the International Tin Research Institute organized a cooperative research project to investigate the environmental behavior of organotin compounds. The Organotin Environmental Project, or ORTEP, is funded by the prinicpal manufacturers of organotin chemicals in the U.S., U.K., Switzerland, Germany and Japan.

Human Exposure. Organotin compounds can be assimilated by inhalation, ingestion through food or water or by absorption through the skin. The lipid soluble derivatives will accumulate in fatty tissue. No data concerning the rates of metabolism degradation or excretion in humans or test animals are available. The total daily intake for tin in the diet has been estimated to be in the range 187 to 8,800 μg/day which corresponds to an average of 2.7 to 126 μg/kg for an adult (153). Tin, presumably as inorganic tin, has been shown to be an essential growth factor in the rat (154).

A harmful physiological effect was first noticed in 1858 for a diethyltin chloride which had a "powerfully pungent odor" and, when heated, produced a vapor that "painfully attacks the skin of the face" and caused fits of sneezing (155). Similar effects were experienced with triethyltin chloride and tetraethyltin (156). The vapors of triethyltin acetate were observed to cause nausea, headache, general weakness, diarrhea and albuminuria, and tetraethyltin caused headache (157). The sternutatory, irritative and lachrimatory properties of triethyltin iodide were studies for possible chemical warfare applications, but none of the effects were considered potent enough and the idea was, mercifully, abandoned (158, 159, 160). Widespread poisoning occurred, however, in France and Algeria in 1954 as a result of taking Stalinon capsules, each of which was stated to contain 15 mg of diethyltin diiodide, for the systematic treatment of staphylococal infections of the skin (161). The incident involved 210 known cases of harmful effects and one hundred deaths. It is estimated that the fatal doses of the organotin were in the range 380 to 750 mg (162-171). The preparation may have also contained tri-, mono- and tetraethyltin (172).

Adverse effects produced by occupational exposure to

triphenyltin acetate used as an agricultural fungicide in Eastern Europe (172-176) include general malaise, headache, loss of conciousness, epigastric pain, vomiting, irritation of the skin, conjunctivae and mucosae, dyspepsia, diarrhea, foggy vision, dizziness, hyperglycemia, glycosuria and damage to the liver shown by increased collagen, some fibrosis and increased serum glutamic-pyruvic transmutase levels. All of these effects were reversible in a few week's time and complete recovery was effected. Organotins have been found to be highly irritating to the skin. Experiments with volunteers showed that undiluted butyltin derivatives produced follicular inflamation and pustulation. The lesions were most severe with tributyltin chloride which produced mild edema and itching (176, 177). In the most severe cases of organotin poisoning, involving peralkylated materials, bradycardia, hypotension and and abrupt variations in the sinus rhythm of the heart were noted (177). The illness lasted 4-10 weeks.

The one fatality in which occupational exposure to organotins has been implicated involved a 29 year old woman drenched with a slurry of tri- and diphenyltin chloride at 175°F (80°C) causing first-degree thermal burns over 90% of the body with erythema and second- and third-degree burns setting in with 85% desquamated skin. Death from renal failure set in 12 days later, but the agent responsible cannot be determined from the available data (179).

An analytical method based upon observation of SnH fluorescence in an H_2-air flame has been applied to organotin compounds present in human urine from males age 25-47. Reduction of the organotins to the corresponding hydrides allows speciation of the tin(IV), mono-, di- and trimethyltin derivatives. Averages of 11 determinations gave concentrations in the ppb range for tin(IV) (0.82 ppb; 82%), mono-(0.090 ppb; 9%), di-(0.073 ppb; 7.3%) and trimethyltin (0.042 ppb; 4.2%). Total tin constituted 1.0 ppb (180). The same techniques were applied to rain (25 pptr total tin, 44% tin(IV), 24% mono-, 30% di- and 0.88% trimethyltin) tap (9.2 pptr total tin, 24% tin(IV), 47% mono-, 14% di- and 16% trimethyltin), fresh ([average of 16 sites around Tampa Bay, Florida] 9.1 pptr total tin, 46% tin(IV), 22% mono-, 15% di- and 16% trimethyltin) estuarine ([average of a dozen sites around Tampa Bay, Florida] 12 pptr total tin, 63% tin(IV), 19% mono-, 14% di- and 3.7% trimethyltin) and saline ([average of a dozen Florida sites] 4.2 pptr total tin, 54% tin(IV), 15% mono-, 33% di- and 12% trimethyltin) waters.

Regulation of Organotins. The U.S. Occupational Safety and Health Act (OSHA) of 1970 requires standards for the protection of workers exposed to hazards at their workplace. The U.S. National Institutes for Occupational Safety and Health (NIOSH) has recommended a standard for occupational exposure to organotin compounds. The recommended threshold limit value (TLV) of 0.1 mg/m^3 on a time-weighted average, calculated as elemental tin has not yet been adopted (57). This standard was itself based upon antecedent documents. The American Conference of Governmental Industrial Hygienists (ACGIH) established a TLV of 0.1 mg/m^3 measured as tin in 1965 for all organotin compounds in the occupational environment (186), and this standard was reaffirmed in 1971 by analogy with the standards for mercury, thallium and selenium because of a lack of pertinent data (187). A separate ACGIH standard was proposed for tricyclohexyltin hydroxide as 1.2 mg/m^3 measured as tin in 1973, and the standard was adopted in 1975 (188). In the German Democratic Republic the maximum allowable concentration of organotins is given as 0.1 mg/m^3 (189), and Romania and Yugoslavia use the same figure for a TLV based upon the 1966 ACGIH value (190). The current U.S. Federal standard, 29 CFR 1910.1000, is a time-weighted average concentration limit of 0.1 mg/m^3 measured as tin, and is based upon the 1965 ACGIH TLV (57). This value equates to an exposure of 0.02 mg/kg/day. The World Health Organization is currently drafting a review of the environmental health aspects of tin (191). The Midwest Research Institute report recommended a maximum daily dosage of 0.02 mg/kg which is the estimated dosage of a person exposed to 0.1 mg/m^3 over an eight-hour period for five days per week (58).

The Council of the European Communities listed organotin compounds as part of a group of substances selected on the basis of their toxicity, persistence and bioaccumulation to which states were directed to apply a system of zero-emission to discharges into ground water, and to require prior authorization by competent authority of the member states for all discharges into the aquatic environment of the Community (192). Russian authors have proposed standards of 20 mg/liter for dibutyltin sulfide (193), 2 mg/liter for dibutyltin dichloride (193), and 0.2 mg/liter for tetraethyltin (195) in resevoir water based upon no-effect doses found in chronic toxicity testing.

Both the NIOSH and Midwest Research Institute reports recognize the strong dependence of the toxicity of organotin compounds upon the number of organic substituents and their nature. The single standard that is recommended for airborne organotins does not reflect the varied toxicity and hazard of these compounds. However, there is lack of adequate data upon which to distinguish quantitatively between different compounds, and a lack of analytical techniques for the speciation of organotins in low environmental concentrations.

Another area for the regulation of organotins is in their use as additives in packaging materials for food and beverages. The Federal Food, Drug and Cosmetic Act was amended in 1963 to allow the introduction of dibutyltin dilaurate in silicone polymer compositions not to exceed 1 part of tin per hundred of polymer (196). A subsequent amendment to the act allowed the use of a di-n-octyltin S,S'-bis(isooctylmercaptoacetate) formulation of ≤15.1-16.1% by weight of tin in PVC. The starting materials for the synthesis of the mercaptoacetate derivative were controlled at not more than 5% of the trichloro- and not less than 95% of the dichlorotin. The Food and Agricultural Organization and the World Health Organization have jointly recommended residue limits of 2 ppm for apples and pears, 0.2 ppm for meat and 0.05 ppm for milk (197), and acceptable daily intake of less than 0.0075 mg/kg of body weight for the pesticide tricyclohexyltin hydroxide (198).

There are currently no specific federal restrictions on the transport or handling of organotin compounds. Some organotins are voluntarily labeled as Corrosive Materials in the U.S. Handling procedures for dealing with accidental spills are not legally specified. Neither the Hazardous Substances List of the Environmental Protection Agency (EPA) nor the Carcinogenic Suspect Agent list of the Occupational Safety and Health Administration (OSHA) contain any organotin compounds. The EPA Office of Pesticides has registered bis(tributyltin) oxide and tributyltin fluoride for use in antifouling paints. As for all pesticides, the labels of the containers of undiluted organotin as well as finished product must show analysis, contents, directions for use, accidental contact remedies and container disposal. Bis(trineophyl)tin oxide and tricyclohexyl- and triphenyltin hydroxides have been registered by the EPA as agricultural chemicals (199). The Food and Drug Administration code permits the use of organotin stabilizers in rigid PVC food containers, and in PVC for food wrap, including di-n-octyltin S,S'-bis(isooctylmercaptoacetate) and maleate polymer and butylthiostannoic acid (200). The EPA allows use in PVC pipe and conduit for potable water. Organotins have also been approved as catalysts and curing agents in the production of polyurethane resins, silicone polymers and PVC products for food packaging, and as adhesive preservatives for food use (200).

Water and Waste Treatment. A U.S. Patent has been issued for removal of inorganic and organolead compounds from the aqueous effluent from the manufacture of tetraalkyllead by treatment with an alkali metal borohydride to form insoluble lead products including hexaalkyldilead and lead metal which can be separated by settling (201). Russian authors have utilized absorption onto activated carbon, extraction into ether and electrochemical oxidation methods for the detoxification of waters polluted by organotin compounds (202, 203). Anodic stripping polarography has been used to determine organotin residues in surface waters (204).

1. E.S. Hedges, "Tin in Social and Economic History," Edward Arnold, London, 1964.
2. Grass, G. "The Tin Drum," transl. by R. Manheim, Pantheon Books, New York, 1962.
3. "The Apocrypha," Ecclesiasticus or the Wisdom of Sirach, 47: 18 (Said of Soloman) "Thou didst gather gold as tin, and multiply (accumulate) silver as lead."
4. Smith, P.J. and Smith, L., Inorg. Chim. Acta Rev. (1973), 7, 11.
5. Kennedy, J.D., and McFarlane, W., in "Reviews on Silicon, Germanium, Tin and Lead Compounds," ed. M. Gielen, Freund, Tel Aviv, Israel (1974), 1, 235.
6. Litzow, M.R. and Spalding, T.R., "Mass Spectrometry of Inorganic and Organometallic Compounds," Elsevier, Amsterdam, 1973.
7. Zuckerman, J.J., Advan. Organometal. Chem. (1970), 9, 21.
8. Tanaka, T., Organomet. Chem. Revs. (1970), A5, 1.
9. Limouzin, Y. and Maire, J.C., "Organotin Compounds: New Chemistry and Applications," J.J. Zuckerman, Ed., Advances in Chemistry Series, No. 157, p. 227, American Chemical Society, Washington, D.C.
10. Zubieta, J.A. and Zuckerman, J.J., Prog. Inorg. Chem. (1978), 24, 251.
11. Geller, S., Matthias, B.T. and Goldstein, R., J. Amer. Chem. Soc. (1955), 77, 1502.
12. Galasso, F.S., "Structure and Properties of Inorganic Solids," Pergamon Press, Oxford, 1970.
13. Vaipolin, A.A., Asmanov, E.O., and Tset'yakov, D.N., Izv. Akad. Nauk SSSR, Neorg. Mater. (1967), 3, 260.
14. Corbett, J.D. and Edwards, P.A., Chem. Commun. (1975), 984.
15. Einstein, F.W.B. and Penfold, B.R., J. Chem. Soc. (A), (1968), 3019.
16. Schlemper, E.O. and Britton, D., Inorg. Chem. (1966), 5, 507.
17. Rush, J.J. and Hamilton, W.C., Inorg. Chem. (1966), 5, 2238.
18. Schlemper, E.O. and Hamilton, W.C., Inorg. Chem. (1966), 5, 995.
19. Forder, R.A. and Sheldrick, G.M., J. Chem. Soc. (A) (1971), 1107.
20. Almenningen, A., Haaland, A., and Motzfeldt, T., J. Organomet. Chem. (1967), 7, 97.
21. Goldberg, D.E., Harris, D.H., Lappert, M.F., and Thomas, K.M., Chem. Commun. (1976), 261.
22. Hulme, R., J. Chem. Soc. (1963), 1524.
23. Nardelli, M., Pelizzi, C., Pelizzi, G., and Tarasconi, P., Z. Anorg. Chem. (1977), 431, 250.
24. Smart, L.E. and Webster, M., J. Chem. Soc. Dalton Trans., (1976), 1924.
25. Brownlee, G.S., Walker, A., Nyburg, S.C., and Szymanski, J.T., Chem. Commun. (1971), 1073.
26. Harrison, P.G., Coord. Chem. Rev. (1976), 20, 1.

27. Neumann, W.P., "The Organic Chemistry of Tin," Wiley-Interscience, New York, 1970.
28. Poller, R.C., "The Chemistry of Organotin Compounds," Academic Press, New York, 1970.
29. Sawyer, A.W., ed., "Organotin Compounds," 3 vols. M. Dekker, New York, 1971-1972.
30. Luijten, J.G.A. and van der Kerk, G.J.M., in "Organometallic Compounds of the Group IV Elements," ed. by A.G. MacDiarmid, vol. 1, part II, p. 91, M. Dekker, New York, 1968.
31. Clark, H.C. and Puddephatt, R.J., in "Organometallic Compounds of the Group IV Elements," ed. by A.G. McDiarmid, vol. 2, part II, p. 71, M. Dekker, New York, 1972.
32. Zuckerman, J.J., ed., "Organotin Compounds: New Chemistry and Applications," Advances in Chemistry Series, No. 157, American Chemical Society, Washington, D.C., 1976.
33. Schumann, H. and Schumann, I., "Gmelin Handbuch der Anorganischen Chemie," Zinn-Organische Verbindungen, Teil 1. Zinntetraorganyle, SnR_4, Springer-Verlag, Berlin, 1975.
34. Schumann, H. and Schumann, I., "Gmelin Handbuch der Anorganischen Chemie," Zinn-Organische Verbindungen, Teil 2, Zinntetraorganyle R_3SnR', Springer-Verlag, Berlin, 1975.
35. Schumann, H. and Schumann, I., "Gmelin Handbuch der Anorganischen Chemie," Zinn-Organische Verbindungen, Teil 3, Zinntetraorganyle R_2SnR_2', $R_2SnR'R''$, $RR'SnR''R'''$, Heterocyclen U. Spirane, Springer-Verlag, Berlin, 1976.
36. Schumann, H. and Schumann, I., "Gmelin Handbuch der Anorganischen Chemie." Zinn-Organische Verbindungen, Teil 4, Organozinnhydride, Springer-Verlag, Berlin, 1975.
37. Seyferth, D. and King, R.B., "Annual Survey of Organometallic Chemistry," covering the year 1964, vol. 1, p. 124, Elsevier, Amsterdam, 1976.
38. Seyferth, D. and King, R.B., "Annual Survey of Organometallic Chemistry," covering the year 1965, vol. 2, p. 161, Elsevier, Amsterdam, 1966.
39. Seyferth, D. and King, R.B., "Annual Survey of Organometallic Chemistry," covering the year 1966, vol. 3, p. 236, Elsevier, Amsterdam, 1967.
40. Luijten, J.G.A., "Tin, Annual Survey Covering the Year 1967," Organometal. Chem. Rev. (1968), B4, 359.
41. Luijten, J.G.A., "Tin, Annual Survey Covering the Year 1968," Organometal. Chem. Rev. (1969), B5, 687.
42. Luijten, J.G.A., "Tin, Annual Survey Covering the Year 1969," Organometal. Chem. Rev. (1970), B6, 486.
43. Bulten, E.J., "Tin, Annual Survey Covering the Year 1970," Organometal. Chem. Rev. (1972), B9, 248.
44. Bulten, E.J., "Tin, Annual Survey Covering the Year 1971,"

J. Organometal. Chem. (1973), 53, 1.

45. Harrison, P.G., "Tin, Annual Survey Covering the Year 1972," J. Organometal. Chem. (1973), 58, 49.
46. Harrison, P.G., "Tin, Annual Survey Covering the Year 1973," J. Organometal. Chem. (1974), 79, 17.
47. Harrison, P.G., "Tin, Annual Survey Covering the Year 1974," J. Organometal. Chem. (1976), 109, 241.
48. Harrison, P.G., "Tin, Annual Survey Covering the Year 1975," J. Organometal. Chem. Library (1977), 4,
49. Abel, E.W. and Stone, F.G.A., "Organometallic Chemistry, A Review of the Literature Published During 1970," Specialist Periodical Report, The Chemical Society, London, 1971.
50. Abel, E.W. and Stone, F.G.A., "Organometallic Chemistry, A Review of the Literature Published During 1971," Specialist Periodical Report, The Chemical Society, London, 1972.
51. Abel, E.W. and Stone, F.G.A., "Organometallic Chemistry, A Review of the Literature Published During 1972," Specialist Periodical Report, The Chemical Society, London, 1973.
52. Abel, E.W. and Stone, F.G.A., "Organometallic Chemistry, A Review of the Literature Published During 1973," Specialist Periodical Report, The Chemical Society, London, 1974.
53. Abel, E.W. and Stone, F.G.A., "Organometallic Chemistry, A Review of the Literature Published During 1974," Specialist Periodical Report, The Chemical Society, London, 1975.
54. Abel, E.W. and Stone, F.G.A., "Organometallic Chemistry, A Review of the Literature Published During 1975," Specialist Periodical Report, The Chemical Society, London, 1976.
55. Abel, E.W. and Stone, F.G.A., "Organometallic Chemistry, A Review of the Literature Published During 1976," Specialist Periodical Report, The Chemical Society, London, 1977.
56. T.W. Lapp, "Study on Chemical Substances from Information Governing the Manufacture, Distribution, Use, Disposal, Alternatives and Magnitude of Exposure to the Environment and Man," Task II - The Manufacture and Use of Selected Alkyltin Compounds, Environmental Protection Agency, Office of Toxic Substances, EPA. 560/6-76-011, PB-251819 (1976).
57. NIOSH Criteria Document No. 77-115, "Occupational Exposure to Organotin Compounds," U.S. Department of HEW, November, 1976.
58. EPA-OTS, Draft Report, "Assessment of the Need for, the Character of, and the Impact Resulting from Limitations on Selected Organotins," Phase I - Assessment of the Need for Limitations on Organotins, July, 1977.

59. Good, M.L., "Abstracts," Workshop on Organotins, New Orleans, 17-19 February, 1978.
60. Rochow, E.G., "Organometallic Chemistry," p. 98, Reinhold, New York, 1964.
61. Gmelin, L., "Karlesruher Zeitung," November, 1838.
62. Fleck, ., Z. Biol. (1872), 8, 444.
63. Gosio, B., Ber. (1897), 30, 1024.
64. Challenger, F., Chem. Ind. (London) 1935, 657.
65. Challenger, F., Chem. Rev. (1945), 36, 315.
66. Challenger, F., Quart. Rev. (1955), 9, 255.
67. Lenhert, P.G. and Crowfoot-Hodgkin, D., Nature (1961), 192, 937.
68. Thayer, J.S., J. Organometal. Chem. (1974), 76, 265; and references therein.
69. Jensen, S. and Jernelöv, A., Nature (1969), 223, 753.
70. Thayer, J.S., J. Chem. Educ. (1971), 48, 806.
71. Thayer, J.S., J. Chem. Educ. (1973), 50, 390.
72. Wood, J.M., Science (1974), 183, 1049.
73. Nriagu, J.O., ed., "Environmental Biogeochemistry," 2 vols., Ann Arbor Science, Ann Arbor, MI.
74. Hutchinson, E.C., ed., "Heavy Metals in the Environment," Univ. Toronto Press, Toronto, Ont., 1976.
75. Wood, J.M., in "Biochemical and Biophysical Perspectives in Marine Biology," ed. by D.C. Malins and J.R. Sargent, Vol. 3, p. 407, Academic Press, New York, 1976.
76. Ridley, W.P., Dizikes, L.J. and Wood, J.M., Science, (1977), 197, 329.
77. Thayer, J.S., J. Chem. Educ. (1977), 54, 604.
78. Thayer, J.S., J. Chem. Educ. (1977), 54, 662.
79. Ayrey, G., Head, B.C. and Poller, R.C., Macromol. Rev. (1974), 8, 1.
80. van der Kerk, G.J.M. in "Organotin Compounds: New Chemistry and Applications," ed. by J.J. Zuckerman, Advances in Chemistry Series No. 157, p. 1, American Chemical Society, Washington, D.C., 1976.
81. Davies, A.G. in "Organotin Compounds: New Chemistry and Applications, ed. by J.J. Zuckerman, Advances in Chemistry Series No. 157, p. 26, American Chemical Society, Washington, D.C.
82. Abraham, M.H. and Johnston, G.E., J. Chem. Soc. (A), 1970, 88.
83. Gielen, M., Accts. Chem. Res. (1973), 6, 198.
84. Deblandre, C., Gielen, M. and Nasielski, J., Bull. Soc. Chem. Belg. (1964), 73, 214.
85. Kuivila, H.G. and Verdone, J.A., Tetrahedron Lett., 1964, 119.
86. Roberts, R.M.G. and Kaissi, F.El., J. Organomet. Chem. (1968), 12, 79.
87. Eaborn, E., Trevorton, J.A. and Walton, D.R.M., J. Organomet. Chem. (1967), 9, 259.

88. Eaborn, C., Hornfield, H.L. and Walton, D.R.M., J. Chem. Soc. (B), 1967, 1036.
89. Bassingdale, A.R., Eaborn, C., Taylor, R., Thompson, A.R. and Walton, D.R.M., J. Chem. Soc. (B), 1971, 1155.
90. D.S. Matteson, "Organometallic Reaction Mechanisms of the Nontransition Elements," Academic Press, New York, 1974.
91. Gielen, M., Dehouck, C., Mokhtar-Jamai, H. and Topart, J., Rev. Silicon, Germanium, Tin, Lead Compd. (1974), 1, 9.
92. Mazaev, V.T., Golovanov, O.V., Igumnov, A.S. and Tsay, V.N., Gig. Sanit. (1976), 3, 17; Chem. Abstr. (1976), 85, 98903z.
93. Barnes, R.D., Bull, A.T. and Poller, R.C., Pestic. Sci. (1973), 4, 305.
94. Chapman, A.H. and Price, J.W., Int. Pest. Control (1972), 14, 11.
95. Freitag, K. and Bock, R., Pestic. Sci. (1974), 5, 731.
96. Engelhart, J.E. and Sheldon, A.W., "Abstr. 15th Marine Coatings Conf.," Point Clear, AL, Feb., 1975.
97. Slesinger, A.E., "Abstr.," Marine Coatings Seminar, National Paint and Coatings Association, Biloxi, MS., March, 1977.
98. Heron, P.N. and Sproule, J.S.G., Indian Pulp Pap. (1958), 12, 510.
99. Beiter, C.G., Engelhart, J.E., Freiman, A. and Sheldon, A., "Abstr. ACS National Meeting," Atlantic City, NJ, August, 1974.
100. Schramel, P.K. , Samsahl, K. and Pavlu, J., Int. J. Environ. Stud. (1973), 5, 37.
101. Jensen, S. and Jernelöv, A., Int. At. Energy Agency Tech. Rep. Ser. No. 137 (1972), p. 43.
102. Wood, J.M. and Brown, D.G., Struct. Bonding, (1972), 11, 47.
103. DeSimone, R.E., Penley, M.W., Charboneau, L., Smith, S., Wood, J.M., Hill, H.A.O., Ridsdale, P. and Williams, R.J.P., Biochim. Biophys. Acta (1973), 304, 851.
104. Jernelöv, A., Lien, E.L. and Wood, J.M. quoted in Wood, J.M., Science (1974), 183, 1049.
105. Huey, C., Brinckman, F.E., Grim, S. and Iverson, W.P., in "Proceedings of the International Conference on the Transport of Persistent Chemicals in Aquatic Ecosystems," Q.N. LeHam, ed., p. 73, National Research Council of Canada, Ottawa, Ontario, 1974.
106. Dizikes, L.J., Ridley, W.P. and Wood, J.M., J. Am. Chem. Soc. (1978), 100, 1010.
107. Pelikan, Z. and Cerny, E., Arch. Toxikol. (1970), 26, 196.
108. Pelikan, Z. and Cerny, E., Arch. Toxikol. (1970) 27, 79.
109. Mazaev, V.T., Korolev, A.A. and Skachkova, I.N., J. Hyg. Epidermiol. Microbiol. Immunol. (1971), 15, 115.

110. Pelikan, Z. and Cerny, E., Arch. Toxikol. (1968), 23, 283.
111. Pelikan, Z., Cerny, E. and Polster, M. Food Cosmet. Toxicol. (1970), 8, 655.
112. Calley, D.J., Guess, W.L. and Autian, J., J. Pharm. Sci. (1967), 56, 240.
113. Jewett, K.L., Brinckman, F.E., and Bellama, J.M., in "Marine Chemistry in the Coastal Environment," ed. by T.M. Church, ACS Symposium Series, No. 18, p. 304, American Chemical Society, Washington, DC, 1975.
114. Jewett, K.L., Brinckman, F.E. and Bellama, J.M., "Abstr. ACS National Meeting," Anaheim, California, March, 1978.
115. Cremer, J.E., Biochem. J. (1958), 68, 685.
116. Casida, J.E., Kimmel, E.C., Holm, B. and Widmark, G., Acta Chem. Scand. (1971), 25, 1497.
117. Bridges, J.W., Davies, D.S. and Williams, R.T., Biochem. J. (1967), 105, 1261.
118. Blair, E.H. Environ. Qual. Saf., Supplement (1975), 3, 406.
119. Fish, R.H., Kimmel, E.C. and Casida, J.E., J. Organomet. Chem. (1975), 93, Cl.
120. Kimmel, E.C., Fish, R.H. and Casida, J.E., J. Agr. Food Chem. (1977), 25, 1201.
121. Fish, R.H., Kimmel, E.C. and Casida, J.E., in "Organotin Compounds: New Chemistry and Applications," ed. by J.J. Zuckerman, Advances in Chemistry Series No. 157, p. 197, American Chemical Society, Washington, D.C., 1976.
122. Fish, R.H., Casida, J.E. and Kimmel, "Chemical Problems in the Environment. Occurence and Fate of the Organoelements," ACS Symposium Series, American Chemical Society, Washington, D.C., 1978.
123. Weisfeld, L.B., J. Paint Technol. (1970), 42, 549.
124. Philips-Dulphar, N.V., as reported in Hopf, H.S., Dunean, J., Beesley, J.S.S., Webley, D.J. and Sturrock, R.F., Bull. WHO (1967), 36, 955.
125. Hopf, H.S., Duncan, J., Beesley, J.S.S., Webley, D.J. and Sturrock, R.F., Bull. WHO (1967), 36, 955.
126. Evans, C.J. and Smith, P.J., J. Oil Colour Chem. Assoc. (1975), 58, 160.
127. Englehart, J., Beiter, C. and Freiman, A., "Abstr. Controlled Release Pesticides Symposium," Akron, Ohio, Sept., 1974.
128. Phillip, A.T., Aust. Oil Colour Chem. Assoc. Proc. News (1975), 58, 160.
129. Frick, L.P. and DeJimenez, W.Q., Bull. WHO (1964), 31, 429.
129a. Weisfeld, Kunststoffe (1975), 65, 298.
130. Klimmer, O.R., Arzneimitt.-Forsch. (1969), 14, 934.
131. Barnes, J. and Stoner, H.B., Brit. J. Ind. Med. (1958), L5, 15.

132. Stoner, H.B., Barnes, J.M. and Duff, J.I., Brit. J. Pharmacol. (1955), 10, 16.
133. Truhaut, R., Chauvel, Y., Anger, J.P., Lieh, N.P., van den Driessche, J., Guesnier, L.R. and Morin, N., Eur. J. Toxicol. Environ. Hyg. (1976), 9, 31.
134. Sheldon, A., J. Paint Technol. (1975), 47, 54.
135. Kimbrough, R.D., Environ. Health Perspect. (1976), 14, 51.
136. Marks, M.J., Winek, C.L. and Shanor, S.P., Toxicol. Appl. Pharmacol. (1969), 14, 627.
137. van der Kerk, G.J.M. and Luyten, J.G.A., J. Appl. Chem. (1954), 4, 314.
137a. Bennett, R.F. and Zedler, R.J., J. Oil Colour Chem. Assoc. (1966), 49, 928.
138. Zedler, R.J. and Beiter, C.B., Soap Cosmet. Chem. Spec. (1962), 38, 75.
139. Hedges, E.S., Research (London) (1960), 13, 449.
140. Sijpesteyn, A.K., Luijten, J.G.A. and van der Kerk, G.J.M., in "Fungicides: An Advanced Treatise," ed. by D.C. Torgeson, p. 331, Academic Press, New York, 1969.
141. Kerr, K.G., Poult. Sci. (1952), 31, 328.
142. Peardon, D.L., Haberman, W.D., Marr, J.E., Garland, F.W., Jr., and Wilcke, H.L., Poult. Sci. (1964), 44, 413.
143. Wilson, H.R., Fry, J.L. and Jones, J.E., Poult. Sci. (1967), 46, 304.
144. McCollister, D.D. and Schober, A.E., Environ. Qual. Saf. (1975), 4.
145. Noel, P.R.B., Heywood, R. and Squires, P.F., quoted in ref. 58, p. 308.
146. Johnson, J.H., Younger, R.L., Witzel, D.A. and Radeleff, R.D., Toxicol. Appl. Pharmacol. (1975), 31, 66.
147. Wolfenberger, D.A., Guerra, A.A. and Loury, W.L., J. Econ. Entomol. (1968), 61, 78.
148. Pieper, G.R. and Casida, J.E., J. Econ. Entomol. (1965), 58, 392.
149. Castel, P., Gras, G., Rioux, J.A. and Vidal, A., Trav. Soc. Pharm. Montpellier (1963), 23, 45.
150. Baker, J.M. and Taylor, J.M., Ann. Appl. Biol. (1967), 60, 181.
151. Kenaga, E.E., J. Econ. Entomol. (1965), 58, 4.
152. Ascher, K.R.S. and Rones, G., Int. Pest Control. (1964), 6, 6.
152a. Anon., Chem. Eng. News, 13 March, 1978, p. 10.
153. Hamilton, E.I. and Minski, M.J., in "The Science of the Environment," p. 375, (1972/1973).
154. Schwartz, K., in "Newer Trace Elements in Nutrition," ed. by W. Mertz and W.E. Cornatzar, Marcel Dekker, New York, 1971.
155. Buckton, G.B., Proc. R. Soc. London (1858), 9, 309.
156. Jolyet, F. and Cahours, A., C.R. Acad. Sci. (Paris) (1869), 68, 1276.

158. McCombie, H. and Saunders, B.C., Nature (1947), 159, 491.
159. Gilman, H., Report No. 6004, National Defense Research Committee, Office of Scientific Research and Development, 1942, p. 1.
160. Glass, H.G., Coon, J.M., Lushbaugh, C.C. and Last, J., Univ. Chicago Toxicology Laboratory Report No. 15, 1942.
161. Anon., Brit. Med. J. (1958), 1, 515.
162. Rouzaud, M. and Lutier, J., Presse Med. (1954), 62, 1075.
163. Rondepierre, J., Truhaut, R., Guilly, P., Hivert, P.E. and Barande, I., Rev. Neurol. (1958), 98, 135.
164. Alajouanine, T.A., Derobert, L. and Thieffry, S., Rev. Neurol. (1958), 98, 85.
165. Gayral, L., Lozorthes, G. and Planques, J., Rev. Neurol. (1958), 98, 143.
166. Druault-Toufesco, M.N., Bull. Soc. Ophtalmol. Fr., 1955, 54.
167. Grossiord, A., Held, J.P., LeCoeur, P.Le., Verley, R. and Drosdowsky, M., Rev. Neurol. (1958), 98, 144.
168. Gruner, J.E., Rev. Neurol. (1958), 98, 109.
169. Pesme, P., Arch. Fr. Pediatr. (1955), 12, 327.
170. Rouzand, M., Rev. Neurol. (1958), 98, 140.
171. Derobert, L., J. Forensic Med. (1961), 7, 192.
172. Lecoq, R., C.R. Soc. Biol. Paris (1954), 239, 678.
173. Guardascione, V., DiBosco, M.N., Lav. Um. (1967), 19, 307.
174. Horacek, V. and Demcik, K., Prac. Lek. (1970), 22, 61.
175. Mijatovic, M., Jugosl. Inostrana Dokumentacija Zastite na Radu (1972), 8, 3.
176. Markicevic, A. and Turko, V., Arh. Hig. Rada. Toksikol. (1967), 18, 355.
177. Lyle, W.H., Br. J. Ind. Med. (1958), 15, 193.
178. Zeman, W., Gadermann, E. and Hardebeck, K., Dtsch. Arch. Klin. Med. (1951), 198, 713.
179. Ref. 57, p. 35.
180. Braman, R.S. and Tompkins, M.A., Anal. Chem. in press.
181. Barnes, J.M. and Magos, L., Organomet. Chem. Rev. (1968), 3, 137.
182. Luijten, J.G.A., in "Organotin Compounds," ed, by A.K. Sawyer, p. 931, Marcel Dekker, New York, 1971.
183. Ivanitskii, A.M., Farmakol. Toksikol. (1963), 26, 629.
184. Gerarde, H.W., Org. Ann. Rev. Pharmacol. (1964), 4, 223.
185. Piver, W.T., Environ. Health Perspect. Exper. Issue (1973), 61.
186. ACGIH, "Threshold Limit Values for 1965," adopted at the 27th ACGIH Annual Meeting, Houston, TX, 1965, p. 15.
187. ACGIH, Committee on Threshold Limit Values: Documentation of the Threshold Limit Values for Substances in Workroom Air, 3rd. ed., pp. 150-51, 224-25, 256, 258, 349-50, Cincinnati, OH, 1971.

188. ACGIH, "Threshold Limit Values for Chemical Substances in the Workroom Air," p. 32, Cincinnati, OH, 1975.
189. Landa, K., Fejfusova, J. and Nedomlelova, R., Prac. Lek. (1973), 25, 391.
190. International Labor Office, "Permissable Levels of Toxic Substances in the Working Environment," Occupational Safety and Health Series, Vol. 20, pp. 239, 353, ILO, Geneva, (1970).
191. Vouk, V.B., private communication, 1977, quoted in ref. 58, p. 344.
192. Off. J. Eur. Commun. (18 May, 1975), 19, 23.
193. Mazaev, V.T. and Shlepnina, T.G., Gig. Sanit. (1973), 38, 10.
194. Mazaev, V.T. and Korolev, A.A., Prom. Zagryaz. Vodoemov (1969), 15.
195. Skachkova, I.N., Gig. Sanit. (1967), 32, 11.
196. Federal Register (31 July, 1963), 28, 7777.
197. Tricyclohexyltin hydroxide, in Joint Meeting of the FAO Working Party of Experts on Pesticide Residues and the WHO Expert Committee on Pesticide Residues: 1973 Evaluations of Some Pesticide Residues in Food - the Monographs, WHO Pesticide Residues Series No. 3, pp.440-52, WHO, Geneva, 1974.
198. Tricyclohexyltin hydroxide, in Joint Meeting of the FAO Working Party of Experts and the WHO Expert Group on Pesticide Residues: 1970 Evaluation of Some Pesticide Residues in Food - The Monographs, pp. 521-42, The FAO of the UN and the WHO, Rome, 1971.
199. Code of Federal Regulations, Title No. 40, Protection of the Environment, SEctions 180.144 and 180.236 (1 July, 1976).
200. Code of Federal Regulations, Title No. 21, Food and Drugs, Sections 121.2514, 121.2520, 121.2522, 121.2566 and 121.2602 (1 April, 1976).
201 Lores, C. and Moore, R.B., U.S. Pat. 3,770,425 (6 Nov., 1973), Chem. Abstr. (1974), 80, 137055.
202. Chernega, L.G., Kozyura, A.S., Kazakova, A.A., Sirak, L.D., Kudenko, N.E. and Linchuk, K.F., Vodosnabzh. Kanaliz. Gidrotekh. Sooruzh. (1971), 13, 75; Chem. Abstr. (1972), 76, 158028.
203. Sirak, L.D., Lutsenko, F.A. and Shkorbatova, T.L., Vodosnabzh. Kanaliz. Gidrotekh. Sooruzh. (1974), 17, 72; Chem. Abstr. (1975), 82, 34748.
204. Woggon, H. and Jehle, D., Nahrung (1975), 19, 271.

Discussion

W. H. RIDLEY (University of Minnesota): What interests me is the toxicity of the organotins. For example, the case of butyltins after they have been hydroxylated by a P450 mechanism that Dr. Fish described. Is there any information on the toxicity of these materials relative to the alkyls? I raise this issue because I think another of the factors controlling the toxicity of these compounds, in addition to steric conditions, is going to be their hydrophobicity, or their partitioning between lipid and aqueous phases.

ZUCKERMAN: I think that's a good question. Dr. Fish, have you put these things back into bugs and seen what the effects are?

R. H. FISH (University of California, Berkeley): We published a paper with Norman Aldridge [*Biochem. Pharmac.* (1977), *26*, 1997] where we examined the inhibition of oxidative phosphorylation with *in vitro* experiments. It was found that the (δ-hydroxybutyl)dibutyltin chloride was a better inhibitor in those systems than tributyltin. However, Ella Kimmell in our laboratory did toxicity studies and found these were slightly less toxic than tributyltin chloride. So there is a dichotomy where they are better inhibitors *in vitro* but less toxic in the whole animal.

ZUCKERMAN: After listening to papers in the symposium, I have learned to ask "More toxic to what?" To comment on your question, the molecules with the terminal hydroxide group would have one end which is lipid-soluble and another end which is water-soluble, like a detergent molecule. Maybe the transport through membranes is more important than the toxicity. Maybe that's a predominimating factor in determining whether it's going to be effective as a toxicant.

FISH: It's interesting to point out that the γ-hydroxy species was less active than the δ-hydroxy compound, perhaps because you get a trigonal bipyramidal interaction as it's exchanging Cl for OH when it goes across the membrane. Both compounds were better than tributyltin, because there was a definite difference between the γ- and δ-hydroxy compounds.

J. S. THAYER (University of Cincinnati): It has been presumed that organotin compounds in the environment will end up as SnO_2. Because of the length of time it takes for this process to occur, relative to redistribution, etc., complete removal of all organic groups might not occur. You might get a tin-oxygen polymer with a number of organic groups remaining (perhaps averaging out to less than one group per tin) but with a high molecular weight and which still retains some bioactive properties.

ZUCKERMAN: The higher the molecular weight, the more complex the particle, the further it's removed from an environment which might be one that cleaves the organo groups. But at the same time, if it became that particle it would be in a sense removed from the environment. If it is detoxification that we are after, maybe that is as effective as making something which analyses as $SnO_2 \cdot H_2O$.

R. COHEN (Bell Laboratories): Have these organotins been subjected to Ames tests to determine their possible mutagenicity? Has there been established any mechanism of concentration along the food chain similar to chlorinated hydrocarbons?

A. CHEH (University of Minnesota): The Ames test does not work well for metals. Citrate and other ions would tie up the metal.

F. E. BRINCKMAN (National Bureau of Standards): A possible source for further information would be Dr. Warren Iverson at our institution, who is concerned with the applications of Ames-type, fast turnover microorganism tests with respect to organometals.

ZUCKERMAN: To turn that question around; at the ONR Organotin Workshop at New Orleans [February, 1978], there was discussion of the possible therapeutic uses of organotin compounds by analogy with the square planar platinum complexes. For example, what would be the analogous tin compound that might be effective in those ways, yet get around some of the bad side-effects of platinum therapy (renal failure, etc.)

Acknowledgment

This manuscript was based in part on NIOSH Criteria Document No. 77-115, and on work done at Midwest Research Institute by R. R. Wilkinson under EPA contract number 68-01-4313. The figures and tables were excerpted from Dr. Wilkinson's draft report by permission of P. Hilgard at EPA and are not subject to U.S. copyright. Our work is supported by the Office of Naval Research.

RECEIVED September 18, 1978.

INDEX

C

D

E

F

I

K

L

M

R

S

T

U

V

W